KT-453-936

BIOCHEMISTRY OF EXERCISE

Mark Hargreaves, PhD
Deakin University
Burwood, Australia

Martin Thompson, PhD
The University of Sydney
Sydney, Australia

Editors

Donation

LIBRARY

ACC No.
36069308
DEPT.

CLASS No.
612.044 HAR

UNIVERSITY
OF CHESTER

Human Kinetics

Library of Congress Cataloging-in-Publication Data

Biochemistry of exercise X / Mark Hargreaves, editor.
      p. cm. -- (International series on sport sciences)
    The 10th International Biochemistry of Exercise Conference, on
which this volume is based, was held in Sydney, Australia on July
15-19, 1997.
    Includes bibliographical references.
    ISBN 0-88011-758-3
    1. Exercise--Physiological aspects--Congresses. 2. Muscles-
-Metabolism--Congresses. I. Hargreaves, Mark, 1961- .
II. International Biochemistry of Exercise Conference (10th : 1997 :
- Sydney, Australia) III. Series.
QP301.B475   1998
612'.044--dc21   98-12383
                                                               CIP

ISBN: 0-88011-758-3

Copyright © 1999 by Human Kinetics

All rights reserved. Except for use in a review, the reproduction or utilization of this work in any form or by any electronic, mechanical, or other means, now known or hereafter invented, including xerography, photocopying, and recording, and in any information storage and retrieval system, is forbidden without the written permission of the publisher.

Permission notices for material reprinted in this book from other sources can be found on page xii.

Acquisitions Editor: Michael S. Bahrke, PhD; **Managing Editor:** Joanna Hatzopoulos; **Assistant Editor:** Jennifer Miller; **Copyeditor:** Donald Amerman; **Proofreader:** Erin Cler; **Graphic Designers:** Keith Blomberg; **Graphic Artist:** Francine Hamerski; **Cover Designer:** Jack Davis; **Photographer (interior):** David Robinson; **Printer:** Edwards Brothers

Printed in the United States of America   10 9 8 7 6 5 4 3 2 1

Human Kinetics
Web site: http://www.humankinetics.com/

United States: Human Kinetics, P.O. Box 5076, Champaign, IL 61825-5076
1-800-747-4457
e-mail: humank@hkusa.com

Canada: Human Kinetics, 475 Devonshire Road Unit 100, Windsor, ON N8Y 2L5
1-800-465-7301 (in Canada only)
e-mail: humank@hkcanada.com

Europe: Human Kinetics, P.O. Box IW14, Leeds LS16 6TR, United Kingdom
(44) 1132 781708
e-mail: humank@hkeurope.com

Australia: Human Kinetics, 57A Price Avenue, Lower Mitcham, South Australia 5062
(088) 277 1555
e-mail: humank@hkaustralia.com

New Zealand: Human Kinetics, P.O. Box 105-231, Auckland 1
(09) 523 3462
e-mail: humank@hknewz.com

This book is dedicated in memory of John Robert Sutton, MD, DSc, FRACP
March 31, 1941—February 6, 1996

Much has been written and spoken of John Sutton since his untimely death in 1996. He was a truly remarkable man and an inspiration to many around the world, but particularly here in Australia where he was our hero. John had a long association with the biochemistry of exercise group, having attended several meetings over the years and served on the organizing committee of the 7th Conference in London, Ontario, in 1988. He was instrumental in bringing the meeting to Sydney, and we dedicate the Proceedings of the 10th Conference to his memory.

Contents

Conference Organization

Organizing Committee

Lyndall Burke, Sydney, Australia
Mark Hargreaves, Melbourne, Australia
Jacques Poortmans, Brussels, Belgium
Martin Thompson, Sydney, Australia

Research Group on Biochemistry of Exercise
(International Council of Sports Sciences and Physical Education)

J.R. Poortmans, Belgium, Chairman
H. Galbo, Denmark, Vice-Chairman
F. Booth, U.S.A.
J.D. Chen, China
C. Guezennec, France
M. Hargreaves, Australia
J. Henriksson, Sweden
J.O. Holloszy, U.S.A.
C.K. Kim, South Korea
R. Maughan, U.K.
K. Nazar, Poland
E.A. Newsholme, U.K.
E. Noble, Canada
D. Pette, Germany
V. Rogozkin, Russia
Y. Sato, Japan
T. Takala, Finland
A. Tsopanakis, Greece
A.J.M. Wagenmakers, The Netherlands

Contributors

David G. Allen
Department of Physiology
University of Sydney
Sydney, Australia

Doron Aronson
Research Division
Joslin Diabetes Center
Department of Medicine
Brigham and Women's Hospital
and Harvard Medical School
Boston, MA, USA

F. Kris Aubrey
School of Kinesiology
University of Western Ontario
London, Ontario, Canada

Christopher D. Balnave
Department of Physiology
University of Sydney
Sydney, Australia

C.J. Barclay
Department of Physiology
Monash University
Clayton, Victoria, Australia

Arend Bonen
Department of Kinesiology
University of Waterloo
Waterloo, Ontario, Canada

Frank W. Booth
Department of Integrative
Biology, Pharmacology,
and Physiology
University of Texas-Houston
Health Science Center
Houston, TX, USA

Marni D. Boppart
Department of Health Sciences
Sargent College of Allied
Health Professions
Boston University
Boston, MA, USA

Anna Casey
Department of Physiology
and Pharmacology
University of Nottingham
Nottingham, United Kingdom

Eva R. Chin
University of Texas
Southwestern Medical Center
Dallas, TX, USA

Michael G. Clark
Division of Biochemistry
Medical School
University of Tasmania
Hobart, Australia

Dumitru Constantin-Teodosiu
Department of Physiology
and Pharmacology
University of Nottingham
Nottingham, United Kingdom

Edward F. Coyle
Department of Kinesiology
and Health Education
University of Texas at Austin
Austin, TX, USA

Michael R. Deschenes
Department of Kinesiology
College of William & Mary
Williamsburg, VA, USA

Kim A. Dora
Division of Biochemistry
Medical School
University of Tasmania
Hobart, Australia

Scott Dufresne
Research Division
Joslin Diabetes Center
Department of Medicine,
Brigham and Women's Hospital
and Harvard Medical School
Boston, MA, USA

Roger A. Fielding
Department of Health Sciences
Sargent College of Allied
Health Professions
Boston University
Boston, MA, USA

Martin Flück
Department of Integrative
Biology, Pharmacology,
and Physiology,
University of Texas-Houston
Health Science Center
Houston, TX, USA

Simon C. Gandevia
Prince of Wales Medical
Research Institute
Sydney, Australia

Laurie J. Goodyear
Research Division
Joslin Diabetes Center
Department of Medicine
Brigham and Women's Hospital
and Harvard Medical School
Boston, MA, USA

Paul L. Greenhaff
Department of Physiology
and Pharmacology
University of Nottingham
Nottingham, United Kingdom

Marc T. Hamilton
Department of Integrative
Biology, Pharmacology,
and Physiology
University of Texas-Houston
Health Science Center
Houston, TX, USA

Tatsuya Hayashi
Research Division
Joslin Diabetes Center
Department of Medicine
Brigham and Women's Hospital
and Harvard Medical School
Boston, MA, USA

Jørn W. Helge
Copenhagen Muscle
Research Center
August Krogh Institute
Copenhagen, Denmark

Michael F. Hirshman
Research Division
Joslin Diabetes Center
Department of Medicine
Brigham and Women's Hospital
and Harvard Medical School
Boston, MA, USA

Harinder S. Hundal
Department of Anatomy
and Physiology
University of Dundee
Dundee, Scotland

Hiroshi Itoh
Department of Physical
Education
Nagoya Institute of Technology
Nagoya, Japan

David E. James
Centre for Molecular
and Cellular Biology
University of Queensland
Brisbane, Australia

Carsten Juel
Copenhagen Muscle
Research Centre
August Krogh Institute
University of Copenhagen
Copenhagen, Denmark

Michael Kjær
Copenhagen Muscle
Research Center and Sports
Medicine Research Unit
Bispebjerg Hospital
Copenhagen, Denmark

Paavo Korge
Cardiovascular Research
Laboratory
Department of Physiology
UCLA School of Medicine
Los Angeles, CA, USA

Søren Kristiansen
Copenhagen Muscle
Research Centre
August Krogh Institute
University of Copenhagen
Copenhagen, Denmark

Graham D. Lamb
School of Zoology
La Trobe University
Bundoora, Victoria, Australia

Michael J. McKenna
Department of Human
Movement, Recreation,
and Performance
Victoria University
of Technology
Footscray, Victoria, Australia

P. Darrell Neufer
John B. Pierce Laboratory
Yale University
New Have, CT, USA

John M.B. Newman
Division of Biochemistry
Medical School
University of Tasmania
Hobart, Australia

Earl G. Noble
School of Kinesiology
University of Western Ontario
London, Ontario, Canada

L. Maureen Odland
Department of Human Biology
& Nutritional Sciences
University of Guelph
Guelph, Ontario, Canada

Robert W. Ogilvie
Department of Cell Biology
and Anatomy
Medical University
of South Carolina
Charleston, SC, USA

Henriette Pilegaard
Copenhagen Muscle
Research Centre
August Krogh Institute
University of Copenhagen
Copenhagen, Denmark

Stephen Rattigan
Division of Biochemistry
University of Tasmania
Hobart, Australia

Michael B. Reid
Department of Medicine
Baylor College of Medicine
Houston, TX, USA

Erik A. Richter
Copenhagen Muscle
Research Centre
August Krogh Institute
University of Copenhagen
Copenhagen, Denmark

Daniel J. Sherwood
Research Division
Joslin Diabetes Center
Department of Medicine
Brigham and Women's Hospital
and Harvard Medical School
Boston, MA, USA

Lawrence L. Spriet
Department of Human Biology
& Nutritional Sciences
University of Guelph
Guelph, Ontario, Canada

Len H. Storlien
Department of Biomedical
Science
Metabolic Research Center
University of Wollongong
Wollongong, Australia

Ronald L. Terjung
College of Veterinary Medicine
University of Missouri
Columbia, MO, USA

Brian S. Tseng
Department of Integrative
Biology, Pharmacology,
and Physiology
University of Texas-Houston
Health Science Center
Houston, TX, USA

Lorraine P. Turcotte
Metabolic Regulation Lab
Department of Exercise Science
University of Southern
California
Los Angeles, CA, USA

Kostas Tzintzas
Department of Physical
Education, Sports Science
and Recreation Management
University of Loughborough
Loughborough,
United Kingdom

Michelle A. Vincent
Division of Biochemistry
Medical School
University of Tasmania
Hobart, Australia

Anton J. M. Wagenmakers
Department of Human
Biology & Stable Isotope
Research Centre
Maastricht University
Masstricht, The Netherlands

Håkan Westerblad
Department of Physiology
and Pharmacology
Karolinska Institute
Stockholm, Sweden

Jørgen F.P. Wojtaszewski
Copenhagen Muscle
Research Centre
August Krogh Institute
University of Copenhagen
Copenhagen, Denmark

H.T. Yang
College of Veterinary Medicine
University of Missouri
Columbia, MO, USA

Preface

The 10th International Conference on the Biochemistry of Exercise, on which this volume is based, was held in Sydney, Australia on July 15–19, 1997. The conference theme was "Membranes, muscle & exercise," and a total of 25 invited lectures were presented. The meeting attracted more than 200 delegates from 26 countries. Two poster sessions were held for the presentation of free communications.

The Honour Award of the Research Group on Biochemistry of Exercise was presented to Prof. Jacques Poortmans for his contributions to exercise biochemistry over many years. It was Jacques who convened the first meeting in 1968, and his efforts in the years since have ensured that this conference remains the most prestigious meeting in the field of exercise biochemistry. The Wander Award for young investigators, sponsored by the Isostar Sports Nutrition Foundation, was awarded to Dr. Labros Sidossis for his paper "Biochemical mechanisms regulating fatty acid oxidation during exercise."

On behalf of the Australian exercise biochemistry community, I would like to thank the Research Group for the opportunity to host the meeting, the invited speakers for their excellent presentations that ensured the success of the meeting, the many delegates from Australia and overseas who attended and, most importantly, Lyndall Burke, who worked tirelessly before and during the meeting to make sure everything ran smoothly.

Mark Hargreaves

Acknowledgments

This conference was made possible by the generous support of a number of organizations, and their contribution to the success of the meeting is gratefully acknowledged by the Organizing Committee.

Deakin University
Experimental and Applied Sciences
Gatorade Sports Science Institute
Isostar Sports Nutrition Foundation
Mars Inc.
New South Wales Department of Sport and Recreation
The University of Sydney

Conference Welcome

At 6:30 a.m. on July 24, 1994, in the rising sun, the Board of the Research Group on Biochemistry of Exercise had a meeting during its 9th Conference in Aberdeen. Everyone was there to hear the three potential candidates for the location of the 10th Conference to be held in 1997. Among them was John Sutton, as fit as a fiddle, eagerly waiting to convince us to choose "his" country. He was talented enough, in his usual way I should say, to show us the benefit of Sydney as the best place. Needless to say, he succeeded elegantly, and we became convinced to go from the "Pom's country" down to "God's own country"! His sudden death on February 6, 1996, stunned all those who knew him well. This was an appalling situation for his family and for the many colleagues who were his best friends.

Australians and Belgians face the same difficulty—in order to succeed we have to leave our country first. So did John Sutton. Besides his verbal skills, John was a man of many talents and professional capabilities. The first time I met him, about 22 years ago, I knew he would become a leading figure in his field of interest. He was a man who could be relied upon, charming with his friends, firm with his opponents. In every circumstance, he would choose the right word to reinforce his view, the correct term to smooth things out. John Sutton also undertook difficult, demanding experiments including those where he exposed himself to extreme environmental conditions. Nevertheless, he was successful in his many attempts to improve our knowledge of humans submitted to physical work. It has been a major loss for the international community, for the Australian scientists, and for the organization of the 10th Conference. John had many reflections on our previous meetings and new ideas to put forward. Unfortunately, the unpredictable has cut, without any notice, the Ariadne's clew, the vital lead that retains each of us on this Earth. His body has gone, but his spirit remains within us.

Mark Hargreaves had a difficult task to replace John. Nevertheless, he succeeded because he is an open-minded person, he had a good supervisor, and he was convinced that the 10th Conference should be of a high level to honor John Sutton's contribution to the sport sciences. For those who attended the Conference it was a real success, and we are convinced that the readers of these Proceedings will share that opinion.

Prof. Jacques Poortmans
Chairman, Research Group on Biochemistry of Exercise

Credits

Figures 3.2, 3.4 Adapted, by permission, from L.J. Goodyear, P.Y. Chang, D.J. Sherwood, S.D. Dufresne, and D.E. Moller, 1996, "Effects of exercise and insulin on mitogen-activated protein kinase signaling pathways in rat skeletal muscle," *American Journal of Physiology* 271: E403-E408.

Figure 6.2 Reprinted, by permission, from L.H. Storlien, A.B. Jenkins, D.J. Chisholm, W.S. Pascoe, and E.W. Kraegen, 1991, "Influence of dietary fat composition on development of insulin resistance in rats: relationship to muscle triglycerine and omega-3 fatty acids in muscle phospholipids," *Diabetes* 40: 280-289.

Figures 7.3, 7.4 Adapted, by permission, from M.J. McKenna, G.J.F. Heigenhauser, R.S. McKelvie, J.D. MacDougall, and N.L. Jones, 1997, "Sprint training enhances ionic regulation during intense exercise in men," *Journal of Physiology* 501: 687-702.

Figure 8.3 Reprinted, by permission, from V.J. Owen, G.D. Lamb, and D.G. Stephenson, 1996, "Effect of low [ATP] on depolarization-induced Ca^{2+} release in skeletal muscle fibres of the toad," *Journal of Physiology* 493: 309-315.

Figure 12.1 Adapted, by permission, from H. Westerblad and D.G. Allen, 1993, "The role of $[Ca^{2+}]_i$ in the slowing of relaxation in fatigued single fibres from mouse skeletal muscle," *Journal of Physiology* 46: 729-740.

Figures 12.2, 12.3 Adapted, by permission, from H. Westerblad, J.D. Bruton and J. Lännergren, 1997, "The effect of intracellular, pH on contractile function of intact, single fibres of mouse declines with increasing temperature." *Journal of Physiology* 500: 193-204.

Figure 14.1 Reprinted, by permission, from S. Kristiansen, M. Hargreaves, and H.A. Richter, 1997, "Progressive increase in glucose transport and GLUT4 in human sarcolammal vesicles during moderate exercise," *American Journal of Physiology* 272: E385-E389.

Figure 14.2 Reprinted, by permission, from S. Kristiansen, M. Hargreaves, and E.A. Richter, 1996, "Exercise-induced increase in glucose transport, GLUT-4 and VAMP-2 in plasma membrane and human muscle," *American Journal of Physiology* 270: E197-E201.

Figure 14.3 Adapted, by permission, from S. Kristiansen, F. Darakhshan, E.A. Richter, and H. Hundal, 1997, "Fructose transport and GLUT-5 protein in human sarcolemmal vesicles," *American Journal of Physiology* 273: E543-E548.

Figure 14.4 Reprinted, by permission, from J.F.P. Wojtaszewski, B.F. Hansen, B. Urso, and E.A. Richter, 1996, "Wortmannin inhibits both insulin- and contraction-stimulated glucose uptake and transport in rat skeletal muscle," *Journal of Applied Physiology* 81: 1501-1509.

Table 14.1 From *Handbook of Physiology*: Section 12: Exercise: Regulation and Integration of Multiple Systems, edited by Loring B. Rowell and John T. Shepherd. Copyright 1996 by The American Physiological Society. Used by permission of Oxford University Press, Inc.

Figure 16.1 Reprinted, by permission, from Loring B. Rowell and John T. Shepherd, 1996, *Handbook of Physiology Section 12: Exercise: Regulation and Integration of Multiple Systems* (New York: Oxford University Press), 952-994.

Figure 16.4   Reprinted, by permission, from L.M. Isola, S-L Zhou, C-L Kiang, D.D. Stump, M.W. Bradbury, and P.D. Berk, 1995, "3T3 fibroblasts transfected with cDNA for mitochondrial aspartate aminotransferase express plasma membrane fatty acid binding protein and saturable fatty acid uptake," *Proceedings of National Academy of Sciences* 92: 9866-9870.

Figures 19.2, 19.3, 19.4, 19.5   Reprinted, by permission, from M.L. Odland, G.J.F. Heigenhauser, D. Wong, M.G. Hollidge-Horvath, and L.L. Spriet, 1998, "Effects of increased fat availability on fat-carbohydrate interaction during prolonged exercise in men," *American Journal of Physiology* 274, E541-E544.

Figure 19.6   Reprinted, by permission, from M.L. Odland, G.J.F. Heigenhauser, G.D. Lopaschuk, and L.L. Spriet, 1996, "Human skeletal muscle malonyl-CoA at rest and during prolonged submaximal exercise," *American Journal of Physiology* 270: E541-E544.

Figure 19.7   Reprinted, by permission, from M.L. Odland, R.A. Howlett, G.J.F. Heigenhauser, E. Hultman, and L.L. Spriet, 9 March 1998, "Skeletal Muscle malonyl-CoA at the onset of exercise at varying power outputs in humans," *APStracts*.

Figures 20.1, 20.2   Reprinted, by permission, from J.A. Romijn, E.F. Coyle, L.S. Sidossis, A. Gastaldelli, J.F. Horowitz, E. Endert, and R.R. Wolfe, 1993, "Regulation of endogenous fat and carbohydrate metabolism in relation to exercise intensity and duration," *American Journal of Physiology* 265: E380-E391.

Figure 20.3   Reprinted, by permission, from J.F. Horowitz, R. Mora-Rodriguez, L.O. Byerley, and E.F. Coyle, 1997, "Lipolytic suppression following carbohydrate ingestion limits fat oxidation during exercise," *American Journal of Physiology* 273: E768-E775.

Figure 21.2   Reprinted, by permission, from D. Constantin-Teodosiu, S. Howell, and P.L. Greenhaff, 1996, "Carnitine metabolism during spring running," *Journal of Applied Physiology* 80: 1061-1064.

Figures 21.3, 21.4   Reprinted from P.W. Greenhaff et al., 1997, "Biochemical aspects of muscle fatigue." *Physiology, stress and malnutrition: Functional correlates, nutritional intervention*, edited by J.M. Kenney and H.N. Tucker (Philadelphia: Lippincott-Raven), 463-481.

Figure 21.5   Reprinted, by permission, from A. Casey, A.H. Short, E. Hultman, and P.L. Greenhaff, 1996, "The metabolic response of type I and type II muscle fibres during repeated bouts of maximal exercise in humans," *American Journal of Physiology* 271: E38-E43.

Figure 22.1   Reprinted, by permission, from P.D. Neufer, G.A. Ordway, G.A. Hand, J.M. Shelton, J.A. Richardson, I.J. Benjamin, and R.S. Williams, 1996, "Continuous contractile activity induces fibre-type specific expression of HSP70 in skeletal muscle," *American Journal of Physiology* 271: C1828-C1837.

Figure 23.1   Adapted, by permission, from D.M. Robinson, R.W. Ogilvie, P.C. Tullson, and R.L. Tenjung, 1994, "Increased peak oxygen consumption of trained muscle requires increased election flux capacity." *Journal of Applied Physiology* 77: 1941-1952.

Figure 23.2   Adapted from H.T. Yang, R.W. Ogilvie, and R.L. Terjung, 1995, "Heparin increases exercise-induced collateral blood flow in rats with femoral artery ligation," *Circulation Research* 76: 448-456.

Figure 23.3   Adapted from H.T. Yang, M.R. Deschenes, R.W. Ogilvie, and R.L. Terjung, 1996, "Basic fibroblast growth factor increases collateral blood flow in rate with femoral arterial ligation," *Circulation Research* 76: 62-69.

Figure 24.2, 24.4   Reprinted, by permission, from L.A. Megeney, P.D. Neufer, G.L. Dohm, M.H. Tan, C.A. Blewtt, G.C.B. Elder, and A. Bonen, 1993, "Effects of muscle activity and fiber composition on glucose transport and GLUT4," *American Journal of Physiology* 264: E583-E593.

Figure 24.5   Reprinted, by permission, from A. Handberg, L.A. Megeney, K.J.A. McCullagh, L. Kayser, and A. Bonen, 1996, "Reciprocal GLUT1 and GLUT4 expression and glucose transport in denervated muscles," *American Journal of Physiology* 271: E50-E57.

Figure 24.6   Reprinted, by permission, from L.A. Megeney, M. Prasad, M.H. Tan, and A. Bonen, 1994, "Expression of the insulin-regulatable transporter GLUT-4 is influenced by neurogenic factors," *American Journal of Physiology* 266: E813-E816.

Figure 24.7   Reprinted, by permission, from L.A. Megeney, R.N. Michel, C.S. Boudreau, P.K Fernando, M. Prasad, M.H. Tan, and A. Bonen, 1995, "Regulation of muscle glucose transport and GLUT-4 by nerve-derived factors and activity-related processes," *American Journal of Physiology* 269: R1148-R1153.

PART I

Keynote Lectures

CHAPTER 1

Mind Over Muscle: The Role of the CNS in Human Muscle Performance

Simon C. Gandevia

Prince of Wales Medical Research Institute, Sydney, Australia

For the opening of the 10th International Meeting on the Biochemistry of Exercise, I was challenged to provide some neurophysiological background to the subsequent deliberations on muscle biochemistry. I will focus not on the peripheral role of skeletal muscles but on the central neural structures that drive the muscles' motoneurons during exercise. Put rather simply, it is the operation of mind rather than the matter! Historically, the debate has long been fought about whether the critical limit to muscle performance lies in the muscle, the cardiorespiratory system, or the central nervous system. However, while some uncertainties remain, the techniques are now evolving to solve the problem for any particular type of exercise (isometric or isotonic; brief or sustained) and any particular type of exerciser (i.e., trained or untrained; healthy subject or a patient limited by disease). Some of the topics mentioned here have been reviewed recently in more detail (8, 10).

Given that the goal is to understand how muscle behaves during voluntary exercise, some refinement of the definitions of fatigue are required, and these clearly bring the central nervous "drive" to motoneurons into focus. Fatigue is best defined as any reduction in the maximal ability to generate force or power during exercise: This will have both a peripheral and a central component when voluntary exercise is considered. Any definition restricting fatigue to the point at which a task no longer can be performed (see 4) ignores the almost immediate changes occurring both in the muscle and at the motoneuron level as exercise begins. Accompanying the newer definition is the tacit implication that methods are available to determine what is the maximal voluntary performance of the muscle.

Twitch Interpolation and Fatigue

As recognized over a century ago by Mosso, Waller, and others, assessment of truly maximal voluntary performance will require artificial electrical stimulation of the muscle. Various techniques that are currently used rely on electrical or "magnetic" stimulation of the motor output, anywhere from the motor cortex to the intramuscular nerves supplying the muscle. Recently, sensitive techniques have been used, including motor nerve and transcranial magnetic stimulation of the motor cortex, to reevaluate the original observations on adductor pollicis by Merton (24, see 16), the originator of the technique of twitch interpolation. He concluded that all motoneurons supplying this intrinsic muscle of the hand could be recruited and discharged sufficiently fast to produce maximal force from the muscle in voluntary isometric efforts. He suggested that this ability remained even when fatigue developed peripherally during exercise under ischemic conditions. More sensitive versions of this technique have indicated that in well-motivated subjects there is some deficit in the ability to recruit and drive motoneurons at the optimal frequency for force generation during isometric exercise and that this deficit increases as force declines with fatigue during isometric contractions (e.g., 10, 12). Hence "central" fatigue occurs: this is an exercise-induced reduction in the voluntary activation of muscle. At one extreme it may produce subjective exhaustion or "task failure" when the peripheral muscle could have reached the necessary force if it had been stimulated artificially. Simplistically, in such a circumstance, the "fatigue" is of central origin.

Although it is difficult to perform the perfect experiment to document this phenomenon, Löscher and colleagues (19) recently have documented failure to produce a target force voluntarily with isometric contractions of the plantar flexor muscles at the ankle when electrical stimulation over the muscles easily produced the force. Another example of task failure with the inspiratory muscles is given below.

Motor Commands, Imagination, and Muscle Performance

Apart from its role in voluntary force production, the central nervous system also exerts critical control over the learning and maintenance of skilled activities. Much of this involves a comparison of the afferent inputs generated by the task (based on visual, proprioceptive, and other information) and the actual goals of the performance. Hence, cognitive neuroscientists recently have been intrigued to determine the extent to which imagination of performing the task improves its execution. Mere imagination of undertaking a task is widely believed by athletes, musicians, ballet dancers, and even psychologists to enhance its performance, a view for which there is strong experimental support (5). The neural mechanisms responsible for this improvement have not been elucidated, but it seems that many cortical structures are brought into play to "design" the spatial and temporal features of the necessary corticospinal output (e.g., 3).

As long suspected (2), overt muscle contractions also occur with the imagination of performing a task, a phenomenon that is sufficient to activate specialized intramuscular receptors (muscle spindle endings and Golgi tendon organs, see 13) and thus to ensure that there is a significant proprioceptive input to aid cortical planning of imagined movements. These small "inadvertent" contractions during imagined performance may provide a critical link between cortical input and peripheral output so that the correct performance can be learned. The improvement of performance associated with mental imagery may involve the association of endogenous cortical commands and inputs from intramuscular receptors activated by the inadvertent contractions.

Imagination of training for maximal strength itself can improve strength of an intrinsic muscle of the hand (28). Whether this simply reflects the fact that these muscles (at least adductor pollicis) are not usually fully activated in "maximal" isometric efforts (see above) is unresolved. Certainly, the ability of imagined training to enhance strength of more proximal muscles is questionable (17).

Another method to access the central nervous system's intrinsic capacity to command motoneurons is to study their recruitment under circumstances in which proprioceptive input has been eliminated. This occurs rarely in neurological diseases (e.g., 6, 25, 26), but it can be studied in experimental animals with different forms of surgical deafferentation (e.g., by dorsal rhizotomy) and in a limited way in healthy subjects by examination of the central control of motoneurons when they have been recorded using microneurography at a site proximal to a complete nerve block (e.g., 9; for review, see 8). Under these circumstances, subjects can recruit the motoneurons innervating particular muscles, grade, and sustain their discharge, with these latter two abilities being improved if some information is given to the subjects about the success of their attempts (e.g., with visual or auditory ancillary feedback). Interestingly, in the absence of feedback from the muscles, maximal voluntary motor unit firing rates are about 30% lower than when the usual muscle feedback reaches the spinal cord and higher centers (22). During a sustained maximal voluntary isometric contraction, the discharge frequency of normally innervated motor units declines with time, perhaps due to altered feedback from the muscle (e.g., 1, 14). These observations suggest that facilitation of motoneurons occurs via muscle spindle feedback during normal isometric contractions and that some input from the muscle may directly or indirectly inhibit and/or disfacilitate motoneurons during fatigue. Presumably in fatigue both the intrinsic properties of motoneurons and the inputs they receive from peripheral and descending sources are adjusted to optimize the actual force output of the whole muscle. Available evidence suggests that the peripheral input from most classes of receptors, especially those innervated by group III and IV afferents, change with fatigue. Descending drive will need to increase to recruit more motoneurons not only because muscle fibers fatigue but because the strength of the fusimotor-muscle spindle servo is not sufficient to maintain spindle discharge during fatigue (23, 1, 15). The fatigue-induced changes in motoneuronal behavior have been reviewed in more detail (7, 10).

Central Factors and Muscle Endurance

To what extent is endurance performance affected by a failure to attain maximal voluntary activation of the muscles? Such failure or "central fatigue" during a task should be measured accurately, for example, with valid "maximal" levels of or, better still, estimates of maximal force with interpolated stimulation. As described earlier for the study of sustained plantar flexion of the ankle, at the time the task is stopped by the subject (i.e., "task failure"), there is failure at high levels in the central nervous system to drive the motoneurons (19). This phenomenon is likely to depend on the type of subjects, the type of "event," and the muscle groups that are used.

Another way to observe a related phenomenon is to check the level of voluntary drive to motoneurons during sustained isometric maximal contractions (12).

Maximal voluntary activation declines with sustained or intermittent contractions, so that any level of motoneuronal activation may be high but not high enough to ensure that optimal force is generated voluntarily by the muscles. This can be observed in the force increases produced by stimulation of the peripheral nerve or motor cortex during maximal fatiguing isometric contractions (12). Many factors contribute to this, not only inhibition (or disfacilitation) at a spinal level, but also effects mediated supraspinally by peripheral afferents (particularly group III and IV afferents) that can lower voluntary activation. Perhaps this is a natural way of conserving the longer-term ability of limb muscles to produce force as more prolonged activation ultimately may risk the critical depletion of ATP.

It is intriguing that during sustained voluntary isometric contractions, changes occur in the responses of the motor cortex to stimulation (particularly its inhibitory intracortical circuitry). Fortunately, these changes recover within about 30 seconds of finishing a prolonged maximal voluntary contraction. Recovery occurs more rapidly than peripheral force or the ability to activate the muscles with maximal voluntary effort (12, 27). Another interesting phenomenon based on the use of various protocols of isometric exercise is that the inability to activate muscles with volition (i.e., the development of central fatigue) loosely parallels the actual decline in peripheral force produced by isometric exercise.

Are all skeletal muscles equal in this decrement in voluntary activation and susceptibility to central fatigue during exercise? The inspiratory muscles are relatively difficult to fatigue at a peripheral level during inspiratory resistive loading (21, 20) when tested with repeated maximal inspiratory efforts or twitches produced by phrenic stimulation (20). Instead, central drive to the muscle and ventilation becomes insufficient, arterial PCO_2 rises, and subjects are unable to continue the task. Here acute ventilatory failure is produced in the laboratory: end-tidal PCO_2 may rise more than 10 torr within 5 to 10 minutes depending on the size of the inspiratory resistive load. Eventually, after remaining capable of driving the inspiratory motoneurons to optimal levels in "test" voluntary contractions and without the peripheral muscles showing fatigue, the CNS decides that "enough is enough." The cause well may be the overwhelming sensation of dyspnea that develops as PCO_2 rises during the loading. Presumably, there must be a survival advantage in such a strategy.

While cause and effect may be obvious in studies of isolated muscle preparations, the separation is more difficult in studies of the neural factors in human muscle fatigue. Hopefully, the examples described above have provided convincing evidence of the need to study central factors in fatigue. However, the results do not contradict the proposition that during most forms of exercise in healthy subjects, the major site for any force loss is within the muscle itself.

Acknowledgments

The author's laboratory is supported by the NHMRC of Australia and the Asthma Foundation of New South Wales.

References

1. Bigland-Ritchie, B., Dawson, N.J., Johansson, R.S., and Lippold, O.C. Reflex origin for the slowing of motoneurone firing rates in fatigue of human voluntary contractions. *Journal of Physiology (London)* 379: 451–459; 1986.
2. Carpenter, W.B. In *Principles of Mental Physiology*, 279–315. New York: Appleton; 1984.
3. Decety, J., Perani, D., Jeannerod, M., Bettinardi, V., Tadary, B., Woods, R., Mazziotta, J.C., and Faxio, F. Mapping motor representations with positron emission tomography. *Nature* 371:600–603; 1994.
4. Edwards, R.H.T. Human muscle function and fatigue. In *Human Muscle Fatigue: Physiological Mechanisms*. Ciba Foundation Symposium No. 82:19–40. London: Pitman Medical; 1981.
5. Feltz, D.L., and Landers, D.M. The effects of mental practice on motor skill learning and performance: a meta-analysis. *Journal of Sport Psychology* 5:25–57; 1983.
6. Forget, R., and Lamarre, Y. Rapid elbow flexion in the absence of proprioceptive and cutaneous feedback. *Human Biology* 6:27–37; 1987.
7. Fuglevand, A.J. Neural aspects of fatigue. *Neuroscientist* 2:203–206; 1996.
8. Gandevia, S.C. Kinesthesia: roles for afferent signals and motor commands. In *Handbook on Integration of Motor, Circulatory, Respiratory and Metabolic Control During Exercise*. Ed. L.B. Rowell and J.T. Shepherd, 128–172. American Physiological Society, New York: Oxford University Press; 1995.
9. Gandevia, S.C., Macefield, G., Burke, D., and McKenzie, D.K. Voluntary activation of human motor axons in the absence of muscle afferent feedback: the control of the deafferented hand. *Brain* 113:1563–1581; 1990.
10. Gandevia, S.C. Neural control in human muscle fatigue: changes in muscle afferents, motoneurones and motor cortical drive. *Acta Physiologica Scandinavica*. In press.
11. Gandevia, S.C., Allen, G.M., and McKenzie, D.K. Central fatigue: critical issues, quantification and practical implications. In *Fatigue: Neural and Muscular Mechanisms*. Ed. S.C. Gandevia, R.M. Enoka, A.J. McComas, D.G. Stuart, and C.K. Thomas, 281–294. New York: Plenum; 1995.
12. Gandevia, S.C., Allen, G.M., Butler, J.E., and Taylor, J.L. Supraspinal factors in human muscle fatigue: evidence for suboptimal output from the motor cortex. *Journal of Physiology (London)* 490:529–536; 1996.
13. Gandevia, S.C., Wilson, L.R., Inglis, J.T., and Burke, D. Mental rehearsal of motor tasks recruits a-motoneurones but fails to recruit human fusimotor neurones selectively. *Journal of Physiology (London)* 505:259-266;1997.
14. Garland, S.J., and McComas, A.J. Reflex inhibition of human soleus muscle during human muscle fatigue. *Journal of Physiology (London)* 401: 547–558; 1990.
15. Hagbarth, K.-E., Bongiovanni, L.G., and Nordin, M. Reduced servo-control of fatigued human finger extensor and flexor muscles. *Journal of Physiology (London)* 485:865–872; 1995.

16. Herbert, R.D., and Gandevia, S.C. Muscle activation in unilateral and bilateral efforts assessed by motor nerve and cortical stimulation. *Journal of Applied Physiology* 80:1351–1356; 1996.

17. Herbert, R.D., Gandevia, S.C., and Dean, C. Effects of real and imagined training on voluntary muscle activation during maximal isometric contractions, Acta Physiologica Scandinavica, in press; 1998.

18. Kirsch, R.F., and Rymer, W.Z. Neural compensation for muscular fatigue evidence for significant force regulation in man. *Journal of Neurophysiology* 57:1893–1910; 1987.

19. Löscher, W.N., Cresswell, A.G., and Thorstensson, A. Central fatigue during a long-lasting submaximal contraction of the triceps surae. *Experimental Brain Research* 108:305–314; 1996.

20. Macefield, G., Gandevia, S.C., Bigland-Ritchie, B., Gorman, R.B., and Burke, D. The firing rates of human motoneurones voluntarily activated in the absence of muscle afferent feedback. *Journal of Physiology (London)* 471:429–443; 1993.

21. Macefield, V.G., Hagbarth, K.-E., Gorman, R., Gandevia, S.C., and Burke, D. Decline in spindle support to a motoneurones during sustained voluntary contractions. *Journal of Physiology (London)* 440:497–512; 1991.

22. McKenzie, D.K., Allen, G.M., Butler, J.E., and Gandevia, S.C. Task failure with lack of diaphragm fatigue during inspiratory resistive loading in human subjects. *Journal of Applied Physiology* 82:2011–2019; 1997.

23. McKenzie, D.K., and Bellamare, F. Respiratory muscle fatigue. In *Fatigue: Neural and Muscular Mechanisms.* Ed. S.C. Gandevia, R.M. Enoka, A.J. McComas, D.G. Stuart, and C.K. Thomas, 401–414. New York: Plenum; 1995.

24. Merton, P.A. Voluntary strength and fatigue. *Journal of Physiology (London)* 123:553–564; 1954.

25. Rothwell, J.C., Traub, M.M., Day, B.L., Obeso, A., Thomas, P.K., and Marsden, C.D. Manual motor performance in a deafferented man. *Brain* 105:515–542; 1982.

26. Sanes, J.N., Mauritz, K.-H., Dalakas, M.C., and Evarts, E.V. Motor control in humans with large-fiber sensory neuropathy. *Human Neurobiology* 4:101–114; 1985.

27. Taylor, J.L., Butler, J.E., Allen, G.M., and Gandevia, S.C. Changes in motor cortical excitability during human muscle fatigue. *Journal of Physiology (London)* 490:519–528; 1996.

28. Yue, G., and Cole, K.J. Strength increases from the motor program: comparison of training with maximum voluntary and imagined muscle contractions. *Journal of Neurophysiology* 67:1114–1118; 1992.

CHAPTER 2

Beyond Exercise Biochemistry 2000: A New Golden Age

Frank W. Booth, Brian S. Tseng, Marc T. Hamilton, and Martin Flück
Department of Integrative Biology, Pharmacology, and Physiology,
University of Texas-Houston Health Science Center, Houston, TX, USA

The biochemistry of exercise is a rapidly evolving field. The First International Symposium on Exercise Biochemistry was held in 1968 (14), which coincided with early publications on adaptive changes in muscle mitochondria and glycogen to training. In the intervening years, molecular and anatomical cell approaches have delineated the plasticity of cells and tissues to exercise training. As the year 2000 approaches, the field of the biochemistry of exercise now includes many more topics than at its inception in the 1960s. Young investigators will have access to the following knowledge that the 1960 pioneers did not: the entire sequence of the human genome; habitual exercise prevents most chronic diseases; the extreme plasticity of gene expression in skeletal muscle; exercise training compensates for many of the physiological declines in aging; lack of exercise contributes to the epidemic in obesity and type II dependent diabetes mellitus; the central nervous system rapidly adapts to changes in the amount of daily physical activity; moderate exercise improves immune function while overtraining depresses immunity, etc. This information provides the framework for a golden age in exercise biochemistry.

Definition of the Biochemistry of Exercise

Biochemistry is the study of the chemistry of living organisms (7). Exercise now includes the disciplines of biochemistry, biophysics, cell biology, developmental biology, endocrinology, genetics, gerontology, immunology, medicine, molecular biology, neurosciences, organic chemistry, pharmacology, physics, physiology, preventive medicine, and public health. If the unanesthetized human is studied at maximal exertion, every field of science can and must be applied to its study.

Human Genome

Only 4% of the 130 billion bases in the human genome have now been sequenced, but by the year 2003 its entire sequence will be known (4). The human genome likely contains 50,000 to 100,000 genes (5). Most of these genes are not known now. Once they are identified, much work will be required to determine their function. Exercise biochemistry will be a necessary discipline needed to delineate the physiology of the new genes to be identified by the human genome project. Questions to be answered start with: a) Which of the new genes is most important for minimizing disruption in cellular and systemic homeostasis? b) Which of the new genes confer improved physical performance? c) What is the exercise duration or intensity required for the cell's gene expression to benefit health, etc.? Answers to these questions will require exercise biochemists to perform physiological, biochemical (including protein chemistry), molecular, anatomical cell biology, and (including inducible transgenics) investigations.

Knowledge of all genes will be a major revolution. It will provide exercise biochemistry with a research tool more powerful than any other technique at its disposal today and will consume many research careers because of the complexity of integration of function of the whole unanesthetized animal responding to the stress of maximal exercise.

Disease Prevention

Epidemiology now has proven that daily exercise decreases the risk of many chronic diseases. There are two classes of genetic diseases: Mendelian disease caused by a single mutant gene (monogenetic), and multigene disorders (oligogenic). Mendelian diseases are rare and are not prevented by exercise training. Chronic diseases, such as cardiovascular diseases, type II dependent diabetes mellitus, hypertension, stroke, and breast and colon cancers are relatively common. The frequency of these chronic diseases is dependent upon the cumulative interaction of many genes at independent loci (oligogenic) and environmental factors (multifactorial diseases) (5). At least 28% of all gene loci harbor polymorphic alleles that vary among individuals (5). It is thought that this polymorphism contributes to the tendency of chronic diseases to "run in families." "Environmental components" such as exercise alter the relative risk of expression of collective interaction of "good" and "bad" genes contributing to the occurrence of a chronic disease. In many cases exercise determines whether an individual crosses a "threshold" of biological significance for clinical symptoms of a chronic disease. It is now clear that many oligogenic diseases are prevented by exercise (3).

For many Americans, this is the first time in the history of human beings that physical activity has been engineered out of their lives. Epidemiological studies now provide convincing data that increasing the physical activity of sedentary populations decreases their risk of acquiring many chronic diseases, such as cardiovascular diseases, type II dependent diabetes mellitus, hypertension, stroke, and breast and colon cancer (3, 13). Physical inactivity contributes to more than one-third of the 500,000 annual heart disease-related deaths in the United States. Each year in the United States 300,000 deaths are a result of physical inactivity and poor diet (12). Neither the molecular and gene mechanism for the preventive effect of exercise nor the proper dosage of exercise to obtain a healthful effect is known. A recent News & Comment in *Science* highlighted a controversy in the United States with its headline that government guidelines say that moderate exercise—walking, housework, gardening, or playing with children—throughout the day is enough to prevent chronic diseases, but that some researchers say that the epidemiological science doesn't support that conclusion (10). In future years exercise biochemists will help resolve this debate by their research into the dose-response relationships for exercise modulation of "healthy" gene expression. Potential questions that can be asked are: a) What are the biochemical changes that are induced by exercise to reduce the risk of these oligogenic diseases? For example, we know that exercise lowers postprandial triglycerides and raises HDL, in part by increasing lipoprotein lipase activity in plasma and in skeletal muscle, but we are uncertain as to how exercise decreases atherosclerosis. b) What are the molecular mechanisms by which exercise increases lipoprotein lipase in skeletal muscle? c) What is the biochemical dose-response curve between a biochemical/molecular change and decreased risk of a certain disease? What is the duration and intensity threshold of exercise needed to produce a healthy effect? What is the minimal threshold to increase signaling pathways to increase lipoprotein lipase in skeletal muscle?

Adaptive Mechanisms

Current approaches of exercise biochemists are to study one gene at a time, termed "gene approach." The selection of this gene requires prior suspicions. After the year 2000, thousands of genes will be able to be studied at the same time, termed "genomics approach." An example of such technology was published recently. A "chip-based" approach involved using one-cm^2 DNA chip with microarrays of cDNA clones as hybridization targets was applied to identify novel heat shock and phorbol ester-regulated genes in human T cells. Using high-speed robotics and performing a highly sensitive two-color hybridization assay, 1,046 human cDNAs were bound to glass or silicone. This technique provides a rapid and efficient method for large-scale gene discovery. Schena et al. (17) wrote that genome analysis will provide insights into growth, development, differentiation, homeostasis, aging, and the onset of diseases. Another usage will be to isolate RNA at various time intervals during and after exercise to see, as Collins (4) describes, how thousands of genes in the cell generate the entire tapestry of its responses. Exercise biochemists will be able to simultaneously monitor the response of thousands of genes during and after a single bout of exercise.

The power of a genetic approach to explaining the functional importance of adaptations to training can be demonstrated by the molecular cause for a benign form of human erythrocytosis. This naturally occurring mutation in a Finnish family introduced an early stopcodon in the erythropoietin receptor (EPO-R) (9). Mutations of the EPO-R gene were autosomal dominant and rendered it to a C-terminal truncated EPO-R, lacking a domain exerting negative control on erythropoiesis in erythroid progenitor cells. Cells expressing the mutant receptor were much more sensitive to erythropoietin and had higher than normal hemoglobin levels. The clinical condition was so mild that many of the affected individuals were not themselves aware of any abnormality, nor do they have any sense of illness. Indeed the life span was unaffected. The family's most famous member, Eero Maentyranta, whose blood carries 25–50% more hemoglobin than the average male's, won three gold medals in cross-country skiing at the 1964 Winter Olympics in Innsbruck, Austria (15). As Lodish, researcher at Whitehead Institute for Biomedical Research, Cambridge, Massachusetts, stated: "This is the first fully characterized mutation that enhances athletic performance."

Compensation for Aging

An exciting development from exercise sciences in recent years has been the observation that some aging is due to physical inactivity. Daily exercise can compensate and prevent aging of the cardiovascular system (10, 18), skeletal muscle (21), and bones (22). Questions that can be posed are: a) What are the biochemical/molecular changes in gene expression for those processes involved with the physical decline with aging? b) Does exercise compensate by an increased expression of a non-aging gene(s) that substitute for the decline in an aging gene or does exercise slow the aging gene? c) What is the dosage of exercise required to maintain appropriate gene expression for those genes involved in the physical decline with aging? d) What growth factors are lost with aging that prevent complete recovery from injury (11)?

Body Weight Control

Recent findings indicate that adipose tissue releases a hormone, leptin, that interacts with a receptor found in high density in the hypothalamus, causing a decrease in neuropeptide Y and a consequent decrease in appetite (food intake) and an increase in physical acitivity. Leptin seems to be conveying information about the amount of body fat to the central nervous system. The more the body fat, the higher the plasma leptin. Exercise reduces body fat and plasma leptin (8).

Questions that need to be answered include: a) A decrease in serum leptin without exercise increases food intake and decreases energy expenditure. Exercise is associated with an increased energy expenditure. How do the differing signals to energy expenditure integrate themselves? b) Does leptin signal a desire to voluntary exercise since it plays a role in energy expenditure?

Brain

The molecular and cellular basis of brain plasticity produced by physical training is largely unexplored. Many individuals exercise daily because it makes them "feel good." The molecular threshold for the very healthful feeling of well-being after exercise is unknown. The central nervous system displays rapid plasticity to resistance training, as most of the increased strength during the first week of training is a result of the greater synchronization of motor units (6). Little is known as to how this occurs. The sympathetic nervous system drive to many peripheral targets is decreased by training (16), but the cellular and molecular basis is little understood.

Integration of Organ Systems

Leptin was identified by positional cloning of ob/ob mice (23). These mice are obese and overeat because their leptin gene is mutated and there is no leptin signal to the hypothalamus. The brain is led to believe that there is no body fat, so appetite increases while energy expenditure decreases. This example illustrates the assistance of molecular biology and genetics in elucidating physiology. Physiological responses can be nonlinear (19). An example is that if you leave out yeast from the bread mix, the bread fails to rise. If you study yeast alone, you might not detect the bread-rising property of yeast. The whole is different than the sum of each individual part. Likewise, molecules, cells, and tissues alone are different than the sum of their parts (19). Exercise biochemists have to integrate many parts to determine the response of the whole unanesthetized animal to exercise. This aspect makes them unique and requires them to explain gene function for new genes found by sequencing the human genome.

Summary

We predict a golden age for exercise biochemistry in the 21st century due to the availability of the sequence of the human genome, new research techniques, and major societal forces. By the year 2003, the entire sequence of the human genome will be found. Consequently, many new genes will be identified. Exercise biochemists will have to assist in the delineation of their function. One new technological advance, DNA chips, will permit the simultaneous study of thousands of genes at once by exercise biochemists. The tapestry of a cell's regulatory processes during and post exercise will be mapped. Some of the societal forces that will foster more research in the field of exercise biochemistry are the increase in chronic diseases as a result of the engineering of physical activity out of the daily lives of many individuals in the United States. It is now clear that the threshold for the development of chronic disease is modulated by exercise's effect on genes. Chronic diseases are multigenic: some genes potentiate a disease while other genes prevent this disease. As the percentage of old people in the population of the United States grows, interest will grow to support research in how exercise compensates for aging of many organ systems. Due to all these events, the future of exercise biochemistry for young investigators is thus very bright. The golden era of exercise biochemistry is upon us if we respond to the challenge.

Acknowledgments

Supported by U.S. Public Health Service grants RO1-AR19393 (FB), RO1-AR44208 (FB), and F32-AR-08455 (MH), National Shriner's Hospital for Crippled Children Research Project #5953 (BT), the MD-PhD Program at the University of Texas Medical School (BT), and the Swiss Government (MF). A special thanks is given to Dr. Peter Davies for his support of the MD-PhD Program.

References

1. Barinaga, M. How much pain for cardiac gain? *Science* 276:11,324-11,327; 1997.
2. Beaudet, A.L., Scriver, C.R., Sly, W.S., and Valle, D. Genetics, biochemistry, and molecular basis of variant human phenotypes. In *The metabolic and molecular bases of inherited disease* (Vol. 1; 7th Ed.). Ed. C.R. Scriver, A.L. Beaudet, W.S. Sly, and D. Valle, 53-118, New York: McGraw-Hill; 1995.
3. Booth, F.W., and Tseng, B.S. America needs to exercise for health. *Med. Sci. Sports Exerc.* 27: 462-465; 1995.
4. Collins, F.S. Sequencing the human genome. *Hosp. Pract.* 32: 35-54; 1997.
5. Green, E.D., Cox, D.R., and Myers, R.M. The human genome project and its impact on the study of human disease. In *The metabolic and molecular bases of inherited disease* (Vol. 1; 7th Ed.). Ed. C.R. Scriver, A.L. Beaudet, W.S. Sly, and D. Valle, 401-436, New York: McGraw-Hill; 1995.
6. Häkkinen, K. Neuromuscular and hormonal adaptations during strength and power training. *J. Sports Med. Physical Fit.* 29: 9-26; 1989.
7. Kent, M. *Oxford Dictionary of Sports Science and Medicine*. New York: Oxford University Press; 1994.
8. Kohrt, W.M., Landt, M., and Birge, S.J. Serum leptin levels are reduced in response to exercise training, but not hormone replacement therapy, in older women. *J. Clin. Endocr. Met.* 81: 3980-3985; 1996.
9. La Chapelle, A., Träskelin, A.L., and Juvonen, E. Truncated erythropoietin receptor causes dominantly inherited benign human erythrocytosis. *Proc. Nat. Acad. Sci. U.S.A.* 90: 4495-4499; 1993.
10. Lakatta, E.G. Cardiovascular regulatory mechanisms in advanced age. *Physiol. Rev.* 73: 413-467; 1993.
11. Marsh, D.R., Criswell, D.S., Carson, J.A., and Booth, F.W. Myogenic regulatory factors during regeneration of skeletal muscle in young, adult, and old rats. *J. Appl. Physiol.* 83: 1270-1275; 1997.
12. McGinnis, J.M., and Foege, W.H. Actual causes of death in the United States. *JAMA* 270: 2207-2212; 1993.
13. Pate, R.R., Pratt, M., Blair, S.N., Haskell, W.L., Macera, C.A., Bouchard, C., Buchner, D., Ettinger, W., Heath, G.W., and King, A.C. Physical activity and public health. A recommendation from the Centers for Disease Control and Prevention and the American College of Sports Medicine. *JAMA* 273: 402-407; 1995.

14. Poortmans, J.R. *Biochemistry of Exercise*. New York: Basel; 1969.

15. Roush, W. An "off switch" for red blood cells. *Sci.* 268: 27-28; 1995.

16. Rowell, L.B., O'Leary, D.S., and Kellogg, D.L. Integration of cardiovascular systems in dynamic exercise. In *Handbook of Physiology. Section 12: Integration of Motor, Circulatory, Respiratory, and Metabolic Control during Exercise*. Ed. L.B. Rowell and J.T. Shepherd, 770-838, New York: Oxford University Press; 1996.

17. Schena, M., Shalon, D., Heller, R., Chai, A., Brown, P.O., and Davis, R.W. Parallel human genome analysis: Microarray-based expression monitoring of 1000 genes. *Proc. Natl. Acad. Sci.* 93: 10,614-10,619; 1996.

18. Schulman, S.P., Fleg, J.L., Goldberg, A.P., Busby-Whitehead, J., Hagberg, J.M., O'Connor, F.C., Gerstenblith, G., Becker, L.C., Katzel, L.I., Lakatta, L.E., and Lakatta, E.G. Continuum of cardiovascular performance across a broad range of fitness levels in healthy older men. *Cir.* 94: 359-367; 1996.

19. Schultz, S.G. Homeostasis, humpty, dumpty, and integrative biology. *News in Physiol. Sci.* 11: 238-246; 1996.

20. Tilghman, S.M. Lessons learned. Promises kept: A biologists eye view of the genome project. *Genome Res.* 6: 773-780; 1996.

21. Tseng, B.S., Marsh, D.R., Hamilton, M.T., and Booth, F.W. Strength and aerobic training attenuate muscle wasting, and improve resistance to the development of disability with aging. *J. Geront.* 50A:113-119; 1995.

22. Vuori, I. Exercise and physical health: musculoskeletal health and functional capabilities. *Res. Quart.* 66: 276-285; 1996.

23. Zhang, Y., Proenca, R., Maffei, M., Barone, M., and Leopold, L. Positional cloning of the mouse obese gene and its human homologe. *Nature* 372: 425-432; 1995.

PART II

Signaling

CHAPTER 3

Intracellular Signaling Pathways in Contracting Skeletal Muscle

Tatsuya Hayashi, Scott D. Dufresne, Doron Aronson, Daniel J. Sherwood, Michael F. Hirshman, Marni D. Boppart**, Roger A. Fielding**, and Laurie J. Goodyear

Research Division, Joslin Diabetes Center, Department of Medicine, Brigham and Women's Hospital, and Harvard Medical School, Boston, MA, USA
**Department of Health Sciences, Sargent College of Health and Rehabilitation Sciences, Boston University, Boston, MA, USA

Physical exercise has major effects on numerous cellular growth and metabolic processes in skeletal muscle. Exercise can be a potent stimulator of skeletal muscle glucose uptake and metabolism, changes that can have significant effects on overall glucose homeostasis (22). Exercise also can have profound effects on glycogen metabolism, with increased rates of glycogenolysis during exercise, followed by a rapid resynthesis of glycogen in the post-exercise state (7). An acute bout of exercise can induce transient increases in skeletal muscle gene transcription (52), which may be a mechanism for chronic adaptations to repeated bouts of exercise (32). Protein metabolism also has been shown to be altered by exercise, with most studies showing decreased rates of synthesis during short-term exercise and increased protein synthesis in the post-exercise state (10). Although exercise is an important regulator of all these biological responses in skeletal muscle, little is known about the molecular signaling events that mediate these alterations. In fact, in recent years there have been few advances in elucidating specific intracellular signaling pathways that may lead to the exercise-induced activation of glucose uptake, glycogen synthesis, gene transcription, and protein synthesis.

Intracellular Signaling Pathways

Studies in mammalian cells have established the existence of numerous intracellular signaling cascades that are critical intermediaries in the regulation of various biological functions (figure 3.1). Over the past few years, our work, and that of others, has shown that many of these signaling proteins are expressed and functional in skeletal muscle. One important cascade of signaling proteins in skeletal muscle is initiated by the hormone insulin. Studies of insulin signaling are potentially relevant to that of exercise signaling because there are several similar biological responses to these two stimuli (e.g., increase in glucose uptake, activation of glycogen synthase). Insulin action is initiated by insulin binding to the extracellular α-subunit of the insulin receptor resulting in the autophosphorylation of tyrosine residues in the receptor β-subunit and activation of a tyrosine kinase intrinsic to the β-subunit (35). The receptor kinase then tyrosine phosphorylates non-receptor proteins, including the insulin receptor substrate-1 (IRS-1) (64). The insulin receptor kinase also can tyrosine phosphorylate IRS-2 (65), but our work has suggested little insulin-stimulated IRS-2 phosphorylation in skeletal muscle (S.D. Dufresne and L.J. Goodyear, unpublished observations). Phosphorylation of IRS-1 by the insulin receptor kinase is known to occur at Tyr-X-X-Met (YXXM) or Tyr-Met-X-Met (YMXM) motifs (60), which then bind with high affinity to *src*-homology 2 (SH2) domains in cellular proteins. Association of SH2 domains in the 85 kD regulatory subunit of phosphatidylinositol 3-kinase (PI 3-kinase) with IRS-1 (6) has been shown to activate the catalytic subunit of this enzyme in insulin-stimulated cultured cells (55), isolated rat adipocytes (36), and intact rat skeletal muscle (14, 23). Activation of PI 3-kinase is thought to be an important component of the intracellular signaling events that lead to one or more of the biological actions of insulin (13).

In addition to PI 3-kinase signaling, a major focus of signaling research during the past several years has been the study of the three parallel mitogen-activated protein kinase (MAP kinase) cascades. These pathways, known as the MAP kinase pathway (also called extracellular-signal-regulated protein kinase or ERK), the c-jun N-terminal kinase (JNK) pathway (also stress activated protein kinase or SAPK), and the p38 kinase pathway, are activated by a variety of growth factors and/ or environmental stresses (12, 18, 46). The defining characteristic of these proteins is that they require dual phosphorylation of a Thr-X-Tyr motif for full enzymatic activation. These signaling cascades have been characterized predominantly in cellular, in vitro systems, and have been shown to be essential for numerous physiological processes. Although all three of these signaling proteins are sometimes referred to as "MAP kinases" in the literature, for purposes of clarity we will refer only to the ERK/MAP kinase as MAP kinase in this manuscript.

The MAP kinase signaling cascade can be stimulated by ras-dependent and independent mechanisms, including G-protein coupled receptors (1, 69) and numerous growth factors, such as growth hormone (2, 11), epidermal growth factor (EGF) (39, 50), FGF (12), and insulin (63). Growth factor activation of the MAP kinase pathway occurs through sequential activation of tyrosine kinase receptors, Shc phosphorylation, and the association of phosphorylated Shc with the adaptor protein Grb2 and the exchange factor SOS (58). The formation of the Shc-Grb2-SOS complex results in the activation of Ras, Raf-1 kinase, the MAP kinase kinases (MEK1 and MEK2), and the two MAP kinase isoforms $p42^{MAPK}$ and $p44^{MAPK}$ (19, 37, 51). The $p42^{MAPK}$ and $p44^{MAPK}$ are proline directed serine/threonine kinases that require dual phosphorylation on a Thr-Glu-Tyr motif for activation (53). MAP kinase can phosphorylate and activate various proteins in the nucleus and cytoplasm. A fraction of the activated MAP kinase

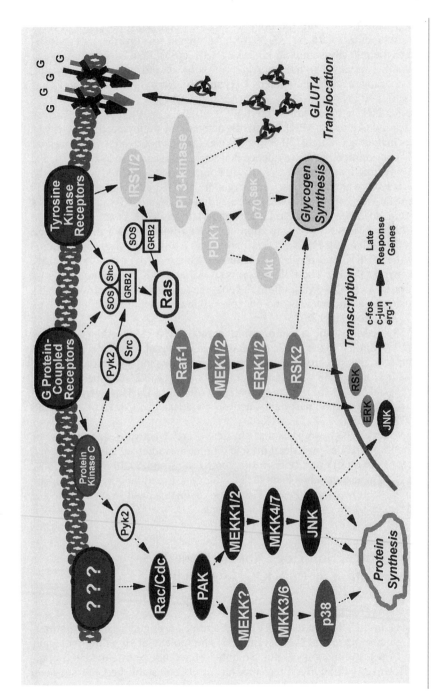

Figure 3.1. Putative and established signaling cascades in skeletal muscle.

population translocates into the nucleus (15, 42) and phosphorylates several transcription factors (9, 25, 31, 33, 59, 68). Another set of substrates are cytosolic, such as the p90 ribosomal S6 kinase 2 (RSK2)(63). RSK2 itself may act as a mediator between the signal transduction pathway and intranuclear events, since it can translocate to the nucleus (15) and phosphorylate c-fos, Elk-1, and CREB (9, 67, 70).

The c-jun NH_2 terminal kinase (JNK) (20, 30, 34, 38, 39), also called stress-activated protein kinase (SAPK), is potently activated in cultured cells by a variety of stresses including UV light (20, 30), cytokines (54), osmotic shock (24), and heat shock (39). In addition, there is evidence that some mitogens such as EGF (30, 39) and NGF (49) can increase JNK activity. In comparison to the MAP kinase pathway, less is known about the upstream activators of JNK, although recent reports have identified a protein with JNK kinase activity named MAP kinase kinase-4 (MKK4) (21, 43, 57). MKK4 in turn is phosphorylated by MEK kinase 1 (MEKK1) (43, 49, 72) and MEKK2 (8). Activation of the Rho-related GTPases Rac1/Rac2 and Cdc42 stimulates the JNK signaling pathway (17, 48), probably through stimulation of the MEKK isoforms by p21 activated kinase (PAK) (45, 74) or a PAK related enzyme (17, 48). The JNK isoforms (JNK1 and JNK2) both require phosphorylation on a Thr-Pro-Tyr motif for full enzyme activation (12). Activated JNKs may translocate to the nucleus and phosphorylate transcription factors, such as c-jun (17, 30) and ATF2 (28).

The p38 kinase cascade is part of a third parallel "MAP kinase-like" signaling pathway (29). Similar to the JNKs, p38 activity is increased when cells are treated with osmotic stress (29), proinflammatory cytokines (54), and other environmental stresses (12, 18). The p38 activity is stimulated by specific MAP kinase kinases, MKK3 (21) and MKK6 (62), but only poorly by MEKK1 (43). However, similar to the JNK pathway, the p38 pathway is activated via Rac and Cdc42 (48, 74), and probably through PAK (74).

We have tested the hypothesis that physical exercise activates one or more of these signaling cascades in skeletal muscle. For these studies we have determined the effects of treadmill exercise or in situ contractions on IRS, PI 3-kinase, JNK, p38, and MAP kinase signaling in rat skeletal muscles. In addition, we measured the effects of bicycle exercise on the MAP kinase signaling pathway in skeletal muscle obtained by needle biopsy from human subjects.

Muscle Contraction Does Not Increase IRS Phosphorylation or PI 3-Kinase Activity

Exercise and insulin are the most potent and physiologically relevant stimulators of glucose uptake in skeletal muscle. The molecular signaling pathways that lead to the stimulation of glucose uptake have not been elucidated completely. However, although a direct link has not been established, studies have suggested that IRS-1 and PI 3-kinase are essential components for insulin-stimulated Glut4 translocation in skeletal muscle (40, 44, 73). We hypothesized that these signaling proteins may be involved in exercise-stimulated Glut4 translocation. For these experiments rats were studied after contraction of hindlimb muscles by electrical stimulation of the sciatic nerve for periods of 15 sec to 20 min or after maximal insulin injection in the absence of contraction (27). Insulin injection resulted in a dramatic increase in tyrosine phosphorylation of the insulin receptor and IRS-1, and IRS-1-associated PI 3-kinase activity. In contrast, contraction alone, at all time points studied, did not increase insulin receptor or IRS-1 phosphorylation. Muscle contraction also had no effect on PI 3-kinase activity measured in IRS-1, insulin receptor, phosphotyrosine, or p85 immune complexes (27), and more recently we have shown that contraction does not increase PI 3-kinase activity associated with the PDGF receptor (M. Boppart, T. Hayashi, and L.J. Goodyear, unpublished observation). In addition to the lack of effect of in situ muscle contractions on these signaling intermediaries, we also have shown that treadmill running of rats does not alter skeletal muscle tyrosine phosphorylation of the insulin receptor, IRS-1, and IRS-2 (S.D. Dufresne and L.J. Goodyear, unpublished observations). These experiments demonstrate that the initial signaling events in insulin action that lead to the activation of Glut4 translocation are not part of the mechanism by which exercise stimulates Glut4 translocation in skeletal muscle.

Effects of Exercise on JNK and p38 Signaling in Rat Skeletal Muscle

JNK and p38 are activated in cultured cells by a variety of environmental stresses (20, 24, 30, 39, 54); however, little is known about the regulation of these signaling proteins in the in vivo context. We determined if physical exercise, a physiological stressor, increased JNK and p38 activities in rat skeletal muscle (26). Animals were studied immediately after treadmill running for 10, 30, 45, or 60 min (20 m/min, 10% grade). This exercise protocol was metabolically effective as indicated by a reduction in soleus muscle glycogen concentrations by approximately 50% at 10 min, and 75% at 30, 45, and 60 min. Mixed hindlimb muscles (soleus, gastrocnemius) were processed and JNK and p38 activities measured using immune complex kinase assays developed for use in skeletal muscle (26). Figure 3.2 shows that exercise significantly increased c-jun phosphorylation in αJNK immune complexes derived from the skeletal muscle. By 10 min of exercise JNK activity was increased by more than twofold above basal, and continued to increase to approximately threefold above basal at 45 min. Exercise also slightly increased p38 kinase activity as indicated by kinase activity assay ($144 \pm 11\%$ of basal; $p<0.007$), and by phospho-specific immunoblotting (data not shown). These studies were the first to show that the stress-activated pathways (JNK and p38) can be activated by a normal physiological stimulus in an intact tissue.

The signaling proteins that lead to the activation of JNK have not been elucidated completely, but the MEKK1/2 and MKK4 are all thought to be upstream signaling molecules in the JNK signaling pathway (figure 3.1). Using a commercially available phospho-specific antibody to MKK4 and by developing in vitro kinase assays for MKK4 and MEKK1, we recently have demonstrated that both

MEKK1 and MKK4 activity are increased in response to contraction induced by sciatic nerve stimulation in rat skeletal muscle (4). Furthermore, the sequential phosphorylation and activation of MEKK1, MKK4, and JNK is temporally related to an increased expression of c-jun mRNA (4). Based on these findings, and given that both of these signaling cascades have been implicated in the regulation gene transcription in various cell types, we believe that exercise-induced activation of these stress-activated signaling cascades also may play a role in transcriptional regulation in skeletal muscle. Studies currently are under way to more directly test this hypothesis.

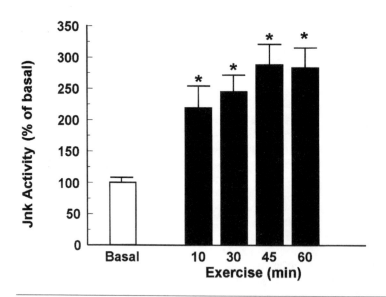

Figure 3.2.   Effects of exercise on JNK activity in rat skeletal muscle. Rats were exercised on a treadmill for the times indicated. Hindlimb muscles were dissected, processed, as described in Methods, and assayed for JNK activity as previously described (26). The upper panel is a representative phosphoimage showing c-jun phosphorylation in two basal rats (B) and three rats exercised for 45 min (E). The bar graph is phosphoimager quantitation of multiple experiments. Data are means ± SE and were analyzed using analysis of variance. * indicates P < 0.05 vs basal, N = 3-8/ group.

Adapted, by permission, from Goodyear et al. 1996.

Exercise Increases MAP Kinase Signaling in Rat Skeletal Muscle

The MAP kinase cascade is an ubiquitously expressed intracellular network of proteins that forms a major signaling system by which cells transduce extracellular cues into intracellular responses (9, 59) (figure 3.1). Initially, this signaling cascade was widely hypothesized to be a regulatory mechanism for insulin-stimulated glycogen synthesis and glucose uptake. More recently, most evidence has suggested that a major biological response to MAP kinase signaling is in the regulation of cell growth. Of particular relevance to studies of exercise regulation in skeletal muscle have been experiments showing that the MAP kinase pathway is critical for mediating the hypertrophic response of cultured cardiac myocytes to stretch (56, 66, 71). Thus, the MAP kinase pathway is a possible candidate for a signaling mechanism that converts the contraction stimulus into intracellular signals in skeletal muscle.

In the first experiments we determined the effects of exercise and insulin on $p42^{MAPK}$ /$p44^{MAPK}$ phosphorylation and RSK2 activity, two key components of the MAP kinase signaling cascade (26). For the exercise treatment, rats were run on a motorized treadmill for 10, 30, 45, or 60 min as described above. To measure MAP kinase phosphorylation, muscle proteins (300 mg) were separated by 10% SDS-PAGE and immunoblotted with a phospho-specific $p42^{MAPK}$/$p44^{MAPK}$ antibody. This antibody recognizes the $p42^{MAPK}$ and $p44^{MAPK}$ isoforms of MAP kinase only when phosphorylated at Thr183 and Tyr185, which are required for full enzymatic activity (3, 53). RSK2 activity was measured in αRSK2 immune complexes using the 3R S6 substrate peptide (RRRLSSLRA) as substrate.

Exercise significantly increased the phosphorylation of the $p42^{MAPK}$ and $p44^{MAPK}$ isoforms (figure 3.3), suggesting that exercise increases MAP kinase activity in skeletal muscle. In subsequent studies we have measured MAP kinase activity by

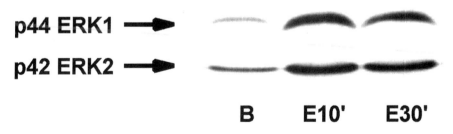

Figure 3.3.   Effects of exercise on MAP kinase phosphorylation in rat skeletal muscle. Rats were exercised on a treadmill for the times indicated. Hindlimb muscles were dissected, processed, and aliquots of muscle proteins (300 mg) were resolved by SDS-PAGE followed by immunoblotting with an antibody that only recognizes the phosphorylated ERK1 and ERK2 species. This image shows that exercise increases both ERK1 and ERK2 phosphorylation.

immune complex and in-gel kinase assays and have confirmed that phospho-specific immunoblotting is a valid technique for assessing MAP kinase activation. Exercise also significantly increased RSK2 activity by 50% to 75% throughout the time course studied (P < 0.01). These data were the first demonstration that exercise can activate MAP kinase signaling in skeletal muscle.

Activation of MAP Kinase Signaling in Human Skeletal Muscle

To understand if these observations in the rat are relevant to exercise in human subjects, we tested the hypothesis that a single bout of exercise activates the MAP kinase signaling pathway in human skeletal muscle (5). Needle biopsies of vastus lateralis muscle were taken from nine subjects at rest (right leg) and after 60 min of cycle ergometer exercise at 70% of $\dot{V}O_2$ max (left leg). Muscle samples were processed and assayed for signaling intermediates as described above for rat skeletal muscle. In all subjects, exercise increased Raf-1 kinase activity and mobility shift, MEK1 activity, p42[MAPK] and p44[MAPK] phosphorylation, and RSK2 activity and mobility shift. Figure 3.4 summarizes the effects of exercise on the MAP kinase signaling cascade in one of the subjects that participated in this study. These studies demonstrate that exercise activates the entire MAP kinase signaling cascade and illustrate the relevance of the animal studies to human physiology.

To determine if the exercise-induced stimulation of the MAPK signaling pathway occurs as a result of a local, tissue-specific mechanism, or as a systemic response to hormones or other factors, we used a one-legged exercise protocol that allowed for matched comparisons within the same subject and controlled for humoral influences and biological heterogeneity between subjects. For these subjects, a basal biopsy was obtained, and 24 hours later the subjects performed 60 min of one-legged exercise. Immediately after the exercise, one muscle sample was biopsied from the exercised leg and another sample biopsied from the nonexercised leg. There was clear stimulation of all components of the MAPK cascade only in the exercised leg (data not shown). It is also important to point out that the exercise-induced stimulation of enzyme activities in the exercising leg of these subjects was of the same magnitude as that observed in subjects who exercised with both legs. These results suggest that activation of the MAPK signaling cascade occurs primarily as a local response to exercise rather than as a systemic response.

Discussion and Future Directions

Exercise has numerous growth and metabolic effects in skeletal muscle, including changes in glycogen metabolism, glucose and amino acid uptake, protein synthesis,

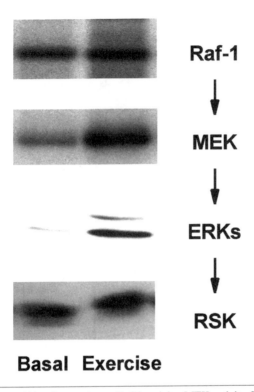

Raf-1

↓

MEK

↓

ERKs

↓

RSK

Basal Exercise

Figure 3.4.   Effects of bicycle exercise on Raf-1 activity, MEK activity, MAP kinase phosphorylation, and RSK2 mobility shift in human skeletal muscle. Needle biopsies of muscle were obtained before and after a 60 min exercise bout at 70% of V̇O₂max. Samples were processed and assayed for the various signaling intermediaries as previously described (5). Adapted, by permission, from Goodyear et al. 1996.

and gene transcription. Despite considerable research in this area, the intracellular signaling mechanisms that transmit the "exercise signals" to the effector molecules are not known. Our laboratory has investigated several intracellular signaling proteins that may be involved in regulating cellular functions in response to exercise. Our studies of IRS and PI 3-kinase signaling have shown that these proteins are not involved in exercise signaling, demonstrating that insulin and exercise stimulate glucose transporter translocation by different mechanisms. Given that PI 3-kinase is activated by so many diverse stimuli in numerous different cell types, and given the plethora of growth and metabolic effects of PI 3-kinase activation, it is fascinating that exercise and muscle contraction per se appear to have no direct effect on the activity of this enzyme. Interestingly, muscle contraction actually can down-regulate insulin-stimulated PI 3-kinase activity in both rat (27) and human skeletal muscle (J.F.P. Wojtaszewski et al., *Diabetes* 46:1775-1781; 1997). Continued work in this area is needed to understand the biological significance of this down-regulation of insulin-stimulated PI 3-kinase signaling after exercise.

Our recent studies described in this manuscript demonstrate that exercise activates multiple intracellular signaling pathways in skeletal muscle. In addition to the studies described above, we also have preliminary data demonstrating that exercise increases JNK and p38 activity in human skeletal muscle (D. Aronson et al., unpublished observations). Thus, exercise in both rats and humans results in a very potent stimulation of MAP kinase and JNK activities, and a modest stimulation of p38 activity in skeletal muscle.

It is likely that a major consequence of the exercise-induced stimulation of the MAP kinase pathways is transcriptional activation. Contractile activity induced by electrical stimulation (47) and exercise (52) have been shown to induce transient increases in c-fos, c-jun, and erg-1 mRNA (47) and GLUT4 gene transcription (52). JNK activation causes phosphorylation of c-jun, resulting in increased AP-1 activity and an alteration in the expression of several genes including c-jun (20,39). Furthermore, both MAP kinase and RSK have been shown to phosphorylate and activate nuclear proteins (15, 25). As mentioned above, we have recently shown that c-fos and c-jun mRNA expression are increased by sciatic nerve induced contractile activity in rat skeletal muscle (S.D. Dufresne and L.J. Goodyear, unpublished observation). The time course of this increased expression follows a similar time course of the activation of downstream signaling components of the MAP kinase and JNK signaling cascades in the same muscle. Although these are only temporal relationships, these results, taken together with previous studies, have led us to speculate that the activation of the JNK and MAP kinase pathways by exercise function to stimulate transcription of multiple genes.

Our human studies demonstrate that a normal "physiological" bout of exercise in human subjects can activate the MAP kinase signaling cascade from Raf-1 to RSK2. How does physical exercise in humans cause the activation of the MAP kinase pathway? One possibility is that the stimulation of the MAP kinase pathway during exercise is mediated by hormones or cytokines released systemically. The results of the one-legged exercise studies suggest that the activation of the MAP kinase signaling cascade occurs primarily as a local response to exercise rather than as a systemic response. Local activation of muscle MAP kinase signaling is fundamental to the hypothesis that this pathway is involved in converting mechanical stimuli into transcriptional responses, since increased gene expression (41, 47) and protein synthesis (16) are restricted to the contracting or overloaded skeletal muscle.

In summary, physical exercise does not stimulate PI 3-kinase signaling in skeletal muscle but does potently activate the MAP kinase, JNK, and p38 signaling cascades. Turning on multiple genes through exercise-induced MAP kinase signaling may lead to the significant and divergent alterations in the biochemical profile of skeletal muscle that occur with repeated bouts of exercise. We hope that the observations described here will provide the basis for several new lines of research that should lead to a better understanding of the molecular signaling mechanisms that regulate the exercise-induced changes in skeletal muscle growth and metabolism.

Acknowledgments

The work described in this manuscript was supported by NIAMS Grant AR42238 and a grant from the Juvenile Diabetes Foundation, International (L.J. Goodyear). Tatsuya Hayashi was supported by a fellowship grant from the Manpei Suzuki Diabetes Foundation. Roger A. Fielding is a Brookdale National Fellow at Boston University.

References

1. Alblas, J., van-Corven, E.J., Hordijk, P.L., Milligan, G., and Moolenaar, W.H. Gi-mediated activation of the p21ras-mitogen-activated protein kinase pathway by alpha 2-adrenergic receptors expressed in fibroblasts. *J. Biol. Chem.* 268: 22,235-22,238; 1993.
2. Anderson, N.G. Growth hormone activates mitogen-activated protein kinase and S6 kinase and promotes intracellular tyrosine phosphorylation in 3T3-F442A preadipocytes. *Biochem. J.* 284: 649-652; 1992.
3. Anderson, N.G., Maller, J.L., Tonks, N.K., and T. W. Sturgill, T.W. Requirement for integration of signals from two distinct phosphorylation pathways for activation of MAP kinase. *Nature* 343: 651-653; 1990.
4. Aronson, D., Dufresne, S.D., and L. J. Goodyear, L.J. Contractile activity stimulates the c-Jun NH2-terminal kinase pathway in rat skeletal muscle. *J. Biol. Chem.* 272(41): 25,636-25,646; 1997.
5. Aronson, D., Violan, M.A., Dufresne, S.D., Zangen, D., Fielding, R.A., and Goodyear, L.J. Exercise stimulates the mitogen-activated protein kinase pathway in human skeletal muscle. *J. Clin. Invest.* 99:1251-1257; 1997.
6. Backer, J. M., Schroeder, G.G., Kahn, C.R., Meyers, M.G., Wilden, P.A., Cahill, D.A., and White, M.F. Insulin stimulation of phosphatidylinositol 3-kinase activity maps to insulin receptor regions required for endogenous substrate phosphoryation. *J. Biol. Chem.* 267:1367-1374; 1992.
7. Bergstrom, J., and Hultman, E. Muscle glycogen synthesis after exercise: an enhancing factor localized to the muscle cells in man. *Nature* 210:309-310; 1966.
8. Blenis, J. Signal transduction via the MAP kinases: proceed at your own RSK. *Proc. Natl. Acad. Sci USA.* 90: 5889-5892; 1993.
9. Booth, F.W., and Watson, P.S. Control of adaptions in protein levels in response to exercise. *Fed. Proc.* 44: 2293-2300; 1985.
10. Campbell, G.S., Pang, L., Miyasaka, T., Saltiel, A.R., and Carter Su, C. Stimulation by growth hormone of MAP kinase activity in 3T3-F442A fibroblasts. *J. Biol. Chem.* 267: 6074-6080; 1992.
11. Cano, E., and Mahadevan, L.C. Parallel signal processing among mammalian MAPKs. *Trends Biochem. Sci.* 20:117-122; 1995.

12. Cheatham, B., Vlahos, C.J., Cheatham, L., Wang, L., Blenis, J., and Kahn, C.R. Phosphatidylinositol 3-kinase activation is required for insulin stimulation of pp70 S6 kinase, DNA synthesis, and glucose transporter translocation. *Molec. Cell. Biol.* 14: 4902-4911; 1994.

13. Chen, K.S., Friel, J.C., and Ruderman, N.B. Regulation of phosphatidylinositol 3-kinase by insulin in rat skeletal muscle. *Am. J. Physiol. Endocrinol. Metab.* 265: E736–E743; 1993.

14. Chen, R.H., Sarnecki, C., and Blenis, J. Nuclear localization and regulation of erk- and rsk-encoded protein kinases. *Molec. Cell. Biol.* 12: 915-927; 1992.

15. Chesley, A., MacDougall, J.D., Tarnopolsky, M.A., Atkinson, S.A., and Smith, K. Changes in human muscle protein synthesis after resistance exercise. *J. Appl. Physiol.* 73:1383-1388; 1992.

16. Coso, O.A., Chiariello, M., Yu, J.C., Teramoto, H., Crespo, P., Xu, N., Miki, T., and Gutkind, J.S. The small GTP-binding proteins rac1 and cdc42 regulate the activity of the JNK/SAPK signaling pathway. *Cell* 81: 1137-1146; 1995.

17. Davis, R. J. MAPKs: new JNK expands the group. *Trends Biochem. Sci.* 19: 470-473; 1994.

18. Dent, P., Haser, W., Haystead, T.A.J., Vincent, L.A., Roberts, T.M., and Sturgill, T.W. Activation of mitogen-activated protein kinase by v-raf in NIH 3T3 cells and in vivo. *Science* 257: 1404-1407; 1992.

19. Derijard, B., Hibi, M., Wu, I.H., Barrett, T., Su, B., Deng, T., Karin, M., and Davis, R.J. JNK1: A protein kinase stimulated by UV light and ha-ras that binds and phosphorylates the c-jun activation domain. *Cell* 76:1025-1037; 1994.

20. Derijard, B., Raingeaud, J., Barrett, T., Wu, I.H., Han, J., Ulevitch, R.J., and Davis, R.J. Independent human MAP kinase signal transduction pathways defined by MEK and MKK isoforms. *Science* 267:682-685; 1995.

21. Devlin, J.T. and Horton, E.S. Effects of prior high-intensity exercise on glucose metabolism in normal and insulin-resistant men. *Diabetes* 34: 973-979; 1985.

22. Folli, F., Saad, M.J.A., Backer, J.M., and Kahn, C.R. Regulation of phosphatidylinositol 3-kinase activity in liver and muscle of animal models of insulin-resistant and insulin-deficient diabetes mellitus. *J. Clin. Invest.* 92:1787-1794; 1993.

23. Galcheva-Gargova, Z., Derijard, B., Wu, I.H., and Davis, R.J. An osmosensing signal transduction pathway in mammalian cells. *Science* 265: 806-811; 1994.

24. Gille, H., Sharrocks, A.D., and Shaw, P.E. Phosphorylation of transcription factor p62TCF by MAP kinase stimulates ternary complex formation at c-fos promoter. *Nature* 358: 414-417; 1992.

25. Goodyear, L.J., Chung, P.Y., Sherwood, D., Dufresne, S.D., and Moller, D.E. Effects of exercise and insulin on mitogen-activated protein kinase signaling pathways in rat skeletal muscle. *Am. J. Physiol.* 271: E403-E408; 1996.

26. Goodyear, L.J., Giorgino, F., Balon, T.W., Condorelli, G., and Smith, R.J. Effects of contractile activity on tyrosine phosphoproteins and phosphatidylinositol 3-kinase activity in rat skeletal muscle. *Am. J. Physiol.* 268: E987-E995; 1995.

27. Gupta, S., Campbell, D., Derijard, B., and Davis, R.J. Transcription factor ATF2 regulation by the JNK signal transduction pathway. *Science* 267: 389-393; 1995.

28. Han, J., Lee, J.D., Bibbs, L., and Ulevitch, R.J. A MAP kinase targeted by endotoxin and hyperosmolarity in mammalian cells. *Science* 265: 808-811; 1994.

29. Hibi, M., Lin, A., Smeal, T., Minden, A., and Karin, M. Identification of an oncoprotein- and UV-responsive protein kinase that binds and potentiates the c-jun activation domain. *Genes & Dev.* 7: 2135-2148; 1993.

30. Hill, C. S. and Tresman, R. Transcriptional r⁻ gulation of extracellular signals: mechanism and specificity. *Cell* 80:199-211; 1995.

31. Holloszy, J.O. and Booth, F.W. Biochemical adaptations to endurance exercise in muscle. *Ann. Rev. Physiol.* 38:273-2911; 1976.

32. Hunter, T. Protein kinases and phosphatases: the yin and yang of protein phosphorylation and signaling. *Cell* 80: 225-236; 1995.

33. Kallunki, T., Su, B., Tsigelny, I., Sluss, H.K., Derijard, B., Moore, G., Davis, R., and Karin, M. JNK2 contains a specificity-determining region responsible for efficient c-jun binding and phosphorylation. *Genes & Dev.* 8: 2996-3007; 1994.

34. Kasuga, M., Karlsson, F.A., and Kahn, C.R. Insulin stimulates the phosphorylation of the 95,000-dalton-subunit of its own receptor. *Science* 215:185-187; 1982.

35. Kelly, K.L., Ruderman, N.B., and Chen, K.S. Phosphatidylinositol-3-kinase in isolated rat adipocytes: activation by insulin and subcellular distribution. *J Biol. Chem.* 267:3423—3428; 1992.

36. Kyriakis, J.M., App, H., Zhang, X., Banerjee, P., Brautigan, D.L., Rapp, U.R., and Avruch, J. Raf-1 activates MAP kinase-kinase. *Nature* 358: 417-421; 1992.

37. Kyriakis, J.M. and Avruch, J. pp54 microtubule-associated protein 2 kinase. *J. Biol. Chem.* 265:17,355-17,363; 1990.

38. Kyriakis, J.M., Banerjee, P., Nikolakaki, E., Dai, T., Rubie, E.A., Ahmad, M.F., Avruch, J., and Woodgett, J.R. The stress-activated protein kinase subfamily of c-jun kinases. *Nature* 369:156-160; 1994.

39. Lank, J.L., Gerwins, P., Elliott, E.M., Sather, S., and Johnson, G. L. Molecular cloning of mitogen-activated protein/ERK kinases (MEKK) 2 and 3. Regulation of sequential phosphorylation pathways involving mitogen-activated protein kinase and c-Jun kinase. *J. Biol. Chem.* 271: 5361-5368; 1996.

40. Lee, A.D., Hansen, P.A., and Holloszy, J.O. Wortmannin inhibits insulin-stimulated but not contraction-stimulated glucose transport activity in skeletal muscle. *FEBS Lett.* 361:51-54; 1995.

41. Lee, D.M., Dawes, N.J., Cox, V.M., Hesketh, J.E., and Goldspink, D.F. Immunolocalization of proto-oncogene expression in mechanically stimulated skeletal muscle. *Biochem. Soc. Trans.* 23: 329S; 1955.

42. Lenormand, P., Sardet, C., Pages, G., L'Allemain, G., Brunet, A., and Pouyssegur, J. Growth factors induce nuclear translocation of MAP kinase (p42^{mapk} and p44^{mapk}) but not of their activator MAP kinase kinase (p45^{mapkk}) in fibroblasts. *J. Cell Biol.* 122: 1079-1088; 1993.

43. Lin, A., Minden, A., Martinetto, H., Claret, F.X., Lange-Carter, C., Mercurio, F., Johnson, G.L., and Karin, M. Identification of a dual specificity kinase that activates the jun kinases and p38-Mpk2. *Science* 268: 286-290; 1995.

44. Lund, S., Holman, G.D., Schmitz, O., and Pedersen, O. Contraction stimulates translocation of glucose transporter GLUT4 in skeletal muscle through a mechanism distinct from that of insulin. *Proc. Natl. Acad. Sci USA.* 92: 5817-5821; 1995.

45. Manser, E., Leung, T., Salihuddin, H., Zhao, Z.S., and Lim, L. A brain serine/threonine protein kinase activated by Cdc42 and Rac1. *Nature* 367: 40-466; 1994.

46. Marshall, C.J. Specificity of receptor tyrosine kinase signaling: transient versus sustained extracellular signal-regulated kinase activation. *Cell* 80:179-185; 1995.

47. Michel, J.B., Ordway, G.A., Richardson, J.A., and Williams, R.S. Biphasic induction of immediate early gene expression accompanies activity-dependent angiogenesis and myofiber remodeling of rabbit skeletal muscle. *J. Clin. Invest.* 90: 277-285; 1994.

48. Minden, A., Lin, A., Claret, F.X., Abo, A., and Karin, M. Selective activation of the JNK signaling cascade and c-jun transcriptional activity by the small GTPases rac and cdc42Hs. *Cell* 81:1147-1157; 1995.

49. Minden, A., Lin, A., McMahon, M., Lange-Carter, C., Derijard, B., Davis, R.J., Johnson, G.L., and Karin, M. Differential activation of ERK and JNK mitogen-activated protein kinases by Raf-1 and MEKK. *Science* 266:1719-1723; 1994.

50. Miyasaka, T., Chao, M.V., Sherline, P., and Saltiel, A.R. Nerve growth factor stimulates a protein kinase in PC-12 cells that phosphorylates microtubule-associated protein-2. *J. Biol. Chem.* 265: 4730-4735; 1990.

51. Moodie, S.A., Willumsen, B.M., Weber, M.J., and Wolfman, A. Complexes of Ras*GTP with Raf-1 and mitogen-activated protein kinase. *Science* 260: 1658-1660; 1993.

52. Neufer, P.D. and Dohm, G.L. Exercise induces a transient increase in transcription of the GLUT-4 gene in skeletal muscle. *Am. J. Physiol.* 265: C1597-C1603; 1993.

53. Payne, D.M., Rossomando, A.J., Martino, P., Erickson, A.K., Her, J.H., Shabanowitz, J., Hunt, D.F., Weber, M.J., and Sturgill, T.W. Identification of the regulatory phosphorylation sites in pp42/mitogen-activated protein kinase (MAP kinase). *EMBO J.* 10: 885-892; 1991.

54. Raingeaud, J., Gupta, S., Rogers, J.S., Dickens, M., Han, J., Ulevitch, R.J., and Davis, R.J. Pro-inflammatory cytokines and environmental stress cause p38 mitogen-activated protein kinase activation by dual phosphorylation on tyrosine and threonine. *J. Biol. Chem.* 270: 7420-7426; 1995.

55. Ruderman, N.B., Kapeller, R., White, M.F., and Cantley, L.C. Activation of phosphatidylinositol 3-kinase by insulin. *Proc. Natl. Acad. Sci. USA* 87:1411-1415; 1990.

56. Sadoshima, J. and Izumo, S. Mechanical stress rapidly activates multiple signal transduction pathways in cardiac myocytes: potential involvement of an autocrine/paracrine mechanism. *EMBO J.* 12:1681-1692; 1993.

57. Sanchez, I., Hughes, R.T., Mayer, B.J., Yee, K., Woodgett, J.R., Avruch, J., Kyriakis, J.M., and Zon, L.I. Role of SAPK/ERK kinase-1 in the stress-activated pathway regulating transcription factor c-jun. *Nature* 372: 794-797; 1994.
58. Schlessinger, J. SH2/SH3 signaling proteins. *Curr. Opin. Cell Biol.* 4: 25-30; 1994.
59. Seger, R. and Krebs, E.G. The MAPK signaling cascade. *FASEB J.* 9: 726-735; 1995.
60. Shoelson, S.E., Chatterjee, S., Chaudhuri, M., and White, M.F. YMXM motifs of IRS-1 define the substrate specificity of the insulin receptor kinase. *Proc. Natl. Acad. Sci. USA* 89: 2027-2031; 1992.
61. Soldati, T., Shapiro, A.D., Dirac Svejstrup, A.B., and Pfeffer, S.R. Membrane targeting of the small GTPase Rab9 is accompanied by nucleotide exchange. *Nature* 369: 76-78; 1994.
62. Stein, B., Brady, H., Yang, M.X., Young, D.B., and Barbosa, M.S. Cloning and characterization of MEK6, a novel member of the mitogen-activated protein kinase cascade. *J. Biol. Chem.* 271:11,427-11,433; 1996.
63. Sturgill, T.W., Ray, L.B., Erikson, E., and Maller, J.L. Insulin-stimulated MAP-2 kinase phosphorylates and activates ribosomal protein S6 kinase II. *Nature (London)* 334: 715-718; 1988.
64. Sun, X.J., Rothenberg, P., Kahn, R.C., Backer, J.M., Araki, E., Wilden, P.A., Cahill, D.A., Goldstein, B.J., and White, M.F. Structure of the insulin receptor substrate IRS-1 defines a unique signal transduction protein. *Nature* 352: 73-77; 1991.
65. Sun, X.J., Wang, L.M., Zhang, Y., Yenush, L.P., Myers, Jr., M.G., Glasheen, E.M., Lane, W.S., Pierce, J.H., and White, M.F. Role of IRS-2 in insulin and cytokine signaling. *Nature* 377:173–177; 1995.
66. Thorburn, J., Frost, J.A., and Thorburn, A. Mitogen-activated protein kinases mediate changes in gene expression, but not cytoskeletal organization associated with cardiac muscle cell hypertrophy. *J. Cell Biol.* 126:1565-1572; 1994.
67. Ward, G.E. and Kirschner, M.W. Identification of cell cycle-regulated phosphorylation sites on nuclear lamin C. *Cell* 61:561-577; 1990.
68. Whitmarsh, A.J., Shore, P., Sharrocks, A.D., and Davis, R.J. Integration of MAP kinase signal transduction pathways at the serum response element. *Science* 269: 403-407; 1995.
69. Winitz, S., Russell, M., Qian, N.X., Gardner, A., Dwyer, L., and Johnson, G.L. Involvement of ras and raf in the Gi-coupled acetylcholine muscarinic m2 receptor activation of mitogen-activated protein (MAP) kinase and MAP kinase. *J. Biol. Chem.* 268: 19,196-19,199; 1993.
70. Xing, J., Ginty, D.D., and Greenberg, M.E. Coupling of the RAS-MAPK pathway to gene activation by RSK2, a growth factor-regulated CREB kinase. *Science* 273: 959-963; 1996.
71. Yamazaki, T., Tobe, K., Hoh, E., Maemura, K., Kaida, T., Komuro, I., Tamemoto, H., Kadowaki, T., Nagai, R., and Yazaki, Y. Mechanical loading activates mitogen-activated protein kinase and s6 peptide kinase in cultured rat cardiac myocytes. *J. Biol. Chem.* 268:12,069-12,076; 1993.
72. Yan, M., Dai, T., Deak, J.C., Kyriakis, J.M., Zon, L.I., Woodgett, J.R., and Templeton, D.J. Activation of stress-activated protein kinase by MEKK1 phosphorylation of its activator SEK1. *Nature* 372: 798-800; 1994.

73. Yeh, J.I., Gulve, E.A., Rameh, L., and Birnbaum, M.J. The effects of wortmannin on rat skeletal muscle. Dissociation of signaling pathways for insulin- and contraction-activated hexose transport. *J. Biol. Chem.* 270: 2107-2111; 1995.
74. Zhang, S., Han, J., Sells, M.A., Chernoff, J., Knaus, U.G., Ulevtich, R.J., and Bokoch, G.M. Rho family GTPases regulate p38 mitogen-activated protein kinase through the downstream mediator pak1. *J. Biol. Chem.* 270: 23,934-23,936; 1995.

CHAPTER 4

Interaction Between Blood Flow, Metabolism, and Exercise

Michael G. Clark, Stephen Rattigan, Kim A. Dora, John M.B. Newman, and Michelle A. Vincent

Division of Biochemistry, Medical School, University of Tasmania, Hobart, Australia

Blood flow to, and distribution between, muscles in exercise has been the subject of intense study. Radioactive, and now more recently, fluorescent microspheres have enabled detailed assessment of flow, particularly in experimental animals, to every muscle at all stages in exercise and during rest. However, the issue of distribution of blood flow between nutritive and non-nutritive pathways of muscle, which confronted investigators in the past, has not been resolved. In this article we wish to present new findings as well as revisit some older studies that confirm the presence of non-nutritive pathways in muscle. Indications are that the non-nutritive route is located in connective tissue of septa and tendons associated with each muscle and is capable of carrying more than half the blood flow at rest. The non-nutritive pathway may have an important function in amplifying nutrient delivery (and waste removal) during exercise.

Regional Blood Flow Within Muscle and Its Influence on Muscle Metabolism at Rest

There appears to be no question that vascular tone can influence muscle metabolism, but the relationship may be complex. Indeed, from a number of studies (e.g., 5) using the constant-flow perfused rat hindlimb, we have found that vasoconstrictors fall into either of two groups (table 4.1). There are those that we have nominally called Type A which, concomitant with their vasoconstrictor activity, increase a variety of metabolic changes, including oxygen uptake, lactate, glycerol, urate and uracil efflux, as well as insulin-mediated glucose uptake. The vasoconstrictor activity and associated metabolic changes of these Type A vasoconstrictors are prevented when Ca^{2+} is omitted from the perfusion medium, or when O_2 is replaced by N_2. In addition vasodilators such as those releasing NO, blocking Ca^{2+} channels or leading to the production of cyclic AMP in vascular smooth muscle, each block both vasoconstrictor and metabolic activity (table 4.1). For any given Type A vasoconstrictor that belongs exclusively to this group (e.g., angiotensin II) there is a proportionality between the extent of vasoconstriction and the increase in metabolism.

Another group of vasoconstrictors behave quite differently. These inhibit metabolism, even though in many cases the pressor effects can be identical in magnitude to Type A vasoconstrictors. These we have called Type B. Type B also differ from Type A by being able to function under low Ca^{2+}, or when O_2 is replaced by N_2

Table 4.1   Vasomodulator Effects on Resting Muscle

Parameter:	% Control	
	Type A*	Type B**
Perfusion pressure	+91	+58.3
Oxygen uptake	+76	−30.3
Lactate efflux	+400	−45
Glycerol efflux	+160	−55
Urate efflux	+200	−44
Uracil efflux	+150	−50
Insulin-mediated glucose uptake	+19	−52
Effect of the following on vasoconstriction and associated changes:		
Removal of perfusate Ca^{2+}	blocked	no effect
Replacement of O_2 by N_2	blocked	
Addition of vasodilators	blocked†	partially blocked#

The data were obtained using the constant flow isolated perfused rat hindlimb.
*Data are for a maximal stimulatory dose of norepinephrine (100 nM); other Type A vasoconstrictors that also increase muscle metabolism during perfusion are epinephrine, phenylephrine, methoxamine, amidephrine, norephedrine, ephedrine, vasopressin, angiotensins I, II, and III, oxytocin, neuropeptide Y, capsaicin, dihydrocapsaicin, gingerols, shogaols, piperine, resinferatoxin, and low frequency sympathetic nerve stimulation.
**Data are for a maximal stimulatory dose of serotonin; other Type B vasoconstrictors that give similar inhibition are norepinephrine at high doses (possibly reflecting nerve synapse concentrations; ≥ 1 μM), high dose vanilloids, and high frequency sympathetic nerve stimulation.
†Includes nitroprusside, nifedipine, isoprenaline, adenosine, AMP, ADP, ATP, and UTP.
#Includes nitroprusside, carbamyl choline, and isoprenaline.
Data are from Clark et al. (1995) and Clark et al. (1994) or references therein.

during hindlimb perfusion conditions. In addition, the types and efficacy of the vasodilators in opposing Type B are not exactly the same ones that are effective against Type A vasoconstrictors (table 4.1).

The differing Ca^{2+} and O_2 dependency of the Type A and Type B vasoconstrictors together with their opposite effects on metabolism suggest that they act at different sites in the muscle vasculature to control the distribution of flow. Further evidence supporting this notion was recently obtained (19). Again, perfusions were conducted at constant flow. Cell-free perfusate was deliberately chosen so that efflux by trapped red blood cells could be detected readily. It was found that although Type A and B vasoconstrictors resulted in similar increases in perfusion pressure, only the Type A vasoconstrictor increased the postequilibrium red blood cell washout. Other experiments in the same study (19) were directed at testing the notion that Type A vasoconstrictors act to access additional nutritive vessels within the hindlimb vasculature. To do this, FITC-labeled dextran (M_r 150,000) was infused over a constant period of 35 min and the Type A vasoconstrictor introduced for 10 min in the middle of the FITC-labeled dextran infusion, the rationale being that if the Type A vasoconstrictor increased perfusate distribution, these newly recruited vessels would be loaded with the dextran and trapped when the vasoconstrictor was withdrawn. A significant new space was indeed recruited by the Type A vasoconstrictor, and this space was emptied by a second exposure to the vasoconstrictor. The space appeared to be perfused more slowly than those regions perfused prior to the vasoconstrictor addition and represented approximately $31.4 \pm 6.0\%$ of the pre-vasoconstrictor vascular volume. There was no evidence that the vasoconstrictor closed off a previously perfused space in conjunction with the recruitment of the new space.

The effects of a Type B vasoconstrictor were different markedly from those of the Type A (19). Thus a Type B vasoconstrictor did not recruit a new vascular space that was filled by labeled dextran but instead resulted in the closing off of a space. Prior loading of this space with labeled dextran was cleared when 5-HT infusion ceased. Estimates suggest that the Type B closed-off space was approximately $13.6 \pm 2.1\%$ of the previously perfused vascular volume. We concluded that the Type B vasoconstrictor acted to close down access to the nutritive pathway (the derecruited space) and increase flow through the non-nutritive pathway that was already perfused. It appeared likely that the Type A vasoconstrictor recruited nutritive flow but not at the expense of non-nutritive flow, which decreased but did not stop.

The advent of laser Doppler flow probes that can be positioned near the muscle nutritive route has enabled one research group (16) to recently assess the relationship between total muscle blood flow (by radioactive microsphere entrapment) and apparent muscle nutritive flow in vivo. Rats were anesthetized and chlorisondamine-induced ganglionic blockade imposed. For phenylephrine or angiotensin II there was a marked increase in muscle vascular resistance, no change in muscle blood flow, and a significant increase in laser Doppler flow (nutritive flow). For the vasodilator, isoproterenol (atenolol was also included), muscle blood flow increased markedly, and muscle vascular resistance and muscle nutritive flow each significantly decreased. Although there may be other explanations for the findings it seems likely that the two Type A vasoconstrictors (phenylephrine and angiotensin II) (5) each has produced effects in vivo involving a selective increase in muscle nutritive flow consistent with our observations using the constant-flow perfused rat hindlimb. The vasodilator, isoproterenol, would appear to have selectively increased flow to the non-nutritive pathway at the expense of nutritive flow. In our experience (6) isoproterenol opposes Type A vasoconstrictor effects in the constant-flow perfused hindlimb by relaxing constricted sites in the vasculature that are preventing access to non-nutritive routes.

We would argue that microspheres do not allow the discrimination between nutritive and non-nutritive routes of muscle, and both sets are excised with an intact muscle when it is removed for microsphere content determination. For example, we have found that the Type B vasoconstrictor, serotonin, did not alter the distribution of microspheres either between different muscles or between muscle and nonmuscle tissue during constant-flow hindlimb perfusion, even though $\dot{V}O_2$ was decreased markedly by this agent (21).

Overall, the data suggest that muscle metabolism at rest is determined in large part by the ratio of nutritive to non-nutritive flow. It is likely, but not yet proven, that access for nutrients such as glucose, amino acids, and lipids thus may be rigorously controlled by this proportion, itself controlled by a set of neuroendocrine inputs.

Regional Blood Flow Within Muscle and Its Influence on Muscle Metabolism in Exercise

From studies described above and made using the constant-flow perfused rat hindlimb at rest, it seems likely that vasoconstrictors that affected metabolism, either positively or negatively, did so by changing nutrient access. It was, therefore, of interest for us to know whether contractile performance of stimulated muscle also was affected by the Type A and B vasoconstrictors under conditions of constant flow. In the first of two studies we used serotonin (5-HT) as a representative Type B vasoconstrictor. Tension development and contraction-induced oxygen uptake by the sciatic nerve-stimulated gastrocnemius-plantaris-soleus muscle group of the perfused rat hindlimb and tension development by electrically stimulated isolated incubated soleus and extensor digitorum longus muscles were examined. In the hindlimb 5-HT increased perfusion pressure, and markedly decreased contraction-induced tension and oxygen uptake (table 4.2), as well as lactate release (7). The release of metabolic vasodilators from exercising skeletal muscle did not appear to affect 5-HT-mediated vasoconstriction; rather, vascular resistance increased during the period of muscle contraction. In contrast, contraction of isolated incubated soleus and extensor digitorum longus muscles was not affected by 5-HT addition to the incubation medium.

The inhibitory effect of 5-HT on aerobic contractility of working muscle in the constant-flow perfused rat hindlimb appeared to have occurred as a result of impaired nutrient delivery associated with vasoconstriction and was evident whether 5-HT was infused before and/or during the stimulation period. There appeared to be no effect of 5-HT on initial tension (anaerobic) development (table 4.2), again consistent with an effect to reduce oxygen delivery for aerobic tension development. In addition, time course studies (7) showed that the sequence of changes induced by 5-HT was increased perfusion pressure (vasoconstriction) closely associated with an inhibition of $\dot{V}O_2$ and lactate release. Maximal impairment of

Table 4.2   Vasomodulator Effects on Contracting Muscle

Infusion	% Control			Comments
	Initial Tension	Tetanic Tension	($\dot{V}O_2$)	
Angiotensin II (3 nM)	126 ± 17 n.s.	175 ± 25*	132 ± 2*	Perfused rat hindlimb at constant flow, using red blood cells at 37°C. Stimulus: sciatic nerve (Rattigan et al. 1996)
Serotonin (0.25 μM)	96 ± 9 n.s.	44 ± 10*	67.9 ± 10*	Similar to above (Dora et al. 1994)
Carbachol (5 μM)		75 ± 3*		Perfused rat hindlimb at constant-flow, no red blood cells at 25°C. Stimulus: sciatic nerve (Dora et al. 1994)
Acetyl choline (2 μg/min)		50		Perfused 'isolated'
Histamine (5 μg/min)		75		gastrocnemius-soleus
Isoproterenol (5 μg/min)		85		muscle group of
ATP (50 μg/min)		85		anaesthetized cats.
Isoxsuprine (500 μg/min)		45		Stimulus: sciatic nerve
Bradykinin (3.5 μg/min)		72		(Hirvonen et al. 1964)

*Significantly different from control.

contraction was somewhat delayed, occurring 1–2 min after the maximal 5-HT-mediated rise in perfusion pressure. This sequence and the lack of an effect in incubated muscles suggest that 5-HT does not impair contractility directly by receptor-mediated effects on skeletal muscle. Rather, the sequence is consistent with our other findings (5) in which the effects of 5-HT in inhibiting $\dot{V}O_2$ in the perfused rat hindlimb have been attributed to functional vascular shunting (increased non-nutritive flow).

In the second similar study angiotensin II (AII), a Type A vasoconstrictor, replaced 5-HT (22). AII infusion alone caused vasoconstriction with increased oxygen and glucose uptake and lactate and glycerol release. AII infusion during muscle contraction gave less vasoconstriction but increased tension development during tetanic stimulation by 80%, contraction-induced oxygen uptake and insulin-mediated 2-deoxyglucose uptake by plantaris and gastrocnemius red and white muscles. In addition, the effect of AII to increase contraction was blocked by the vasodilator carbachol, suggestive that the AII effects were mediated via the vasculature and not a direct effect on muscle. Whereas the stimulatory effect of AII on muscle contraction may be due to the redirection of flow from noncontracting and type I fiber muscles to the type II fiber contracting muscles in the hindlimb, we believe that it is due to increased nutritive flow within the contracting muscles. This is consistent with the variety of metabolic effects seen in perfused muscle at rest and discussed above (table 4.1).

A study using the "isolated" constant-flow gastrocnemius-soleus preparation of the cat (12) showed over thirty years ago that contractile force was depressed by a variety of intra-arterial vasodilator agents (table 4.2). Muscle force was reduced even when flow was increased or unchanged. In addition, a decrease occurred in the ^{131}I clearance rate from the gastrocnemius parallel to the reduction in muscle force from the vasodilator, acetyl choline, while a much smaller, or delayed reduction in clearance rate occurred when force was reduced to the same extent by neuromuscular blockade (d-tubocurarine) (12). From our studies using vasodilators in the constant-flow perfused rat hindlimb (6) it would appear likely that the effects seen by Hirvonen et al. (12) were due to the vasodilator agents redirecting flow to the non-nutritive route.

The Notion of Functional Vascular Shunts (Non-Nutritive Flow) in Muscle

An authoritative review of muscle blood flow in 1973 by Hudlicka (14) suggested that the connection between the arterial and venous vascular beds may be of two kinds. Either a network of true capillaries exists that might be open only partially under resting conditions or there are thoroughfare channels from which true capillaries branch. However, despite an intense search for anatomical shunts in muscle directly by visualization, or indirectly by measuring passage of various size microspheres (14 and references therein), a convincing case for their presence could not be made. This contrasted with the functional evidence which was considerable. Indeed many workers had come to the conclusion that there were two circulation systems in skeletal muscle and that total blood flow to muscle was not simply an indicator of nutrient delivery. Thus there were situations where total blood flow could increase with either an increase or no change to nutritive flow (14). Most notably among these was exercise where both increased together (1) and emotional stress where blood flow rose without change in apparent nutritive flow (3).

Attempts to measure the proportion of total flow that was nutritive appear to date from Friedman (8) who for the dog hindlimb calculated a value of 26%. However, this calculation for the whole hindlimb did not take into account flow in skin. Other estimates of nutritive to total flow are shown in table 4.3. Clearly, there is a wide range of values that may reflect the different methods used. Apart from one very high value of 97%, the average estimated proportion of total flow entering the nutritive route in resting muscle appears to be less than 50%.

A major stumbling block in the acceptance of non-nutritive flow in skeletal muscle has been the anatomical identity of a discrete route associated with muscle that cannot be discriminated by microspheres and that essentially denies nutrient access but is capable of high flow. Vessels of muscle-associated connective tissue (septa and tendons) are possible candidates (2, 9) although not without controversy (10). Recently, our laboratory has measured flow to connective tissue in the constant-flow perfused rat hindlimb (20). Exposed tibial tendon vessels of the

Table 4.3   Calculations of Nutritive: Total Blood Flow Ratio in Skeletal Muscle at Rest

Muscle	Nutritive: Total (%)	Methods	Reference
Cat	60.1	Hydrogen washout	Nakamura et al. (1972)
Cat gastrocnemius	59.3	^{133}Xe clearance/ drop counter	Kjellmer et al. (1965)
Cat gastrocnemius	97.3	^{133}Xe clearance/ drop counter	Sejrsen & Tønnesen (1968)
Chicken ALD	42.0	^{133}Xe clearance from	Hudlická (1969)
Chicken PLD	61.0	intramuscular injection/ drop counter	Hudlická (1969)
Rabbit tenuissimus			
• low surrounding PO$_2$	54.0*	Intravital microscopy	Lindbom & Arfors (1984)
• high surrounding PO$_2$	5.0*	Intravital microscopy	Lindbom & Arfors (1984)
Dog sartorius	16.0	Local hydrogen clearance and microflow	Harrison et al. (1990)

*The authors conclude that the likely *in vivo* average is 80-90% (Lindbom & Arfors 1984). ALD, anterior latissimus dorsi, PLD, posterior latissimus dorsi.

biceps femoris muscle of the perfused leg were positioned either under a surface fluorometer probe to monitor signal strength when pulses of fluorescein isothiocyanate dextran were infused, or over the objective lens of an inverted microscope for photography when pulses of india ink were infused. Measurements were conducted under steady state with vehicle, norepinephrine (a Type A vasoconstrictor, table 4.1) or serotonin (a Type B vasoconstrictor, table 4.1) infusion. Norepinephrine increased perfusion pressure and oxygen uptake, but decreased fluorescence signal from the tendon vessels (figure 4.1). Photomicroscopy of the india ink-filled vessels confirmed that the tendon vessels had generally decreased in diameter. Serotonin, although increasing perfusion pressure like norepinephrine had quite the opposite effect on the other parameters. Oxygen uptake was decreased, and the fluorescence signal from the tendon vessels increased (figure 4.1). For serotonin, the vessels had clearly increased in diameter. Analysis of data for a range of concentrations of norepinephrine as well as serotonin showed that a reciprocal relationship exists between resting muscle metabolism (reflected by oxygen uptake) as controlled by vasoconstrictors and flow through muscle tendon vessels. Overall, such findings heighten the possibility that vessels supplying septa and tendons are the functional shunts, or the non-nutritive flow route, for muscle as proposed by others (2, 9) several years ago. In addition, the data of figure 4.1 would imply that flow through the tendon vessels does not cease even when nutritive flow is high. This is consistent with an earlier study from our laboratory (19) where serotonin was found to close off a vascular space without recruiting a new space.

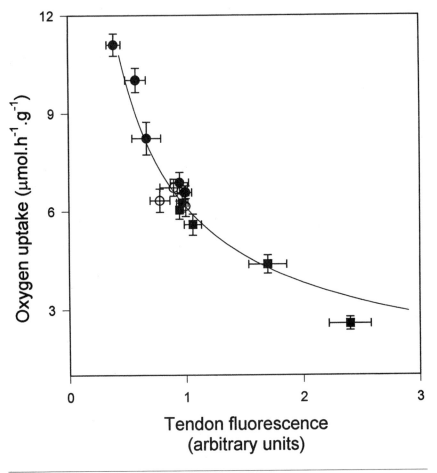

Figure 4.1.   Plot of resting muscle oxygen uptake as a function of tendon vessel flow. Experiments were conducted using the isolated constant-flow perfused rat hindlimb. Oxygen uptake was calculated from arteriovenous difference and flow. Tendon vessel flow was determined from fluorescence signal strength of infused fluorescein isothiocyanate labeled dextran (M_r 150,000) over the tibial tendon of the biceps femoris muscle. Variations to both parameters were achieved by infusing increasing doses of either norepinephrine (●) or serotonin (■). Basal conditions are represented by (○). Data are from Newman et al. (20).

Further evidence for nutritive and non-nutritive routes in muscle comes from studies where microsphere infusion has been used to selectively occlude either route in perfused hindlimb. In that study microspheres (11.9 ± 0.1 μm, mean $\pm$ SE diameter) were injected during angiotensin II (a Type A vasoconstrictor), serotonin or vehicle infusions and the effects on hindlimb (4.7 ± 0.1 g muscle) oxygen uptake and indices of energy status CrP/Cr, CrP/ATP and energy charge of the calf muscle group assessed. Quite unexpectedly, microsphere (1.5×10^6) injection during vehicle or serotonin infusion increased oxygen uptake ($P < 0.05$) and marginally increased (not significant) energy status but did not reduce venous outflow. During angiotensin infusion when oxygen uptake was stimulated, the same number of microspheres inhibited oxygen uptake and lowered CrP/ATP but again, did not reduce venous outflow. When all of the data were considered it was clear that change in oxygen uptake correlated positively with CrP/Cr ($r = 0.68$, $P < 0.0001$) but not EC ($r = 0.08$, $P = 0.59$) (figure 4.2). Additional microspheres ($> 3.0 \times 10^6$) progressively inhibited oxygen uptake, increased perfusion pressure, and reduced flow regardless of agonist present. Thus microsphere injection into the constant-flow perfused rat hindlimb reversed the stimulatory and inhibitory effect of the vasoconstrictors angiotensin and serotonin on metabolism, respectively, by further increasing pressure but with marginal effects on venous flow. We believe this further supports the notion of two circulatory systems in skeletal muscle that can be selectively blocked by 11.9 μm microspheres if the preexisting distribution of flow is itself selectively induced by specific vasoconstriction.

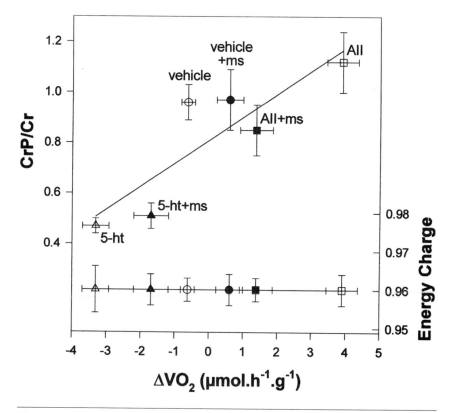

Figure 4.2.   Relationships between V̇O₂ and CrP/Cr and V̇O₂ and EC as affected by vasoconstrictors and microsphere infusions. Data (unpublished) are from constant-flow perfused rat hindlimb at 37°C. Additions were vehicle (O), 1 μM 5-HT (Δ), 5 nM AII (□), vehicle + 1.5 × 10⁶ 11 μm microspheres (●), 5-HT + microspheres (▲) and 5 nM AII + microspheres (■).

Shunts as Flow Reserves for Exercising Muscle

Ultimately, the question arises as to the significance of functional shunts (or non-nutritive flow routes) in muscle. It is likely that one of the roles of shunts is to provide a flow reserve and therefore potential amplification for nutrient delivery during exercise (11). For example, if the ratio of nutritive/total flow in muscle at rest is 50% (table 4.3) this affords an amplification factor approaching 2 when non-nutritive flow is reduced by mild sympathetic nervous system activity accompanying exercise. Combined with a threefold increase in cardiac output and thereby total blood flow to muscle, this would mean an overall amplification of sixfold. Of course, if the ratio of nutritive to total flow is 16%, as predicted by Harrison (11, table 4.3) then the amplification effect is approximately sixfold alone and eighteenfold when combined with a threefold increase in total blood flow to muscle.

In summary, recent new data confirm that functional shunts are associated with skeletal muscle, allowing high flow rates with little or no nutrient delivery. Anatomical candidates for these shunts may be vessels of the muscles' connective tissue. The proportion of flow distributed between capillaries (nutritive) and functional shunts appears to be tightly regulated by endocrine and neuroendocrine inputs and can control muscle metabolism and work performance by regulating nutrient delivery and product removal. A role for these shunts is likely to be in providing a flow reserve for amplifying nutrient delivery as muscle begins to work.

Acknowledgments

Our research was generously supported by grants from the Australian National Health and Medical Research Council, Australian Research Council, Diabetes Australia, and the National Heart Foundation of Australia.

References

1. Barcroft, H., and Dornhorst, A.C. The blood flow through the human calf during rhythmic exercise. *Journal of Physiology London* 10: 402-411; 1949.
2. Barlow, T.E., Haigh, A.L., and Walder, D.N. Evidence for two vascular pathways in skeletal muscle. *Clinical Science* 20: 367-385; 1961.
3. Brod, J., Hejl, Z., and Ulrych, M. Metabolic changes in the forearm muscle and skin during emotional muscular vasodilatation. *Clinical Science* 25:1-10; 1963.
4. Clark, M.G., Colquhoun, E.Q., Dora, K.A., Rattigan, S., Eldershaw, T.P.D., Hall, J.L., Matthias, A., and Ye, J.M. Resting muscle: A source of thermogenesis controlled by vasomodulators. In *Temperature Regulation: Recent Physiological and Pharmacological Advances*. Ed. A.S. Milton, 315-320. Basel: Birkhauser Verlag; 1994.

5. Clark, M.G., Colquhoun, E.Q., Rattigan, S., Dora, K.A., Eldershaw, T.P.D., Hall, J.L., and Ye, J. Vascular and endocrine control of muscle metabolism, *American Journal of Physiology* 268: E797-E812; 1995.
6. Colquhoun, E.Q., Hettiarachchi, M., Ye, J.M., Rattigan, S., and Clark, M.G. Inhibition by vasodilators of noradrenaline and vasoconstrictor-mediated, but not skeletal muscle contraction-induced oxygen uptake in the perfused rat hindlimb; implications for non-shivering thermogenesis in muscle tissue. *General Pharmacology* 21:141-148; 1990.
7. Dora, K.A., Rattigan, S., Colquhoun, E.Q., and Clark, M.G. Aerobic muscle contraction impaired by serotonin-mediated vasoconstriction. *Journal of Applied Physiology* 77:277-284; 1994.
8. Friedman, J.J. Total, non-nutritional and nutritional blood volume in isolated dog hindlimb. *American Journal of Physiology* 210: 151-156; 1966.
9. Grant, R.T., and Payling Wright, H. Anatomical basis for non-nutritive circulation in skeletal muscle exemplified by blood vessels of rat biceps femoris tendon. *Journal of Anatomy* 106: 125-133; 1970.
10. Hammersen, F. The terminal vascular bed in skeletal muscle with special regard to the problem of shunts. In *Capillary Permeability: The Transfer of Molecules and Ions Between Capillary Blood and Tissue*. Ed. C. Crone and N.A. Lassen, 351-365. Copenhagen: Munksgaard; 1970.
11. Harrison, D.K., Birkenhake, S., Knauf, S.K., and Kessler, M. Local oxygen supply and blood regulation in contracting muscle in dogs and rabbits. *Journal of Physiology* 422: 227-243; 1990.
12. Hirvonen, L., Korobkin, M., Sonnenschein, R.R., and Wright, D.L. Depression of contractile force of skeletal muscle by intra-arterial vasodilator drugs. *Circulation Research* 4: 525-535; 1964.
13. Hudlická, O. Resting and post contraction blood flow in slow and fast muscles of the chick during development. *Microvascular Research* 1: 390-402; 1969.
14. Hudlická, O. In *Muscle Blood Flow: Its Regulation to Muscle Metabolism and Function*. Amsterdam: Swets & Zeittinger B.V; 1973.
15. Kjellmer, I., Lindbjerg, I., Prevovský, I. & Tønnesen, H. The relation between blood flow in an isolated muscle measured with Xe-133 clearance and direct recording technique. *Acta Physiologica Scandinavica* 69: 69-78; 1967.
16. Kuznetsova, L.V., Tomasek, N., Sigurdsson, G.H., Banic, A., Erni, D., and Wheatley, A.M. Dissociation between volume blood flow and laser-Doppler signal from rat muscle during changes in vascular tone. *American Journal of Physiology* 274: H1248-H1254; 1998.
17. Lindbom, L., and Arfors, K.-E. Non-homogeneous blood distribution in the rabbit tenuissimus muscle. *Acta Physiologica Scandinavica* 122: 225-233; 1984.
18. Nakamura, T., Suzuki, T., Tsuiki, K., and Tominaga, S. Non-nutritional blood flow in skeletal muscle determined with hydrogen gas. *Tohoku Journal of Experimental Medicine* 106:135-145; 1972.
19. Newman, J.M.B., Dora, K.A., Rattigan, S., Edwards, S.J., Colquhoun, E.Q., and Clark, M.G. Norepinephrine and serotonin vasoconstriction in rat hindlimb control different vascular flow routes. *American Journal of Physiology* 270: E689-E699; 1996.

LIBRARY, UNIVERSITY OF CHESTER

20. Newman, J.M.B., Steen, J.T., and Clark, M.G. Vessels supplying septa and tendons as functional shunts in perfused rat hindlimb. *Microvascular Research* 54: 49-57; 1997.
21. Rattigan, S., Appleby, G.J., Miller, K.A., Steen, J.T., Dora, K.A., Colquhoun, E.Q., and Clark, M.G. Serotonin inhibition of 1-methylxanthine metabolism parallels its vasoconstrictor activity and inhibition of oxygen uptake in perfused rat hindlimb. *Acta Physiologica Scandinavica* 161: 161-169; 1997.
22. Rattigan, S., Dora, K.A., Tong, A.C.Y., and Clark, M.G. Perfused skeletal muscle contraction and metabolism improved by angiotensin II mediated vasoconstriction. *American Journal of Physiology*, 271: E96-E103; 1996.
23. Sejrsen, P., and Tønnensen, K.H. Inert gas diffusion method for measurement of blood flow using saturation techniques. *Circulation Research*, 22: 679-693; 1968.

CHAPTER 5

Neuroendocrine Regulation During Exercise

Michael Kjær

Copenhagen Muscle Research Centre and Sports Medicine Research Unit,
Bispebjerg Hospital, Copenhagen, Denmark

The exercise-induced rise in energy turnover results in increased metabolic activity both in skeletal muscle and liver and fat tissue, and changes in the neuroendocrine activity play an important role for this regulation. Changes in autonomic nervous activity and hormone secretion are coupled to both feedforward and feedback mechanisms, and an interaction between neuroendocrine activity and substrate mobilization and utilization is present during muscular contraction (figure 5.1). The present text will not cover a full overview of the field but rather pinpoint examples of models that can be used in humans to study the interaction between hormones and metabolism.

Neuroendocrine Responses

Motor Center Control—"Central Command"

During dynamic exercise, sympathoadrenergic activity increases, as determined by changes in electrical activity in superficial sympathetic nerves or changes in circulating catecholamines (16, 31). Arterial plasma concentrations of norepinephrine and epinephrine increase linearly with exercise duration and exponentially with intensity (16). In earlier experiments using tubocurarine administration in healthy subjects and thereby weakening skeletal muscle force development and presumably increasing the motor center activity needed to produce a given force output, it has been documented that catecholamine responses were higher when a given workload was performed during partial neuromuscular blockade as compared with control experiments with saline administration (13). This indicates that central motor activity can directly stimulate sympathoadrenergic activity during exercise in humans. These findings are supported by findings in paralyzed cats, in whom electrical stimulation of their subthalamic motor centers elicited a hormonal response similar to that seen in response to exercise in the awake, running cat (36).

Sympathetic activity during exercise results in several other hormonal responses. Epinephrine release from the adrenal medulla increases (12), insulin release from

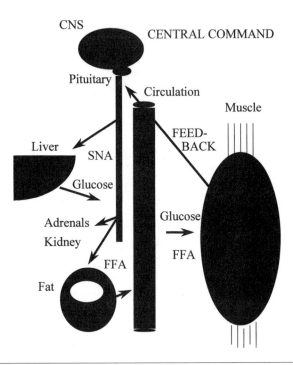

Figure 5.1.   Parts of the neuroendocrine system that are dealt with in the present text, with major emphasis on sympathoadrenergic activity. In addition to the topics present on this figure, pancreatic secretion of insulin and glucagon plays a major role in metabolic regulation during exercise.

the pancreas is inhibited (4, 10), and renin release from the kidneys is increased. The evidence for sympathetic activity being involved in the exercise-induced rise in plasma renin activity has been drawn from both animal experiments (44) and recently also from experiments in kidney-transplanted humans in whom the exercise-induced increase in renin activity was abolished compared with healthy control subjects (19).

Blood-borne Feedback and Afferent Reflex Mechanisms

In addition to central command, feedback mechanisms from contracting muscle also are active during exercise. From experiments with tubocurarine experiments it can be shown that when muscle force was reduced to around 40% MVC, even when individuals worked with maximal effort, the catecholamine responses still were markedly lower than responses in control experiments when maximal exercise was performed (5). Furthermore, during exercise with single muscle groups that worked maximally, only very small catecholamine responses were found (30).

The role of peripheral neural feedback from contracting muscle for the neuroendocrine response to exercise has been studied in man using lumbar epidural anesthesia, where doses of local anesthesia were administered, aimed at a specific block of the impulses in thin afferent nerves while maintaining impulses in larger motor nerves. For pituitary hormones adrenocorticotrophin (ACTH) and beta-endorphin, responses were abolished during submaximal exercise when epidural anesthesia was administered (14). The abolished response of beta-endorphin with epidural anesthesia also was present during hypoxic exercise (23). In animal experiments (cats), direct stimulation of afferent nerves from muscles increased plasma catecholamine and beta-endorphin levels and decreased plasma levels of insulin (38).

Another model to study feedback mechanisms is the use of patients with inborn errors of metabolism, e.g., myophosphorylase deficiency (McArdle's disease), who lack the ability to perform intramuscular glycogenolysis. It has been demonstrated that during exercise in these individuals, a higher than normal hormonal response as well as an increased mobilization of extramuscular fuels (especially glucose) compensated for the impaired glycogenolysis (37). As neither plasma glucose nor circulating FFA levels were lower during exercise in the McArdle patients compared with healthy control individuals, the higher hormonal response to exercise in the patients is most likely due to neural feedback from metabolism in working skeletal muscle. Interestingly, in patients with phosphofructokinase deficiency and with mitochondrial myopathy, an excessive neuroendocrine response and exaggerated mobilization of glucose was found during exercise (39, 40). Although these responses contribute to an augmented ATP synthesis, they also cause a progressive accumulation of nonoxidized substrates in the working muscle. The responses most likely are coupled to neural feedback and illustrate that neuroendocrine responses to exercise in this situation are related more closely to muscle oxidative demands and not so closely to the total oxidative capacity in working muscle.

Hormones and Metabolism

Muscle Glycogenolysis and Glucose Uptake

Although muscle contractions per se can increase glucose uptake, humoral factors can modify this markedly (29). It is well documented that insulin has a synergistic effect on glucose uptake with contractions (28), and epinephrine infusion in the running dog decreased glucose clearance (8). Furthermore, in humans intra-arterial infusion of epinephrine into one of two active legs during cycling resulted in a local decrease in glucose uptake (9). It is at present concluded that epinephrine decreases muscle glucose uptake and utilization by beta-adrenergic mechanisms and that these mechanisms involve increased lipolytic activity and muscle glycogenolysis (24). In a recent experiment using adrenalectomized individuals, it has been shown that the exercise-induced rise in glucose uptake during 45 min of bicycle exercise at 50% $\dot{V}O_2$max and 15 min at 85% $\dot{V}O_2$max was significantly decreased when these individuals performed the exercise bout with epinephrine infusion to substitute epinephrine levels during exercise (Howlett et al., unpublished observation). As plasma levels of blood glucose increased markedly during exercise with epinephrine infusion, the glucose clearance was even more diminished by epinephrine infusion than if data were expressed as glucose uptake. In accordance with the hypothesis that the decreased glucose uptake in experiments with epinephrine infusion was caused by an enhanced glycogenolysis, intramuscular concentrations of glucose-6-phosphate and glucose were higher in experiments with epinephrine than without, indicating a decreased activity of hexokinase. Furthermore, blood lactate rose more during exercise when epinephrine was infused in adrenalectomized individuals. However, somewhat surprising, muscle glycogen levels in vastus lateralis muscle did not diminish more during exercise with epinephrine infusion compared with both a control situation without infusion or healthy control subjects with intact adrenal glands. This is similar to findings in a study where epinephrine was infused in physiological doses and where phosphorylase a fraction was increased during exercise whereas no significant effect could be demonstrated on glycogen breakdown rates (2). Earlier human studies have shown an enhanced muscle glycogenolysis associated with the increase in phosphorylase a fraction during aerobic exercise with epinephrine infusion, although it has to be acknowledged that in these studies supraphysiological concentrations of epinephrine were obtained in relation to the exercise intensity studied (9, 34). Therefore, it has to be concluded that epinephrine stimulates glycogenolysis in human exercising muscle, but that human studies do not always show a significant effect on the glycogen breakdown as determined from muscle biopsies. In spinal cord patients with impaired sympathetic control of the adrenal medulla, and therefore no rise in plasma epinephrine during electrical induced involuntary exercise, glycogen breakdown was pronounced and similar to that seen in control subjects during electrical exercise where plasma epinephrine rose fourfold (20, 21). This indicates that although epinephrine stimulates glycogenolysis in muscle, glycogen breakdown is present even in the absence of any exercise-induced increase in epinephrine.

Hepatic Glucose Production

Pancreatic hormones repeatedly have been shown to be involved in the exercise-induced release of glucose from the liver (41). In exercising dogs, a decrease in insulin and an increase in glucagon is important for the exercise-induced rise in hepatic glucose production. Primarily the effect on glycogenolysis is exerted by increased levels of glucagon and diminished insulin, but during prolonged exercise gluconeogenesis also was stimulated, mainly as a result of increased amounts of gluconeogenic precursor delivery, which is most likely to be a result of an increasing concentration of glucagon (42). Although this may be the case in dogs, in humans glucagon does not increase until exercise has been going on for a prolonged period, and in experiments, in which plasma insulin and glucagon levels were maintained at constant levels, hepatic glucose production still increased (6, 43). This indicates that other factors are involved, and several authors have found a correlation between exercise-induced increase in catecholamines and splanchnic glucose output (15, 25, 32). Also in exercising dogs and rats, epinephrine may play some role in regulation of hepatic glucose production during intense or prolonged exercise (26, 33). In humans the role of epinephrine and sympathetic liver nerves has been addressed by application of local anesthesia of the sympathetic coeliac ganglion innervating the liver, the pancreas, and the adrenal medulla. Pancreatic hormones were controlled both at rest and during exercise by infusion of somatostatin and pancreatic hormones (17). During coeliac blockade, the exercise-induced epinephrine response was reduced by 40–90% without any diminished response of glucose production during contraction. This indicates that neither sympathetic liver nerves nor the circulating catecholamines are responsible for the exercise induced rise in splanchnic glucose output. Further support for this hypothesis comes from experiments in liver-transplanted patients in whom biopsy-determined norepinephrine content in liver tissue indicated a very low amount of norepinephrine, supporting that liver denervation was present (18). The exercise-induced increase in glucose production in liver-transplanted patients was identical to that of healthy control individuals and that of kidney-transplanted individuals matched for intake of medication (19). Furthermore, recent studies using high intensity of exercise in dogs have concluded that catecholamines do not participate in the stimulation of glucose production (3). Although hepatic glucose production is not stimulated directly by the sympathetic nervous system, potentially indirect pathways could be active. In a recent study, the importance of the renin-angiotensin axis was investigated. Administration of an angiotensin converting enzyme inhibitor did not influence the exercise-induced increase in hepatic glucose production, indicating that the renin-angiotensin system is not important for the rise in hepatic glucose production during exercise (22).

In addition to the hormonal influence on glucose production, some unidentified factor produced by contracting muscle may also directly increase hepatic glucose production during exercise. In support of this, when rat liver was perfused with an extract from contracting muscle, hepatic glucose production increased (7).

Fat Metabolism

Mobilization of FFA from triglyceride stored in adipose tissue is increased during exercise in parallel with increases in plasma glycerol and catecholamines as studied by microdialysis of subcutaneous abdominal tissue (1). As the addition of a nonselective beta-adrenoceptor blocker almost completely prevented the exercise-induced increase in the dialysate level of glycerol, this suggests that exercise-induced lipid mobilization is coupled to an increased sympathoadrenergic activity (1). The relative role of increased sympathetic nervous activity and circulating norepinephrine/epinephrine, respectively, in regulation of lipolysis, is, however, is not solved. Infusion of epinephrine at rest repeatedly has shown an increased lipolytic activity, an effect that can be desensitized by a previous increase in epinephrine (35). Recently, the role of direct sympathetic nerves to adipose tissue has been adressed during dynamic exercise. Spinal cord injured individuals (SCI) are known to have disrupted sympathetic nerve traffic below the spinal lesion level, and when the injury is located above the T5 level the functional block includes the adrenal medulla. In SCI who had an injury level of T2-T4 the segmental level of sympathetic output was determined during exercise by the use of ion-starch to indicate the cutaneous area in which sweating was intact or absent. Microdialysis of subcutaneous fat tissue was undertaken in the clavical, and the umbilical area and results were compared with matched controls. It was found that during voluntary arm cycling at 50–60% $\dot{V}O_2$max for 1 hour, the glycerol output increased more in areas with than without sympathetic innervation (Stallknecht et al., unpublished observations). This supports findings in a recent study where an increase in umbilical glycerol release was absent in SCI but present in healthy control subjects during handgrip exercise (11). It has to be noted, however, that the authors were unable to detect any significant increase in lipolysis at the clavicular level in any of the groups with this very moderate type of stress. Our recent findings point to an important role for direct sympathetic nerves to adipose tissue for lipolysis during exercise. In addition to the role of sympathetic nerves it cannot be excluded that circulating epinephrine can modify lipolysis during exercise, as the increase in circulating glycerol during exercise in adrenolectomized subjects was higher when epinephrine was infused to restore normal increments in epinephrine seen during exercise (Howlett et al., unpublished observations).

Regulation of triglyceride metabolism in exercise includes not only lipoprotein lipases but also a hormone-sensitive lipase (HSL) (27). HSL has been identified in adipose tissue and recently also in skeletal muscle, and it is likely that HSL is partly responsible for the mobilization of intramuscular triglyceride during muscular activity. Several hormones (catecholamines, growth hormone, and cortisol) have been suggested to activate HSL in the resting organism. Electrically induced contractions in rat muscle have been shown to cause a transient increase in HSL activity within the first minutes of contraction, and it has been demonstrated that epinephrine stimulates HSL activity potently in incubated rat soleus muscle (Langfort et al., unpublished observations). In adrenalectomized humans, HSL activity recently has been shown not to increase during 60 min of exercise at 50–80% $\dot{V}O_2$max, whereas a marked increase in HSL was observed in both matched control individuals who had a normal epinephrine response to exercise, as well as in the adrenalectomized persons who received an epinephrine infusion during cycling that matched the increase in plasma epinephrine seen in controls (Kjær et al., unpublished observations). Interestingly, when comparing responses for HSL and those of glycogen phosphorylase activity in adrenalectomized individuals, epinephrine infusion caused a parallel increase in these two parameters. This indicates that muscular activity resulting in an increase in epinephrine simultaneously activates breakdown of triglyceride and glycogen in a feedforward manner, and that the choice of substrate for combustion during muscular work must take place at another cellular level.

References

1. Arner, P., Kriegholm, E., Engfeldt, P., and Bolinder, J. Adrenergic regulation of lipolysis in situ at rest and during exercise. *J. Clin. Invest.* 85: 893-898; 1990.

2. Chesley, A., Hultman, E., and Spriet, L.L. Effects of epinephrine infusion on muscle glycogenolysis during intense aerobic exercise. *Am. J. Physiol.* 268: E127-E134; 1995.

3. Coker, R.H., Krishna, M.G., Lacy, D.B., Bracy, D.P., and Wasserman, D.H. Effect of selective hepatic adrenergic blockade on glucose production during heavy exercise in dogs. *Med. Sci. Sports Ex.* 29: S44; 1997.

4. Galbo, H., Christensen, N.J., and Holst, J.J. Catecholamines and pancreatic hormones during autonomic blockade in exercising man. *Acta Physiol. Scand.* 101: 428-437; 1977.

5. Galbo, H., Kjær, M., and Secher, N.H. Cardiovascular, ventilatory and catecholamine responses to maximal dynamic exercise in partially curarized man. *J. Physiol.* 389: 557-568; 1987.

6. Hoelzer, D.R., Dalsky, G.P., Clutter, W.E., Shah, S.D., Holloszy, J.O., and Cryer, P.E. Glucoregulation during exercise: hypoglycemia is prevented by redundant glucoregulatory systems, sympathocromaffin activation and changes in islet hormone secretion. *J. Clin. Invest.* 77: 212-221; 1986.

7. Hua, X. Studies on the regulation of exercise induced muscular uptake and hepatic production of glucose. Ph.D. thesis, University of Copenhagen, Denmark; 1993.

8. Issekutz, B. Effect of epinephrine on carbohydrate metabolism in exercising dogs. *Metabolism* 34: 457-464; 1985.

9. Jansson, E., Hjemdahl, P., and Kaijser, L. Epinephrine-induced changes in muscle carbohydrate metabolism during exercise in male subjects. *J. Appl. Physiol.* 60: 1466-1470; 1986.

10. Karlsson, A.K., Elam, M., Friberg, P., Biering-Sørensen, F., Sullivan, L., and Lønnroth, P. Regulation of lipolysis by the sympathetic nervous system: a microdialysis study in normal and spinal cord injured subjects. *Metabolism* 46: 388-394; 1997.

11. Karlsson, S., and Ahren, B. Insulin and glucagon secretion in swimming mice: effects of autonomic receptor antagonism. *Metabolism* 39: 724-732; 1990.

12. Kjær, M., Christensen, N.J., Sonne, B., Richter, E.A., and Galbo, H. Effect of exercise on epinephrine turnover in trained and untrained male subjects. *J. Appl. Physiol.* 59: 1061-1067; 1985.

13. Kjær, M., Secher, N.H., Bach, F.W., and Galbo, H. Role of motor center activity for hormonal changes and substrate mobilization in exercising man. *Am. J. Physiol.* 253: R687-R695; 1987.

14. Kjær, M., Secher, N.H., Bach, F.W., Sheikh, S., and Galbo, H. Hormonal and metabolic responses to exercise in humans: effect of sensory nervous blockade. *Am. J. Physiol.* 257: E95-E101; 1989.

15. Kjær, M., Kiens, B., Hargreaves, M, and Richter, E.A. Influence of active muscle mass on glucose homeostasis during exercise in humans. *J. Appl. Physiol.* 71: 552-557; 1991.

16. Kjær, M. Regulation of hormonal and metabolic responses during exercise in humans. *Exercise and Sport Science Reviews.* Ed. J.O. Holloszy, 20:161-184; 1992.

17. Kjær, M., Engfred, K., Fernandes, A., Secher, N.H., and Galbo, H. Regulation of hepatic glucose production during exercise in man: role of sympathoadrenergic activity. *Am. J. Physiol.* 265: E275-E283; 1993.

18. Kjær, M., Jurlander, J., Keiding, S., Galbo, H., Kirkegaard, P., and Hage, E. No reinnervation of hepatic sympathetic nerves after liver transplantation in human subjects. *J. Hepatol.* 20: 97-100; 1994.

19. Kjær, M., Keiding, S., Engfred, K., Rasmussen, K., Sonne, B., Kirkegård, P., Galbo, H. Glucose homeostasis during exercise in humans with a liver or kidney transplant. *Am. J. Physiol.* 268: E636-E644; 1995.

20. Kjær, M., Pollack, S.F., Mohr, T., Weiss, H., Gleim, G.W., Bach, F.W., Nicolaisen, T., Galbo, H., and Ragnarsson, K.T. Regulation of glucose turnover and hormonal responses during exercise: electrical induced cycling in tetraplegic humans. *Am. J. Physiol.* 2711: R191-R199; 1996.

21. Kjær, M., Secher, N.H, Bangsbo, J., Perko, G., Horn, A., Mohr, T., and Galbo, H. Neural regulation of hormonal and metabolic responses to exercise studied by electrically induced cycling during epidural anesthesia. *J. Appl. Physiol.* 80: 2156-2162; 1996.

22. Kjær, M., Bergeron, R., Bülow, J., Rørdam, L., Skovgaard, D., Howlett, K., and Galbo, H. Splanchnic glucose production and blood flow during exercise: the renin-angiotensin system. *Med. Sci. Sports Ex.* 29: S43; 1997.

23. Klokker, M., Kjær, M., Secher, N.H., Hanel, B., Worm, L., Kappel, M., and Pedersen, B.K. Natural killer cell response to exercise in humans: effect of hypoxia and epidural anesthesia. *J. Appl. Physiol.* 78: 709-716; 1995.

24. Marker, J.C., Hirsch, I.B., Smith, L.J., Parvin, C.A., Holloszy, J.O., and Cryer, P.E. Catecholamines in the prevention of hypoglycemia during exercise in humans. *Am. J. Physiol.* 260: E705-E712; 1991.

25. Marliss, E.B., Simantirakis, E., Miles, P.D.G., Purnon, C., Gougeon, R., Field, C.J., Halter, J.B., and Vranic, M. Glucoregulatory and hormonal responses to repeated bouts of intense exercise in normal male subjects. *J. Appl. Physiol.* 71: 924-933; 1991.

26. Moates, J.M., Lacy, D.B., Goldstein, R.E., Cherrington, A.D., and Wasserman, D.H. The metabolic role of the exercise induced increment in epinephrine in the dog. *Am. J. Physiol.* 255: E428-E436; 1988.

27. Oscai, L.B., Essig, D.A., and Palmer, W.K. Lipase regulation of muscle triglyceride hydrolysis. *J. Appl. Physiol.* 69: 1571-1577; 1990.

28. Ploug, T., Galbo, H., and Richter, E.A. Increased muscle glucose uptake during contractions: no need for insulin. *Am. J. Physiol.* 247: E726-E731; 1984.

29. Richter, E.A. Glucose utilization. In *Handbook of Physiology 1997.* Ed. L.B. Rowell and J.T. Shepherd, 912-951. Columbia, MD: Bermedica Production; 1996.

30. Savard, G., Richter, E.A., Strange, S., Kiens, B., Christensen, N.J., and Saltin, B. Norepinephrine spillover from skeletal muscle during exercise in humans: role of muscle mass. *Am. J. Physiol.* 257: H1812-H1818; 1989.

31. Seals, D.R., Victor, R.G., and Mark, A.L. Plasma norepinephrine and muscle sympathetic discharge during rhythmic exercise in humans. *J. Appl. Physiol.* 65: 940-944; 1988

32. Sigal, R., Fisher, S.F., Halter, J.B., Vranic, M., and Marliss, E.B. The roles of catecholamines in glucoregulation in intense exercise as defined by the islet cell clamp technique. *Diabetes* 45:148-156; 1996.

33. Sonne, B., Mikines, K.J., Richter, E.A., Christensen, N.J., and Galbo, H. Role of liver nerves and adrenal medulla in glucose turnover of running rats. *J. Appl. Physiol* 59: 1640-1646; 1985.
34. Spriet, L.L., Ren, J.M., and Hultman, E. Epinephrine infusion enhances glycogenolysis during prolonged electrical stimulation. *J. Appl. Physiol.* 64: 1439-1444; 1988.
35. Stallknecht, B., Bülow, J., Frandsen, E., and Galbo, H. Desensitization of human adipose tissue to adrenaline stimulation studied by microdialysis. *J. Physiol.* 500: 271-282; 1997.
36. Vissing, J., Iwamoto, G.A., Rybicki, K.J., Galbo, H., and Mitchell, J.H. Mobilization of glucoregulatory hormones and glucose by hypothalamic locomotor centers. *Am. J. Physiol.* 257: E722-E728; 1989.
37. Vissing, J., Lewis, S.F., Galbo, H., and Haller, R.G. Effect of deficient muscular glycogenolysis on extramuscular fuel production in exercise. *J. Appl. Physiol.* 72: 1773-1779; 1992.
38. Vissing, J., Iwamoto, G.A., Fuchs, I.E., Galbo, H., and Mitchell, J.H. Reflex control of glucoregulatory exercise responses by group III and IV muscle afferents. *Am. J. Physiol.* 266: R824-R830; 1994.
39. Vissing, J., Galbo, H., and Haller, R. Paradoxically enhanced glucose production during exercise in humans with blocked glycolysis due to muscle phosphofructokinase deficiency. *Neurology* 47: 766-771; 1996.
40. Vissing, J., Galbo, H., and Haller, R.G. Exercise fuel mobilization in mitochondrial myopathy: a metabolic dilemma. *Ann. Neurol.* 40: 655-662; 1996.
41. Vranic, M., Kawamori, R., Pek, S., Kovacevic, N., and Wrenshall G. The essentiality of insulin and the role of glucagon in regulating glucose utilization and production during strenuous exercise in dogs. *J. Clin. Invest.* 57: 245-256; 1976.
42. Wasserman, D.H., and Cherrington, A.D. Regulation of extramuscular fuel sources during exercise. In *Handbook of Physiology.*, Ed. L.B. Rowell and J.T. Shepherd, 1036-1074. Columbia, MD: Bermedica Production; 1996.
43. Wolfe, R.R., Nadel, E.R., Shaw, J.H.F., Stephenson, L.A., and Wolfe, M.H. Role of changes in insulin and glucagon in glucose homeostasis in exercise. *J. Clin. Invest.* 77: 900-907; 1986.
44. Zambraski, E.J., Tucker, M.S., Lakas, C.S., Grassl, S.M., and Scanes, C.G. Mechanism of renin release in exercising dog. *Am. J. Physiol.* E71-E78; 1984.

CHAPTER 6

Muscle Membranes, Diet, and Exercise

Jørn W. Helge and Len H. Storlien

Copenhagen Muscle Research Centre, August Krogh Institute, Denmark,
and Metabolic Research Centre, Dept. Biomedical Science,
University of Wollongong, Australia

Skeletal muscle plays a central role in whole-body energy metabolism. It contains significant stores of energy both as glycogen and as triacylglycerol, and it is the major tissue for exercise glucose and lipid utilization. This paper focuses on the possible role of the membrane lipid composition on various functions that are implicated in muscle homeostasis and fuel utilization. A schematic overview is provided in figure 6.1. In the case of dietary modification of membrane lipids and subsequent effects on fuel utilization there is an advanced literature. However, the interactions between diet, acute exercise, and regular exercise training on muscle membrane lipid composition and, in turn, performance is at an embryonic stage.

Membrane Lipids

The currently accepted view of cell membranes is that they exist as a lipid bilayer that provides a flexible structure in which are located proteins important for, among other things, transport of nutrients and ions into and out of cells and cell organelles. The physical properties of the membrane are greatly influenced by the fatty acid composition of lipids in the bilayer. The carbon to carbon bonds in saturated fats are unconstrained. Consequently, packing of phospholipids with mainly saturated fats is tight, and the membrane is relatively rigid. As double bonds are introduced, flexibility of the carbon-carbon bonds is restricted, and packing is less dense. This increase in membrane fluidity is most pronounced with the introduction of one, two, or three double bonds (29). As well as the fatty acid composition, the ratio of phospholipid to cholesterol and the nature of the proteins contained in the lipid bilayer will determine the fluidity of the membrane. Fluidity is a global membrane feature, but it is equally true that the fatty acid composition of the membrane is likely to determine other features of membrane and cellular function, including the expression and activity of membrane proteins that control many aspects of cellular energetics.

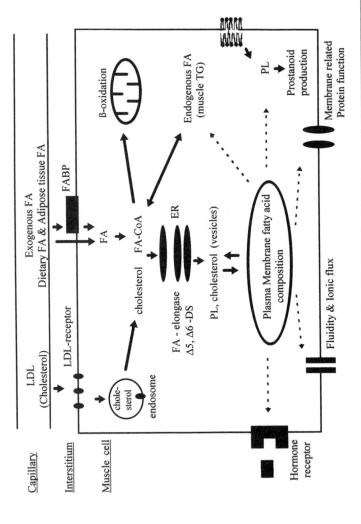

Figure 6.1.   A schematic overview showing pathways leading to the incorporation of cholesterol and fatty acids into muscle plasma membrane (solid lines). In addition, sites are indicated where muscle membrane fatty acid composition have some of their effects (dotted lines) on muscle cell function and homeostasis.

Abbreviations: PL: phospholipids; ER: endoplasmatic reticulum; FA: fatty acid; Δ5, Δ6 -DS: Δ5, Δ6 - fatty acid desaturase enzymes; FA-elongase: fatty acid elongase enzyme; FABP: fatty acid binding proteins.

The regulation of membrane lipid profile is not entirely clear, however dietary fatty acid supply and some obvious steps in the transport and supply of precursors are very important in the regulation of fatty acid composition. (The effect of dietary fatty acid supply will be discussed in the following section.) One step is the fatty acid binding proteins (FABP), which are suggested to be important for the cellular uptake and transport of free fatty acids (10). There is a family of FABPs located in the plasma membrane and the cytosol, and the tissue distribution of these subtypes may favor uptake and intracellular transport of certain fatty acids. Regulatory steps also exist in the conversion of fatty acids to CoA derivatives and their carbon backbone elongation and desaturation. Here one must remember that the Unsaturation Index (UI— an overall measure of degree of fatty acid desaturation, calculated as the number of double bonds per 100 fatty acyl groups) of an average "Western" diet is about 80 (with a preponderance of 18 carbon fatty acids), while the UI of skeletal muscle phospholipids is in the range of 165-170 (see 4). Thus a substantial part of the exogenous fatty acids must be desaturated and elongated before they are incorporated into muscle structural lipid, which emphasizes the importance of the Δ6 and Δ5 fatty acid desaturase and the fatty acid elongase activities. To this point, it is interesting that the apparent activity of the Δ5 fatty acid desaturase enzyme is closely linked with insulin action over a range of human populations (4, 35).

Role of Diet in Altering Membrane Fatty Acid Composition

There is now clear evidence that the fatty acid profile of the diet can significantly change the fatty acid composition of skeletal muscle membrane structural lipids. The evidence is very strong in experimental animals, particularly rodents, which have been well studied. Tight control of the fatty acid profile of the diet is simple in experimental animals, and turnover is rapid. Significant changes in muscle membrane fatty acid composition occur within days, and remodeling approaches plateau within three to four weeks (Pan and Storlien, unpublished observations). An important point is that incorporation of specific classes of fatty acids is strongly influenced by the overall dietary fatty acid profile. This is clearly illustrated in figure 6.2 (also referred to below in relation to effects on insulin action). The two groups indicated by arrows contained identical amounts and types of n-3 fatty acids. However, the diet of the group to the left also contained large amounts of n-6 polyunsaturated fatty acids that are known to compete for enzymes of desaturation and elongation, but based on the evidence in figure 6.2, also clearly compete for incorporation into muscle membrane structural lipid. It is also worth noting that the group whose mean percentage of 20 plus 22-carbon n-3s was highest had a diet almost devoid of those long-chain n-3 fatty acids. It did, however, have high levels of α-linolenic 18:3 n-3 which, in the rat, is rapidly desaturated and elongated before incorporation into phospholipid. That skeletal muscle itself has such capacity is demonstrated in studies on L6 myocytes (from a murine muscle cell line) that were incubated in medium containing α-linolenic n-3 fatty acid. Some α-linolenic is seen to be incorporated in the phospholipid, but what is striking is the increase in

20:5 and 22:6 n-3, reflecting both desaturation by Δ6 and Δ5 fatty acid desaturase and elongation by the fatty acid elongase enzyme occurring within the myocytes themselves (Pan and Storlien, unpublished observations). Data from humans are less easy to obtain, as control of human dietary fatty acid composition is difficult. However, a recent study in infants has provided interesting data (2). Skeletal muscle biopsies were obtained from solely breast-fed or solely formula-fed infants undergoing surgery. These were otherwise healthy infants, and the surgery was of a minor corrective nature. Breast milk contains small but significant amounts of 20 and 22 carbon PUFAs (3). In contrast, the formula diets consumed by that group of infants had no fatty acids of carbon length greater than 18. There were substantial differences in skeletal muscle phospholipid fatty acid composition between breast- and formula-fed infants with the former having 97% more docosahexaenoic 22:6 n-3 fatty acid and overall 21% more 20 and 22 carbon fatty acids. This is a clear demonstration in humans of the effect of diet on skeletal muscle membrane structural lipid fatty acid composition.

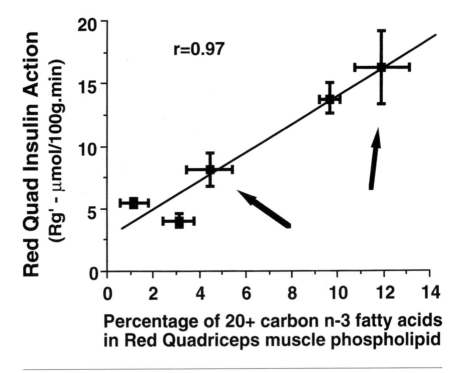

Figure 6.2.   Relationship between insulin-stimulated glucose metabolism (Rg') and percentage of long-chain (20 or more carbon) n-3 fatty acids in the major membrane structural lipid, phospholipid, in red quadriceps (Red Quad) hindlimb muscle (adapted from Storlien et al.[45]). The arrowed groups were fed diets that contained identical amounts and type of n-3 fatty acids but with a different overall dietary lipid profile.

Reproduced with permission of the American Diabetes Association, Inc.

Having evidence that dietary lipid profile does markedly change skeletal muscle membranes, does that change have physiologically relevant effects? The following discussion will focus only on the issues relevant to muscle morphology and fuel supply and in particular as variables linkable to exercise and muscle performance.

Role of Dietary Fatty Acid Profile in Altering Fuel Supply

Factors that influence the balance between fats and carbohydrates as metabolic fuels play a major role in exercise endurance capacity, as well as in propensity to diseases such as NIDDM and obesity. Dietary lipid profile has the capacity to strongly influence fat/carbohydrate balance, although the mechanisms as yet are not completely understood. While a full review of this aspect of the issue is beyond this paper, following are some relevant examples. Changing the fatty acid profile of the diet toward a more unsaturated profile is capable of inducing a fourfold increase in insulin-stimulated glucose uptake in rats without changing the amount of total GLUT4 protein (the insulin regulable glucose transporter) (49). This suggests that these diet-induced changes in membrane lipids might affect glucose transporter intrinsic activity, a concept for which there is now preliminary support (Storlien, unpublished observations). Similarly, there is now evidence that the intrinsic activity of ion transport molecules such as the Na+K+ pump is dependent upon the lipid milieu in which the pump exists (Dr. Paul Else, personal communication). The fatty acid profile of the diet also has a profound effect on insulin action. In studies where rats were fed isocaloric diets identical in all respects except the fatty acid composition of the lipid component, major differences in insulin action were found (45). As can be seen from figure 6.2, those differences in insulin action in individual muscles were closely related to the incorporation of long-chain, highly unsaturated n-3 fatty acids into the membrane phospholipid of those muscles. The design of those studies emphasized dietary inclusion of polyunsaturated n-3 fatty acids, so it is not clear whether the relationships are specific to n-3 fatty acids or are related to unsaturation in general. However, other work in humans has suggested, that among polyunsaturated fatty acids (PUFAs) a high n-6/n-3 ratio is deleterious for insulin action (40, 46).

Dietary lipid profile also exerts its effect on fat/carbohydrate balance through a direct effect on the fat supply, which is a function of both exogenous (dietary) and endogenous (lipogenesis) sources. Feeding rats a highly saturated fat diet for eight weeks has been shown to decrease lipoprotein lipase activity in both fasted and fed rats in a range of tissues including soleus muscle (27). In relation to the latter, n-3 fats have long been known to reduce circulating plasma triacylglycerol and VLDL-cholesterol levels in man (15, 38), an effect that could relate to above mentioned increased endogenous lipase activity and/or to a direct suppression of hepatic lipogenesis (50), the latter perhaps via control at the level of gene expression of enzymes such as fatty acid synthase (6, 21). Also adrenergic receptor binding is altered in parallel with the P/S ratio and membrane fluidity changes resulting from feeding either a highly saturated or highly polyunsaturated diet to rats (26, 28). This would have the effect of altering both sympathetic nervous system-driven lipolytic activity, which potentially could influence the supply of lipids, and uptake of metabolic fuels during exercise. In addition to the effect on circulating lipid supply, the impact of dietary fat profile can be seen in different levels of stored muscle triacylglycerol in rats, where the more saturated fats induce an increased storage (45). In man, there is evidence of a relationship between muscle phospholipid fatty acid composition and the accumulation of storage triacylglycerol in the Pima Indians (34). The more saturated the membrane lipids, the more triacylglycerol. In this population the triacylglycerol is related to insulin resistance, but any causal nature of the relationship between membrane lipids and triacylglycerol storage has not been demonstrated. In this regard an important issue may be the location of those intramuscular triacylglycerols which appears to be influenced by exercise training (18, Helge & Storlien, unpublished observations). The location of intramuscular triacylglycerol in relation to mitochondria and, in turn, the relative ability of this lipid to be used for fuel is undoubtedly a major issue for the etiology of disease as well as in the area of exercise metabolism.

Energy Storage: Relation to Fatty Acid Profile and Fiber Types

On the issue of overall energy balance, there is evidence from experimental animals that the predisposition to obesity following exposure to any particular high-fat diet appears to be markedly influenced by the fatty acid profile of that diet (12, 37, 41 and see 33 for a recent review). These effects have some support from human studies (7, 19) with possible mechanisms including alterations of sympathetic nervous system activity (26) and/or changes in membrane fluidity and/or ion/proton "leakiness" (5, 8). However, evidence also links a relatively high proportion of Type 2b fibers in humans with insulin resistance and obesity (17, 22, 25, 32, 48). This makes intrinsic sense as fat balance is a major factor in development of insulin resistance and obesity (9), and individuals with a high proportion of Type 2b fibers will have limited capacity in the skeletal musculature for using fats as fuel. Recently, it has been shown that the phospholipid fatty acid composition differs between muscles of differing fiber types (23). There was an increased proportion of unsaturated fatty acids, particularly 20 and 22 carbon n-3 fatty acids, in the more insulin-sensitive Type 1 and 2a fiber-predominant muscles compared with a muscle with mainly Type 2b fibers. Similar relationships have been observed in humans (22). Thus muscle fiber type composition and muscle membrane fatty acid composition are interacting factors which, in conjunction with dietary fatty acid profile, affects whole-body energy balance and substrate utilization.

Exercise training over extended periods in both experimental animals and humans has been shown to be capable of altering the proportions of fiber types in muscles, and it is thus possible that this would affect muscle membrane fatty acid composition. The most easily achievable transition is from Type 2b to Type 2a with a further limited capacity for a Type 2a to Type 1 transition (1, 14, 23). In female rats spontaneously exercising in a running wheel more than 45 days, highly significant changes in muscle fiber types were achieved in the extensor digitorum longus (EDL), a mixed hindlimb muscle (23). The percentage of Type 1 fibers significantly increased from 2.9 to 5.7% and Type 2a from 46.8 to 55.7% with a commensurate reduction in Type 2b fibers. This major change in fiber mix was not accompanied by a significant change in muscle membrane phospholipid fatty acid composition (23), which suggests that exercise-induced transitions in muscle fiber types are not of a magnitude sufficient to alter muscle membrane fatty acid composition. However, preliminary evidence, from nerve cross-innervation studies, suggest the possibility of a coordinate regulation of fiber type and membrane lipid composition in skeletal muscle (Kriketos et al., unpublished observations).

Role of Exercise in Altering Membrane Fatty Acid Composition

Regular exercise modifies substrate oxidation toward a higher fat oxidation during exercise. In addition there is cumulating evidence both at rest and during exercise to suggest that some (particularly unsaturated) fatty acids are more readily oxidized for fuel following ingestion in both rats (24, 36) and man (11, 20), and that more unsaturated fats are more easily mobilized by lipolytic stimuli and hence available for use as energy (13, 31, 39). Thus there is reason to suggest that exercise could exert an effect on muscle membrane fatty acid composition through a selective supply and depletion of the fatty acid-CoA precursor pool.

Although limited, there is some evidence to suggest that regular training also might influence muscle phospholipid fatty acid composition. A small but significant decrease in the content of palmitate and a borderline significant increase in the sum of C18-C20 was observed when endurance-trained and sedentary males were compared (47), however the lack of a strict dietary control necessitates some caution. Recent work by Sjodin and colleagues demonstrated that six weeks of low-intensity exercise training resulted in significant changes in muscle phospholipid fatty acid composition, but not in muscle TG fatty acid composition (43). Furthermore, preliminary data obtained from vastus lateralis in untrained males after four weeks' dietary adaptation (B. Kiens, E. Skytte, V. Ainsworth-Zink, and J.W. Helge, unpublished observations) and from red vastus lateralis in male rats subjected to four weeks adaptation to diet suggest that regular endurance training does induce significant changes in muscle phospholipid composition (16), which at least in rats were independent of diet. In addition to the observed effect on muscle phospholipid membrane fatty acid composition, regular training also has been demonstrated to decrease muscle membrane cholesterol content (30). This finding is supported in a cross-sectional study, where a significantly lower membrane cholesterol content was demonstrated in endurance-trained compared to sedentary males (47). Regular exercise decreases plasma cholesterol content at rest, however it remains to be addressed whether this has any influence on membrane cholesterol content. Evidence from studies in cell cultures indicates that major changes in cholesterol content does affect the fatty acid composition, whereas the converse effect does not occur (44).

Future Directions

The ensuing question is obviously whether performance is influenced by the dietary fatty acid composition. So far there are no in vivo studies that directly demonstrate performance-enhancing effects of supplying different fatty acids; on the contrary, we demonstrated that endurance performance in trained and sedentary rats was similar after consumption of two high-fat diets containing predominantly either monounsaturated or saturated fatty acids (authors, unpublished observations). However, it is clear that the changes in fat/carbohydrate balance and utilization and the capacity to maintain homeostasis could be of crucial significance for endurance performance. At present we are still at an embryonic stage in our understanding of dietary and possibly exercise-induced modification of individual cell membranes and subsequently cell organelles or indeed modification of particular phospholipid subclasses and the related issue of asymmetrical changes in the lipid bilayer. Finally, the role of, and interaction among, various other membrane constituents such as cholesterol, cardiolipin, sphingolipids, and the phospholipid fatty acid composition in determining membrane fluidity as well as providing an optimal anchoring for membrane proteins is an exciting new area of further investigation (42).

Acknowledgments

The authors' experimental work referenced herein was supported by NHMRC Australia and the Danish Research Academy (V940098 & 970501-10).

References

1. Adams, G.R., Hather, B.M., Baldwin, K.M., and Dudley, G.A. Skeletal muscle myosin heavy chain composition and resistance training. *Journal of Applied Physiology* 74: 911-915; 1993.
2. Baur, L.A., O'Connor, J., Pan, D.A., Kriketos, A.D., and Storlien, L.H. The fatty acid composition of skeletal muscle membrane phospholipid: its relationship with the type of feeding and plasma glucose levels in young children. *Metabolism* 47: 106-112; 1998.
3. Bitman, J., Wood, L., Hamosh, M., Hamosh, P., and Mehta, N.R.. Comparison of the lipid composition of breast milk from mothers of term and preterm infants. *American Journal of Clinical Nutrition* 38:300-312; 1983.
4. Borkman, M., Storlien, L.H., Pan, D.A., Jenkins, A.B., Chisholm, D.J., and Campbell, L.V. The relationship between insulin sensitivity and the fatty acid composition of phospholipids of skeletal muscle. *New England Journal of Medicine* 328:238-244; 1993.
5. Brand, M.D., Couture, P., Else, P.L., Withers, K.W., and Hulbert, A.J. Evolution of energy metabolism. Proton permeability of the inner membrane of

liver mitochondria is greater in mammal than in a reptile. *Biochemical Journal* 275: 81-86; 1991.

6. Clarke, S.D., and Jump, D.B. Regulation of gene transcription by polyunsaturated fatty acids. *Progress in Lipid Research* 32: 139-149; 1993.

7. Couet, C., Delarue, J., Ritz, P., Antoine, J.M., and Lamisse, F. Effect of dietary fish oil on body fat mass and basal fat oxidation in healthy adults. *International Journal of Obesity* 21: 637-643; 1997.

8. Else, P.L., and Hulbert, A.J. Evolution of mammalian endothermic metabolism: "leaky" membranes as a source of heat. *American Journal of Physiology* 253: R1-R7; 1987.

9. Flatt, J.P. McCollum Award Lecture: Diet, lifestyle, and weight maintenance. *American Journal of Clinical Nutrition* 62: 820-836; 1995.

10. Glatz, J.F.C., and Wusse, G.J. Cellular fatty acid-binding proteins: their function and physiological significance. *Progress in Lipid Research* 35:243-282; 1996.

11. Hagenfeldt, L., and Wahren, J. Human muscle forearm metabolism during exercise II. *Scandinavian Journal of Clinical Laboratory Investigation* 21:263-276; 1968.

12. Hainault, I., Carlotti, M., Hajduch, E., Guichard, C., and Lavau, M. Fish oil in high lard diet prevents obesity, hyperlipemia and adipocyte insulin resistance in rats. *Annals of the New York Academy of Sciences* 683: 99-101; 1993.

13. Halliwell, K.J., Fielding, B.A., Samra, J.S., Humphreys, S.M., and Frayn, K.N. Release of individual fatty acids from human adipose tissue in vivo after an overnight fast. *Journal of Lipid Research* 37:1842-1848; 1996.

14. Harridge, S.D.R. The muscle contractile system and its adaptation to training. In *Human muscular function during dynamic exercise.* Ed. P. Marconnet, B. Saltin, P. Komi, and J. Poortmans, pp. 82-94. Basel: S. Karger; 1996.

15. Harris, W.S., Lu, G., Rambj¢r, G.S., Walen, A.I., Ontko, J.A., Qi, C., and Windsor, S.L. Influence of n-3 fatty acid supplementation on the endogenous activities of plasma lipases. *American Journal of Clinical Nutrition* 66: 254-260; 1997.

16. Helge, J.W., Ayre, K.J., Hulbert, A.J., Kiens, B., and Storlien, L.H. Training modifies the phospholipid fatty acid membrane composition of rat muscle irrespective of diet intake. *ECSS Frontiers in Exercise Science Book of Abstracts 1:* 258-259 (abstract); 1996.

17. Hickey, M.S., Carey, J.O., Azevedo, J.L., Houmard, J.A., Pories, W.J., Israel, R.G., and Dohm, G.L. Skeletal muscle fiber composition is related to adiposity and in vitro glucose transport rate in humans. *American Journal of Physiology* 268: E453-E457; 1995.

18. Hoppeler, H. Exercise-induced ultrastructural changes in skeletal muscle. *International Journal of Sports Medicine* 7:187-204; 1986.

19. Jones, P.J.H., and Schoeller, D.A. Polyunsaturated:saturated ratio of diet fat influences energy substrate utilization in the human. *Metabolism* 37:145-1511; 1988.

20. Jones, P. J. H., Pencharz, P.B., and M. T. Clandinin, M.T. 1985. Whole body oxidation of dietary fatty acids: implications for energy utilization. *American Journal of Clinical Nutrition* 42: 769-777; 1985.

21. Jump, D.B., Clarke, S.D., Thelen, A., Liimatta, M., Ren, B., and Badin, M. Dietary polyunsaturated fatty acid regulation of gene transcription. *Progress in Lipid Research* 35: 227-241; 1997.

22. Kriketos, A.D., Pan, D.A., Lillioja, S., Cooney, G.J., Baur, L.A., Milner, M.R., Sutton, J.R., Jenkins, A.B., Bogardus, C., and Storlien. L.H. Inter-relationships between muscle morphology, insulin action, and adiposity. *American Journal of Physiology* 270: R1332-1339; 1996.

23. Kriketos, A.D., Pan, D.A., Sutton, J.R., Hoh, J.F.Y., Baur, L.A., Cooney, G.J., Jenkins, A.B., and Storlien, L.H. Relationships between muscle membrane lipids, fiber type and enzyme activities in sedentary and exercised rats. *American Journal of Physiology* 269: R1154-R1162; 1995.

24. Leyton, J., Drury, P.J., and Crawford, M.A. Differential oxidation of satu-rated and unsaturated fatty acids in vivo in the rat. *British Journal of Nutri-tion* 57: 383-393; 1987.

25. Lillioja, S., Young, A.A., Culter, C.L., Ivy, J.L., Abbott, W.G.H., Zawadzki, J.K., Jki-Jarvinen, H., Christin, L., Secomb, T.W., and Bogardus, C. Skeletal muscle capillary density and fiber type are possible determinants of in vivo insulin resistance in man. *Journal of Clinical Investigation* 80: 415-424; 1987.

26. Matsuo, T., Shimomura, Y., Saitoh, S., Tokuyama, K., Takeuchi, H., and Suzuki, M. Sympathetic activity is lower in rats fed a beef tallow diet than in rats fed a safflower oil diet. *Metabolism* 44:934-939; 1995.

27. Matsuo, T., and Suzuki, M. Beef tallow diet decreases lipoprotein lipase activities in brown adipose tissue, heart, and soleus muscle by reducing sympathetic activities in rats. *Journal of Nutitional Science and Vitaminol-ogy* (Tokyo) 40: 569-581; 1994.

28. Matsuo, T., and Suzuki, M. Brain beta-adrenergic receptor binding in rats with obesity induced by a beef tallow diet. *Metabolism* 46:18-22; 1997.

29. McMurchie, E.J. Dietary lipids and the regulation of membrane fluidity and function. In *Physiological Regulation of Membrane Fluidity*. Ed. by R.C. Aloia, 1189-1237. New York: Alan R. Liss Inc.; 1988.

30. Morgan, T.E., Short, F.A., and Cobb, L.A. Effect of long-term exercise on skeletal muscle lipid composition. *American Journal of Physiology*, 216: 82-88; 1969.

31. Mougios, V., Kotzamanidis, C., Koutsari, C., and Atsopardis, S. Exercise-induced changes in the concentration of individual fatty acids and triacylglycerols of human plasma. *Metabolism* 44: 681-688; 1995.

32. Nyholm, B., Qu, Z., Kaal, A., Pedersen, S.B., Gravholt, C.H., Andersen, J.L., Saltin, B., and Schmitz, O. Evidence of an increased number of type IIb muscle fibers in insulin-resistant first-degree relatives of patients with NIDDM. *Diabetes* 46: 1822-1828; 1997.

33. Pan, D.A., Hulbert, A.J., and Storlien, L.H. Dietary fats, membrane phospho-lipids and obesity. *Journal of Nutrition* 124:1555-1566; 1994.

34. Pan, D.A., Lillioja, S., Kriketos, A.D., Milner, M.R., Baur, L.A., Bogardus, C., Jenkins, A.B., and Storlien, L.H. Skeletal muscle triacylglycerol levels are inversely related to insulin action. *Diabetes* 46: 983-988; 1997.

35. Pan, D.A., Lillioja, S., Milner, M.R., Kriketos, A.D., Baur, L.A., Bogardus, C., and Storlien, L.H. Skeletal muscle membrane lipid composition is related

to adiposity and insulin action. *Journal of Clinical Investigation* 96: 2802-2808; 1995.

36. Pan, D.A., and Storlien, L.H. Dietary lipid profile is a determinant of tissue phospholipid fatty acid composition and rate of weight gain in rats. *Journal of Nutrition* 123: 512-519; 1993a.

37. Pan, D.A., and Storlien, L.H. Effect of dietary lipid profile on the metabolism of w-3 fatty acids: implications for obesity prevention. In *Omega-3 fatty acids: Metabolism and biological effects.* Ed. C.A. Drevon, I. Baksaas, and H. E. Krokan, pp 97–106. Basel: Birkhauser Verlag; 1993b.

38. Phillipson, B.E., Rothrock, D.W., Connor, W.E., Harris, W.S., and Illingworth, D.R. Reduction of plasma lipids, lipoproteins, and apoproteins by dietary fish oils in patients with hypertriglyceridemia. *New England Journal of Medicine* 312:1210-1216; 1985.

39. Raclot, T., and Groscolas, R. Differential mobilization of white adipose tissue fatty acids according to chain length, unsaturation, and positional isomerism. *Journal of Lipid Research* 34:1515-1526; 1993.

40. Raheja, B.S., Sadikot, S.M., Phatak, R.B., and Rao, M.B. Significance of the n-6/n-3 ratio for insulin action in diabetes. *Annals of the New York Academy of Sciences* 683: 258-271; 1993.

41. Shimomura, Y., Tamura, T., and Suzuki, M. Less body fat accumulation in rats fed a safflower oil diet than in rats fed a beef tallow diet. *Journal of Nutrition* 120:1291-1296; 1990.

42. Simons, K., and Ikonen, E. Functional rafts in cell membranes. *Nature* 387: 569-572; 1997.

43. Sjodin, A., Andersson, A., Olsson, R., and Vessby, B. Effects of exercise on fatty acid composition in skeletal muscle in relation to insulin sensitivity. *Biochemistry of Exercise, Poster Abstracts, 1.19,* 10 (abstract); 1997.

44. Spector, A.A., and Yorek, M.A. Membrane lipid composition and cellular function. *Journal of Lipid Research* 26:1015-1035; 1985.

45. Storlien, L.H., Jenkins, A.B., Chisholm, D.J., Pascoe, W.S., Khouri, S., and Kraegen, E.W. Influence of dietary fat composition on development of insulin resistance in rats. Relationship to muscle triacylglycerol and ω-3 fatty acids in muscle phospholipids. *Diabetes* 40: 280-289; 1991.

46. Storlien, L.H., Pan, D.A., Kriketos, A.D., O'Connor, J., Caterson, I.D., Cooney, G.J., Jenkins, A.B., and Baur, L.A. Skeletal muscle membrane lipids and insulin resistance. *Lipids* 31: S261-S265; 1995.

47. Thomas, T. R., Londeree, B.R., Gerhardt, K.O., and Gehrke, C.W. Fatty acid profile and cholesterol in skeletal muscle of trained and untrained men. *Journal of Applied Physiology* 43: 709-713; 1977.

48. Wade, A.J., Marbut, M.M., and Round, J.M. Muscle fibre type and aetiology of obesity. *Lancet* 335: 805-808; 1990.

49. Wake, S.A., Sowden, J.A., Storlien, L.H., James, D.E., and Clark, P.W. Effects of exercise training and dietary manipulation on insulin-regulatable glucose-transporter mRNA in rat muscle. *Diabetes* 40: 275-279; 1991.

50. Wong, S.H., Nestel, P.J., Trimble, R.P., Storer, G.B., Illman, R.J., and Topping, D.L. The adaptive effects of dietary fish and safflower oil on lipid and lipoprotein metabolism in perfused rat liver. *Biochimica et Biophysica Acta* 792:103-109; 1984.

PART III

Excitation-Contraction

CHAPTER 7

Role of the Skeletal Muscle Na+,K+-Pump During Exercise

Michael J. McKenna

Department of Human Movement, Recreation and Performance,
Victoria University of Technology, Footscray, Victoria, Australia

The marked Na^+ and K^+ disturbances that occur in contracting skeletal muscle and in blood with exercise essentially already were documented six decades ago. For example, Fenn and Cobb (21) reported that electrically stimulated frog and rat muscles lost K^+ and gained Na^+, Cl^-, and water, with all of these changes being reversible with recovery. One of the earliest studies in exercising humans in this field was by Keys (48), who reported that venous $[K^+]$ was increased by 25% during exhaustive short-term exercise, and then fell precipitously in recovery. Our understanding of these cation changes was enhanced greatly with the discovery of the Na^+,K^+-ATPase enzyme (Na^+,K^+-pump) some 20 years later (85). Since then, a tremendous amount of information has been gained on the Na^+,K^+-pump, including its ubiquitous nature; discovery of different isoforms and their organ-, developmental- and tissue-specific expression; determination of the numerous factors influencing both short- and long-term pump regulation; and identification of the important tissue-specific functions of the pump. This paper examines the role of the Na^+,K^+-pump in skeletal muscle during exercise, with an emphasis where possible on exercising humans. The first section examines the basic structure, location, subunit isoforms, and quantification of the Na^+,K^+-pump. The second section covers the electrical, ionic, and hormonal factors responsible for the dramatic acute activation of the pump during exercise. The third section investigates the activity of the pump in human skeletal muscle at rest and during exercise, as assessed from measurements of enzyme activity in biopsy samples and from K^+ differences across exercising limbs. The fourth section reviews the importance of the Na^+,K^+-pump in skeletal muscle in transmembrane Na^+/K^+ exchange and maintenance of the membrane potential. The preservative action of the Na^+,K^+-pump in retarding skeletal muscle fatigue is described, as well as a possible acute inhibition of the pump activity with fatigue. Finally, the effects of chronic alterations in physical activity on the concentration and activity of the Na^+,K^+-pump in skeletal muscle, and the functional consequences of these changes are described.

Structure, Composition, and Location of Na⁺,K⁺-Pumps in Skeletal Muscle

Structure and Subunit Isoforms

Functional $\alpha\beta$ Complex. The Na⁺,K⁺-ATPase enzyme is a transmembranous protein expressed in all excitable cells. The enzyme comprises an α (~112kDa) and a glycosylated β (~55kDa) subunit, present in an $\alpha\beta$ complex, which is the minimum functional Na⁺,K⁺-ATPase unit (44, 57, 60). The α subunit spans the membrane 10 times, with a single span for the β subunit (20). The α subunit contains binding sites for each of the substrates (Na⁺, K⁺, Mg^{2+}, and ATP), as well as the cardiac glycosides digoxin and ouabain, and is usually referred to as the catalytic subunit of the enzyme (44, 60). The functions of the β subunits are less clear, but they are essential for enzyme activity, and are involved in the normal processing of Na⁺,K⁺-ATPase enzymes, as well as their transport to, and insertion into, the plasma membrane (25).

Isoforms. Three isoforms each of the Na⁺,K⁺-pump α and β subunit have been identified, designated α_1, α_2, α_3, β_1, β_2, and β_3; each of these isoforms except β_3 are expressed in skeletal muscle, dependent upon species. In skeletal muscle it has been proposed that the α_1 isoform is responsible for basal Na⁺,K⁺-ATPase activity, whereas the α_2 isoform may be recruited from an intracellular pool for further gain in Na⁺,K⁺-ATPase activity, although this is dependent upon the muscle fiber type (see below). The α_2 isoform is the dominant α subunit in skeletal muscle (60, 87). Several differences in isoforms have been described. The most pronounced is that of ouabain sensitivity, which is lowest in the α_1 isoform in rats, but similar among the α isoforms in primates (57, 87). In HeLa cells transfected with α isoforms from rat muscle, the Na⁺ affinity was similar for α_1 and α_2 isoforms, and higher than in the α_3 isoform; K⁺ affinity was similar in all three α isoforms (42). It is not clear whether these differences also occur in skeletal muscle, since the affinities of these isoforms to Na⁺ and K⁺ varies among different cells (57).

Quantification of Na⁺,K⁺-Pumps in Skeletal Muscle

Ouabain Binding Site Density. The concentration of functional Na⁺,K⁺-pumps in skeletal muscle is best quantified by the density of vanadate-facilitated ³[H]-ouabain binding sites, and this has been widely measured in rat and in human muscle (9, 34). In cut muscle pieces, vanadate ensures ouabain binding to all functional Na⁺,K⁺-pumps (9). Human vastus lateralis muscle typically yields values of around 300 pmol·g ww⁻¹, but with a large individual range reported from ~175 to ~500 pmol·g ww⁻¹ (4, 9, 18, 67, 70). Higher values are obtained after physical training. Although earlier studies indicated no gender differences in skeletal muscle Na⁺,K⁺-pump concentration (70), a recent study reported an 18% higher concentration in men compared with women (18).

Fiber Type Dependence.   The concentration of Na+,K+-pumps was ~20–30% higher in extensor digitorum longus (EDL) than in soleus muscles in rats (14, 70). A later study in rats found no significant differences in the concentration of Na+,K+-pumps between soleus, EDL, and red vastus muscles, but the concentrations in these muscles were 51%, 53%, and 69% higher than in the white vastus muscle (8). A moderate positive relationship was found between the total concentration of Na+,K+-pumps and the oxidative potential in rat muscles (8). In humans, no clear muscle fiber type dependence has been shown (4, 15, 70).

K+-Dependent Phosphatase Activity.   An alternate method of quantifying the pump concentration in muscle homogenates or membrane fractions is through the measurement of K+-dependent phosphatase activity, an enzyme forming a partial reaction of the Na+,K+-pump (34, 44, 72)). This assay represents the terminal step in ATP hydrolysis and is specific to the Na+,K+-pump, since activity is both K+-dependent and inhibited by ouabain. The K+-dependent phosphatase assay utilizes either 3-O-methyl fluorescein phosphate (3-O-MFP), or p-nitrophenyl phosphate (p-NPP) as artificial substrates, and the Na+,K+-pump concentration then can be estimated using the molar activity of the enzyme. In human muscle, K+-stimulated 3-O-MFPase activity of ~1200 to 1400 pmol · min^{-1} · mg protein^{-1} has been measured (23, 24). Assuming a molar activity of 620 min^{-1} (72), this corresponds to a Na+,K+-pump concentration of ~330 pmol.g ww^{-1}, within the normal range quantified by the vanadate-facilitated ^{3}[H]-ouabain binding site method (9).

Location of Na+,K+-Pumps

Ouabain Binding Sites.   In skeletal muscle the Na+,K+-pumps are located principally in the plasma (80%) and t-tubular (20%) membranes (89). Since the extensive t-tubular network in muscle occupies a fourfold to fivefold greater area than the sarcolemma, the t-tubular Na+,K+-pump concentration was estimated at only ~4% that of the sarcolemma (89). The resultant low capacity for Na+/K+ exchange in the t-tubules, together with ionic diffusional limitations, predisposes the t-tubules to greater increases in [K+] and reductions in [Na+] than might occur at the sarcolemma. This may lead to blockade of t-tubular action potential transmission during muscle contraction, which is probably an important mechanism in fatigue induced by high-frequency stimulation (2, 43, 92).

Specific Isoform Locations.   Several recent studies have explored the locations of Na+,K+-pump isoforms in muscle, using biochemical membrane fractionation procedures and labeling with antibodies specific for the different isoforms. Unfortunately, their findings have not been entirely consistent, and this is probably due to variability caused by the very low membrane recoveries inherent in membrane fractionation (35). Although several uncertainties exist, a schematic representation of the known locations of the Na+,K+-pump subunit isoforms in rat and in human skeletal muscle is shown in figure 7.1.

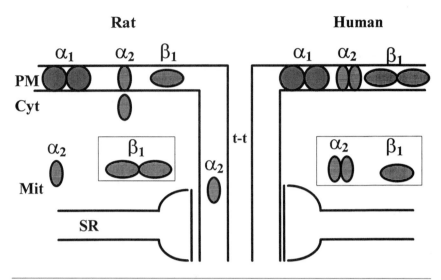

Figure 7.1.   Schematic representation of Na⁺,K⁺-pump isoform locations in rat (left) and human (right) skeletal muscle. The α and β isoforms are shown separately rather than as a functional $\alpha\beta$ complex for simplicity and since the relative abundance of different $\alpha\beta$ complexes (i.e., comprising different isoforms) has not been determined. The number of isoforms shown indicate the relative abundance of a given isoform at different locations; the relative overall abundance of different isoforms is not shown. In rat muscle, the β_1 isoform is expressed predominantly in oxidative muscles and β_2 (not shown) in glycolytic muscles. Human muscle also expresses the α_3 isoform (not shown). Abbreviations: PM, plasma membrane; Cyt, cytoplasm; Mit, mitochondria; t-t, t-tubules; SR, sarcoplasmic reticulum. The unlabeled box indicates an intracellular membrane fraction exclusive of t-tubular and SR membranes.

Fiber Type-Specific Isoform Expression.   The expression of these α and β isoforms is fiber type-specific in rat muscle (39). The α_1 and α_2 isoform mRNA was present in both oxidative and glycolytic muscles, with slightly higher expression in oxidative muscles; however, no differences in α_1 or α_2 protein contents were found in mixed red and white muscles (39). In contrast, marked differences were found between rat muscles in β isoform mRNA and protein contents. Both β_1 isoform mRNA and protein contents (~5-fold) were substantially higher in soleus than in the fast twitch muscle EDL; in contrast, β_2 isoform mRNA and protein contents (~3-fold) were much higher in EDL muscle (39).

α-Isoform Locations.   Analysis of membrane fractions in rat hindlimb muscles indicated that the Na⁺,K⁺-pump α_1 subunit was located almost exclusively at the plasma membrane in both red and white muscle (38, 39). Biochemical studies yielded conflicting results on the α_2 isoform, being principally located within intracellular membrane fractions in one study (38), but more abundant in plasma membranes in another, in both red and white muscle (39). This discrepancy was resolved using immunogold labeling and electron microscopy in rat hindlimb muscles, with the α_2 isoforms found to be located principally in the surface membrane, but also present within t-tubules in triadic regions and in subsarcolemmal regions in vesicles (54, 62). A greater amount of α_2 subunits was found in the plasma membrane in white than in red muscles (54).

β-Isoform Locations.   Similar discrepancies exist in the reported β-isoform locations. About 80% of the β_1 subunit was reported within intracellular membrane fractions in one study, but subsequently reported mainly at the surface membrane (38, 39). The β_2 isoforms were found to be more abundant intracellularly (38), but later to be more diffuse in muscle (39).

Isoforms in Human Skeletal Muscle.   To date only a single study has investigated Na⁺,K⁺-pump isoforms in human muscle, measured using large amounts (20–30 g) of soleus muscle obtained from amputated limbs (40). Several important differences in isoform expression exist between rat and human muscle. Unlike rats, adult human soleus muscle expresses the α_3 but does not express the β_2 isoform, which also was absent in human tibialis anterior and gastrocnemius muscles (40). Thus, in human muscle the activity of Na⁺,K⁺-ATPase α_1, α_2, and α_3 isoforms is supported only by the β_1 isoform (40). Their biochemical membrane fractionation analyses indicated: (a) the α_1 isoform was located almost exclusively in the plasma membrane; (b) the α_2 isoform was present both at the plasma membrane and at intracellular sites; and (c) the β_1 isoform was present mainly at the plasma membrane, but with about one-third present in the intracellular fraction (40).

It is now apparent that these locations of Na⁺,K⁺-pump isoforms are not fixed, since insulin and, most likely, exercise can promote translocation of Na⁺,K⁺-pump isoforms to the plasma membrane in rat muscle.

Intracellular-to-Sarcolemmal Na⁺,K⁺-Pump Translocation

Insulin.   Insulin stimulates translocation of Na⁺,K⁺-pumps to the plasma membrane in frog muscle, with ouabain binding increased in plasma membrane fractions and reduced in internal membrane fractions (73). Later studies in rats demonstrated that this comprised translocation of α_2 and β_1 Na⁺,K⁺-pump isoforms to the surface membrane, but this occurred only in oxidative muscles (38, 54, 62). This translocation had functional benefits, with a corresponding increase in Na⁺,K⁺-pump activity in the plasma membrane fraction (54). These findings are analogous to the insulin- and exercise-induced translocation of GLUT4 (glucose transporter) to the surface membrane in rat muscle, resulting in increased glucose transport (38, 53). Although both the Na⁺,K⁺-pump α_2 subunit and GLUT4 are translocated to the surface membrane by insulin, they do not arise from a common intracellular pool (1).

Exercise.   The only study to have examined whether exercise induces translocation of Na⁺,K⁺-pumps to the plasma membrane subjected rats to one hour of running on a treadmill (88). This resulted in an increased appearance of the α_1 and α_2 isoforms at the plasma membrane, in red (64% and 43%) and white (55% and 94%) hindlimb muscles (88). These changes could not be attributed to translocation of α_1

and α_2 Na$^+$,K$^+$-pump isoforms to the surface membrane, since no corresponding decrease was found in α isoforms in intracellular membrane fractions. However, it is possible that translocation did occur and that these isoforms arose from an unmeasured intracellular pool. A further explanation is that these changes may be due to an increased retention of α subunits at the plasma membrane. After exercise a greater abundance of β subunits also was found at the plasma membrane, suggesting an increase in functional (i.e., $\alpha\beta$ complex) Na$^+$,K$^+$-pumps with exercise (88). It is unknown whether Na$^+$,K$^+$-pump translocation occurs in exercising human muscle, nor has the time course of any such change with exercise been documented. Of interest and possible relevance, is that GLUT4 translocation in human muscle occurred after only ~5 min of exercise (53). Increased Na$^+$,K$^+$-pump abundance at the surface membrane with exercise might attenuate muscle K$^+$ loss during contraction and facilitate K$^+$ reuptake during recovery, thereby minimizing the exercise-induced K$^+$ loss component of fatigue. Translocation of Na$^+$,K$^+$-pumps to the muscle membrane with exercise may also contribute to the progressive decline (narrowing) in the arteriovenous K$^+$ difference reported across contracting muscles. Any increase in pump activity due to translocation of Na$^+$,K$^+$-pumps to the sarcolemma also might act to counterbalance the direct inhibition of membrane Na$^+$,K$^+$-pumps with fatigue. The effects of insulin and exercise on translocation of Na$^+$,K$^+$-pump isoforms to the plasma membrane are shown schematically in figure 7.2.

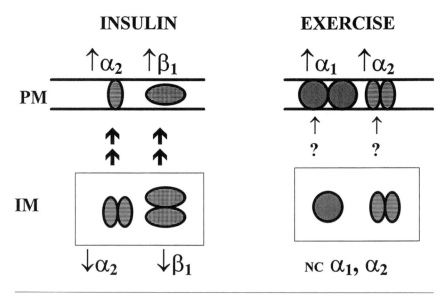

Figure 7.2.   Schematic representation of the effects of insulin (left) and exercise (right) on translocation of the Na$^+$,K$^+$-pump subunit isoform in rat skeletal muscle. Plasma membrane and intracellular membrane fractions are indicated. Solid arrows indicate translocation; dashed arrows indicate that translocation cannot yet be confirmed. Abbreviations: PM, plasma membrane; IM, intracellular membrane fractions; NC, no change.

Activation of the Na⁺,K⁺-Pump in Contracting Muscle

The Na⁺,K⁺-pump operates at low levels under resting conditions, but is rapidly activated during exercise by a combination of electrical, ionic, and hormonal factors. Their relative importance in the acute activation of the Na⁺,K⁺-pump are reviewed here. For long-term regulation of the pump, readers are referred elsewhere (9, 11). It is apparent that electrical events induce a rapid and extensive activation of the pump and most likely form the major mechanism for pump activation during exercise. Increases in muscle intracellular [Na⁺] and in circulating catecholamines are probably also important activators of the Na⁺,K⁺-pump during intense contractions.

Electrical Activity-Induced Activation

Excitation of the muscle membrane rapidly and dramatically increases the activity of the Na⁺,K⁺-pump in isolated rat soleus muscle, to such an extent that the maximal theoretical pump activity may be achieved. Electrical stimulation at frequencies as low as 2 Hz induces a marked activation of the Na⁺,K⁺-pump in rat soleus muscle, despite unchanged $[Na^+]_i$ or $[K^+]_e$ (16, 17). Activation of the Na⁺,K⁺-pump was proportional to stimulation frequency, with 10 s stimulation at 30, 60, 90, and 120 Hz inducing ~28%, 47%, 66%, and nearly 100% of the theoretical maximal Na⁺,K⁺-pump mediated K⁺ uptake (16). Even very brief periods of stimulation strongly activate the Na⁺,K⁺-pump. After only 1 s stimulation at 60 Hz, the rate of muscle ²²Na⁺ efflux over the subsequent 5 min was increased by 22%; after 10 s stimulation this was elevated by 57%, and after 60 s stimulation by 50% (16).

The activation mechanism of the Na⁺,K⁺-pump was linked with opening of the Na⁺ channels, rather than membrane depolarization, since pump activation was blocked by tetrodotoxin (Na⁺ channel blocker) and also since greater pump activation occurred in the presence of veratridine, which maintains Na⁺ channels in the open state, than with a depolarizing $[K^+]_e$ of 50 mM (16). These findings also could reflect a rapid localized increase in subsarcolemmal [Na⁺], consistent with the notion of a subsarcolemmal "fuzzy space" (55, 82).

Ionic Activation

An important acute activator of the Na⁺,K⁺-pump in skeletal muscle is an increased intracellular [Na⁺] and in the physiological range, pump activity increases linearly with $[Na^+]_i$ (81). However, pump activity can be increased substantially in the absence of elevated $[Na^+]_i$ (e.g., 16). The contribution of increased extracellular [K⁺] to pump activation during exercise is probably small, due to the low K_m of the pump for K⁺ (81). The theoretical maximal in-vivo pump activation cannot be achieved during exercise solely on the basis of the observed increases in muscle

$[Na^+]_i$ and $[K^+]_e$. Twofold or even greater increases in $[Na^+]_i$ and $[K^+]_e$ have been reported in stimulated rat, frog, and murine muscles (2, 45, 91), as well as in contracting muscle in exercising humans (31, 68, 84). However, these increases fall far short of the nonphysiologically high $[Na^+]_i$ and $[K^+]_e$ required to fully activate the Na^+,K^+-pump in rat muscle (125 and 85 mM, respectively) (13). There may well be ionic inhibition of Na^+,K^+-pump activity. For example, the maximal in-vitro activity in isolated kidney Na^+,K^+-ATPase enzymes was depressed by a decline in pH and an increased phosphate concentration (37). These possibilities do not appear to have been investigated in skeletal muscle.

3.3 Hormonal Activation

Numerous hormones increase the activity and/or concentration of the Na^+,K^+-pump in skeletal muscle, including adrenaline, noradrenaline, insulin, aldosterone, thyroid hormones, calcitonin gene related peptide, amylin, and insulin-like growth factor. The role of these hormones in the short- and long-term regulation of the skeletal muscle Na^+,K^+-pump, as well as the tissue-specific nature of this regulation, have been covered extensively elsewhere and will not be discussed here (9, 11, 19). Probably the most important hormones involved in extrarenal potassium regulation during exercise are the catecholamines. This section briefly examines their role in activation of the skeletal muscle Na^+,K^+-pump during exercise, including recent information on muscle K^+ efflux with β-adrenergic blockade in exercise.

Catecholamines.   Elevated catecholamines increase the activity of the Na^+,K^+-pump in resting skeletal muscle, and in rats this effect was greater in soleus than in EDL muscles (17). However, these effects were not additive to the stimulatory effects of electrical stimulation on pump activity in contracting muscles (17), which questions the role of catecholamines in Na^+,K^+-pump activation during exercise. This has been assessed through studies employing β-blockade and measuring K^+ fluxes during exercise. In rat muscle, β-blockade inhibited the Na^+,K^+-pump under resting conditions (17). In exercising humans, plasma $[K^+]$ is higher after β-blockade, indicating impaired K^+ regulation during exercise and implicating catecholamines in the regulation of K^+ during exercise (7, 31, 47). Katz et al. (47) first showed that the greater hyperkalaemia with β-blockade was not caused by exacerbated K^+ loss from contracting muscles, since the arteriovenous $[K^+]$ difference across exercising leg muscles was not greater after β-blockade. They concluded that the greater hyperkalaemia must have resulted from reduced K^+ redistribution to inactive muscle and/or other tissues. A series of elegant studies using a K^+-selective electrode inserted into the femoral vein of exercising humans recently confirmed that β-blockade did not have major effects on muscle K^+ release from, and thus, Na^+,K^+-pump activity in, contracting muscles (28, 31). However, they reported that β-blockade did increase the initial muscle cellular K^+ efflux, reflecting an increased lag time of Na^+,K^+-pump activation (28, 31). In addition, a slower post-exercise decline in plasma $[K^+]$ was found after β-blockade, indicating a slowed component of the post-exercise Na^+,K^+-pump activity in exercised muscle (31). Thus, impaired K^+ redistribution in other tissues, a lag in Na^+,K^+-pump activity in contracting muscles, and a subsequent diminished post-exercise K^+ reuptake all contribute to the greater hyperkalaemia after β-blockade. These studies in rats and in humans therefore indicate that catecholamines do play an important role in K^+ regulation during exercise, via activation of the Na⁺,K⁺-pump. Studies with β-adrenergic agonists do not show reduced muscle K^+ release, although this is probably due to greater blood flow (32,77). Catecholamine activation of the pump then would minimize muscle interstitial K^+ accumulation during exercise via (a) transiently reduced K^+ release from contracting muscles, and (b) increased K^+ uptake by resting muscles, thereby maintaining an interstitial-to-plasma [K^+] gradient, and facilitating K^+ removal from muscle.

Insulin.   The insulin-induced activation of the Na⁺,K⁺-pump and consequent lowering action on plasma [K^+] are well known (11). Although both insulin and acute exercise induce an increased appearance of Na⁺,K⁺-pump isoforms at the plasma membrane, these do not involve a common mechanism, and thus the role of insulin in Na⁺,K⁺-pump activation in contracting muscle during exercise is not clear. One argument against an important role of insulin in skeletal muscle activation during exercise is the relatively slow time-course and lesser effect of insulin compared to catecholamine action on Na⁺,K⁺-pump activation (16). In addition, plasma insulin concentration falls during prolonged exercise, although on the other hand, it could also be argued that insulin delivery to contracting muscles is greatly elevated due to the large increase in muscle blood flow.

Effects of Exercise on Na⁺,K⁺-Pump Activity in Human Skeletal Muscle

Numerous problems constrain our ability to measure the activity of the Na⁺,K⁺-pump in contracting human skeletal muscle. These limitations arise due to methodological difficulties with isolation of living cells, use of radio-labeled ion infusion, measurement of intracellular versus extracellular ions, and accurate determination of fluid spaces. It is not possible to detect Na⁺,K⁺-pump activity in small samples of human muscle using standard end-point assays (e.g., P_i liberation) due to the large amount of nonspecific ATPase activity (e.g., 4). One recent approach to this problem has therefore been to measure the K^+-dependent phosphatase activity in skeletal muscle obtained by needle biopsies (4, 23) or in large samples (20–30 g) from limb amputations (40). Alternately, the activation of the skeletal muscle Na⁺,K⁺-pump may be estimated from arteriovenous (a-v) K^+ difference measurements across contracting muscles (31, 90).

Na⁺,K⁺-Pump Activity in Human Skeletal Muscle

Resting Values.   Values reported for K^+-dependent 3-O-MFPase activity in human vastus lateralis muscle range from ~382 pmol·min⁻¹·mg protein⁻¹, in unspecified patients (4), to ~1212 pmol·min^{-1}·mg protein^{-1} for untrained young men (23, 24). A very high K$^+$-dependent p-NPP activity of ~2500 pmol·min^{-1}·mg protein^{-1} was measured in soleus muscle in amputated lower limbs (40). The reasons for the differences between K$^+$-dependent 3-O-MFPase and p-NPP activities are not clear but may reflect differences in subject/patient source as well as methodology.

K$^+$ Differences Across Contracting Muscle

K$^+$ Release From and Reuptake by Contracting Muscle.   Numerous studies have demonstrated K$^+$ loss from contracting muscles and subsequent K$^+$ reuptake during recovery, through measurements of the arteriovenous K$^+$ difference ([K$^+$]$_{a-v}$) across the contracting limb muscles, during brief high-intensity exercise (46, 59, 66, 84, 90), incremental exercise (28, 31), as well as during prolonged submaximal static contractions (83). Venous plasma [K$^+$] is usually between 0.2 and 1.0 mM higher than arterial plasma [K$^+$] during exercise, indicating net K$^+$ release from contracting muscles. The [K$^+$]$_{a-v}$ appears to be influenced by the exercise intensity and duration, fluid shifts, muscle mass, mode of contraction, and training status, some of which are described below.

Increased Pump Activity During Exercise.   It is well established that Na$^+$,K$^+$-pump activity is markedly increased above resting levels in contracting skeletal muscle in humans. This is clear from changes in the arteriovenous K$^+$ difference, which is positive during exercise, indicating muscle K$^+$ release, and is rapidly reversed upon cessation of exercise to a positive value, indicating muscle K$^+$ uptake (31, 46, 66, 83, 84, 90). This rapid reversal from K$^+$ release to reuptake in recovery must be due to a substantial elevation in Na$^+$,K$^+$-pump activity. The relevant question is to what extent these pumps are activated with exercise.

Submaximal Activation of the Muscle Na$^+$,K$^+$-Pumps During Exercise.   Everts and Clausen (16) recently demonstrated full activation of Na$^+$,K$^+$-pumps in isolated electrically stimulated rat muscles due to events associated with membrane excitation. They had argued previously that muscle K$^+$ loss during exercise indicated that the capacity of the Na$^+$,K$^+$-pump must be exceeded by the rate of K$^+$ release (16). An alternate explanation is that activation of the pumps in contracting muscles during exercise is insufficient to balance the cellular K$^+$ efflux, leading to K$^+$ release (31). This problem has been addressed in several studies using an indwelling K$^+$ selective microelectrode inserted into the femoral vein, allowing continuous monitoring of femoral venous [K$^+$] and with simultaneous sampling of arterial blood to calculate [K$^+$]$_{a-v}$ (31, 33, 90). The [K$^+$]$_{a-v}$ is influenced by the rate of K$^+$ efflux from contracting cells, the rate of K$^+$ reuptake by contracting muscle, the rate of change in the interstitial volume and plasma blood flow (33). At the end of exercise it was assumed that K$^+$ efflux ceases suddenly with the end of electrical activity, that the rate of change in the interstitial volume is zero, and therefore the main determinants of [K$^+$]$_{a-v}$ will be the rate of K$^+$ reuptake and blood flow. Under these conditions, over a brief time period, the rate of change in [K$^+$]$_{a-v}$ will reflect the activity of the Na$^+$,K$^+$-pump (31, 33, 90). The immediate post-exercise Na$^+$,K$^+$-pump activity was estimated at only ~15% of the maximal pump capacity, indicating a substantial Na$^+$,K$^+$-pump reserve pumping rate (33). In subsequent studies, the calculated pump activity in the first few seconds following exercise was estimated at only ~15–50% of maximal theoretical activity (28, 31, 32, 90). Their conclusion was that K$^+$ loss from contracting muscles was due to a lag in pump activation, rather than inadequate capacity. Several aspects of their model need to be further explored. These include the effects of K$^+$ efflux mediated by ATP-dependent K$^+$ channels that are activated in fatigued muscle (76), as well as the effects of the rapid post-exercise decline in muscle blood flow on muscle interstitial volume and K$^+$ loss. Despite this, their conclusion of submaximal pump activation during exercise seems reasonable. Consistent with this, the magnitude of pump activation during maximal sprint cycling in humans estimated from arteriovenous whole blood [K$^+$] differences in untrained men (66) was ~16% of the maximal theoretical pump activity. Thus, in exercising humans, activation of the Na$^+$,K$^+$-pump in contracting muscles is well below the theoretical maximal level. The most likely reason for the difference between isolated stimulated rat muscles and contracting human muscles is the stimulation rate. In humans, the motor unit discharge rate is probably only around 30 Hz and rarely above 50 Hz (5, 43), well below the frequency required to fully activate the Na$^+$,K$^+$-pumps in rat muscles (16).

Interpretation of Arteriovenous K$^+$ Differences Across Contracting Muscles.   The plasma [K$^+$]$_{a-v}$ is immediately established at the onset of contraction, is greatest early in exercise, and progressively declines thereafter (28, 31, 32, 83, 84, 86, 90). During submaximal cycling exercise, [K$^+$]$_{a-v}$ returns to zero after 10–15 min but remains negative during exercise at intensities exceeding V̇O$_2$max (31, 90). In contrast, during submaximal exercise with a smaller muscle mass, the [K$^+$]$_{a-v}$ declined but remained above zero at all times, suggesting continuous K$^+$ release throughout exercise (28). The favored explanation by these authors is that muscle K$^+$ loss falls gradually during exercise at the same power output due to an initial lag period followed by a progressive rise in Na$^+$,K$^+$-pump activity, with a time constant of ~3 min. Another possibility is that a reflex slowing of motor unit discharge rate may occur with fatiguing exercise (5), which might result in reduced cellular K$^+$ efflux. A further intriguing possibility is that the declining [K$^+$]$_{a-v}$ with exercise also may be due to recruitment of additional Na$^+$,K$^+$-pumps to the muscle membrane (88). However, it is also clear that the venous [K$^+$] and therefore the [K$^+$]$_{a-v}$ are strongly influenced by the fluid shift from arterial plasma into contracting muscles. Thus, it is possible that a declining [K$^+$]$_{a-v}$ during exercise may also reflect a reduced hemoconcentration effect on venous [K$^+$] due to a smaller fluid shift. Unfortunately, few of the above studies have investigated this effect on K$^+$ homeostasis (28, 77). Rolett et al. (77) reported similar magnitudes of increase in femoral venous [K$^+$] and the decline in plasma volume (ΔPV_{a-v}) during leg exercise. Gullestad et al. (28) reported a net K$^+$ release of 0.13 mM and a ΔPV_{a-v} of 3.6%, after 6 min of high-intensity, two-legged knee extension exercise. Corrections of their data for fluid shifts indicate an actual K$^+$ uptake into contracting muscle. This also is demonstrated by findings of K$^+$ uptake by contracting muscles during the final seconds of maximal cycling sprint exercise (66). At the end of 30 s maximal exercise, arterial and femoral venous plasma [K$^+$] had risen to 7 and 8 mM, respectively. This indicates earlier substantial K⁺ loss and suggests a large ongoing net K⁺ release from the leg muscles. However, a 14% decline in plasma volume occurred across the leg at this time, which effectively returned the a-v plasma [K⁺] difference to zero (66). Further, measurement of the arteriofemoral venous whole K⁺ difference, corrected for fluid shifts, indicated net K⁺ uptake by the contracting leg (66). These findings are shown in figures 7.3

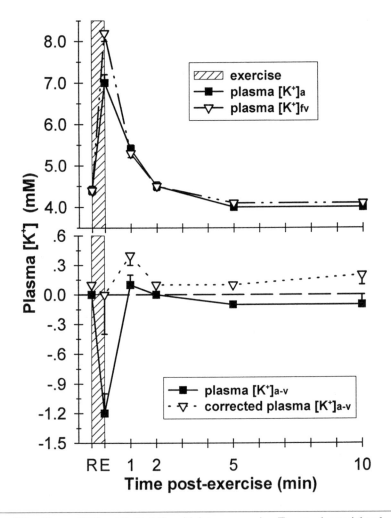

Figure 7.3.   Plasma [K⁺] and fluid shifts with intense exercise. Top panel: arterial and femoral venous plasma [K⁺]. Bottom panel: the arteriovenous plasma [K⁺] difference ([K⁺]$_{a-v}$) and the [K⁺]$_{a-v}$ corrected for the arteriovenous decline in plasma volume. Data collected at rest (R), during the final seconds of a 30 s maximal sprint bout (E) and in recovery. Figure modified from McKenna et al. (66).

and 7.4. This suggests that muscle K⁺ loss earlier during exercise had reversed to a net muscle K⁺ uptake in the final seconds of exercise when fatigue had induced a decline in power output by 40-60% (66). Therefore interpretation of anteriovenous [K⁺] differences in the absence of fluid shifts may result in erroneous conclusions. Nonetheless, it is clear that the magnitude of activation of the pump is far from maximal during exercise in humans.

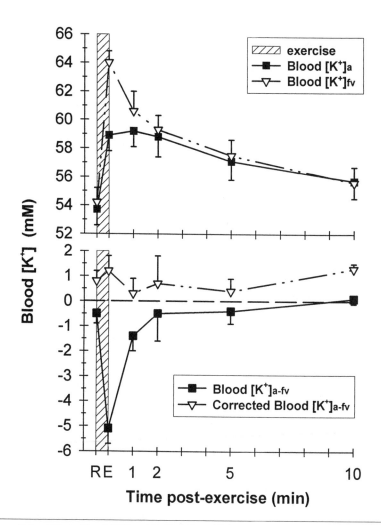

Figure 7.4.   Whole blood [K⁺] and fluid shifts with intense exercise. Top panel: arterial and femoral venous whole blood [K⁺]. Bottom panel: the arteriovenous whole blood [K⁺] difference ([K⁺]$_{a-v}$) and the [K⁺]$_{a-v}$ corrected for the arteriovenous decline in blood volume. Data collected at rest (R), during the final seconds of a 30 s maximal sprint bout (E) and in recovery. Figure modified from McKenna et al. (66).

Na⁺,K⁺-Pump Function During Muscle Contraction

The most important function of the Na⁺,K⁺-pump in skeletal muscle is to regulate Na⁺ and K⁺ fluxes across muscle membranes, thereby maintaining the muscle membrane potential and excitability.

Muscle Na⁺ and K⁺ Fluxes

The rapid and dramatic shifts in water, electrolytes, and metabolites with muscle contraction have been reviewed extensively elsewhere (22, 63, 69). Briefly, a small cellular Na⁺ influx and K⁺ efflux occur with each action potential, and these fluxes are amplified during contraction due to the high rates of membrane excitation. In humans undertaking fatiguing knee extensor exercise, muscle intracellular [Na⁺] rose from ~13 to 23 mM, and intracellular [K⁺] fell from 169 to 125 mM (84). Even greater changes were seen in frog muscles stimulated at high frequency, where intracellular [Na⁺] rose from 16 to 49 mM, and intracellular [K⁺] fell from 142 to 97 mM (2). The calculated decline in membrane potential was 13 mV in human muscle, and the decline measured in frog muscle was 9 mV (2, 84). These ionic and membrane potential changes have been implicated widely as causative factors in skeletal muscle fatigue, although there is not compelling evidence that this is the case. The argument is that K⁺ shifts cause t-tubular membrane depolarization and inactivation of fast Na⁺ channels, resulting in impaired action potential development and propagation, preventing activation of the dihydropyridine receptors, resulting in reduced sarcoplasmic reticulum Ca²⁺ release, smaller tetanic [Ca²⁺], and therefore reduced force. The role of each of these steps in fatigue has been extensively reviewed recently (22). One major difficulty in testing the above concept is our inability to measure t-tubular ion concentrations and membrane potential in intact muscles. However, the low density of Na⁺,K⁺-pumps in the t-tubules makes them susceptible to a failure of membrane excitability. One way to assess the role of the Na⁺,K⁺-pump and therefore ion fluxes in fatigue is through manipulation of the Na⁺,K⁺-pump activity. This is difficult in contracting human muscles although inhibition of Na⁺,K⁺-pumps with digoxin has been used. Therefore this question has been studied largely in isolated animal preparations.

Role of the Na⁺,K⁺-Pump in Skeletal Muscle Fatigue

Na⁺,K⁺-Pump Activation Attenuates Fatigue During Contraction.   When skeletal muscles are incubated in high K⁺ solutions (10-12.5 mM), muscle tetanic force declines steadily, with recovery upon restoration of low extracellular [K⁺] (10, 12). This decline in force was accelerated and recovery delayed when the Na⁺,K⁺-pump was inhibited in a dose-dependent manner by ouabain. The rate of fatigue development was delayed and force recovery increased by the β-adrenergic agonist salbutamol, and the hormones adrenaline and insulin (10, 12). Tetrodotoxin abolished the force recovery with salbutamol, indicating that this was dependent upon functional Na^+ channels (10). Increased cytosolic $[Ca^{2+}]$ could have contributed to the salbutamol-induced force recovery, but not that induced by insulin, providing further evidence that activation of the Na⁺,K⁺-pump was responsible for force restoration and thus plays an important facilitative role in skeletal muscle contraction. Therefore, activation of the Na⁺,K⁺-pump in contracting muscles most likely plays an important role in delaying muscular fatigue.

Depressed Maximal Pump Activity With Fatigue.   Counter to the facilitative role of the Na⁺,K⁺-pump for muscle function are recent findings that intense fatiguing exercise depressed the maximal in-vitro activity of the Na⁺,K⁺-pump (3-0-MFPase activity) in human muscle by 17% (23). This fatigue effect has since been confirmed in a larger group of individuals, including untrained, strength- and endurance-trained men (24). This depression was analogous to the reduced maximal activity of the sarcoplasmic reticulum Ca^{2+}-ATPase enzyme in human muscle with fatigue, which was most likely due to irreversible structural alterations to the Ca^{2+}-ATPase enzyme (6, 58, figure 7.5). Since muscle sarcoplasmic reticulum Ca^{2+}-ATPase activity is inhibited to a greater extent in vivo than in-vitro (6), it is possible that similar greater inhibitory in-vivo effects also occur for the Na⁺,K⁺-pump. The role of other factors such as an intracellular acidosis and phosphate accumulation also may directly effect the activity of the Na⁺,K⁺-pump during intense contractions. The net effect of these changes would be to reduce the available pump reserve capacity for regulating Na^+ and K^+ exchange. If full activation (i.e., the theoretical maximal activity) of the Na⁺,K⁺-pump could be achieved during exercise (16), any impairment in maximal in vivo Na⁺,K⁺-pump activity would have profound augmentative effects on muscle K^+ loss. But since the activation of the Na⁺,K⁺-pump is submaximal during exercise in humans, the significance of depressed maximal in-vitro Na⁺,K⁺-pump activity with fatigue is not yet clear. A possible role is however, indicated by the interesting finding that a larger (i.e., more negative) arteriovenous K^+ difference across the contracting muscles was reported at fatigue during prolonged exercise (78).

Muscle Na⁺ Gain and K⁺ Loss: Na⁺,K⁺-Pump Inadequacy or a Regulated Phenomenon?   The Na⁺,K⁺-pumps are activated only submaximally during exercise, with a consequent gain in Na^+ and loss of K^+ from muscle. This may be interpreted as a failure in ionic homeostasis, or more likely, as a well-regulated phenomenon, with transient perturbations in muscle and extracellular ions with exercise important in maintaining other cellular and systemic functions. For example, elevated extracellular $[K^+]$ plays important regulatory roles in exercise through local muscle vasodilatation, increased arterial blood pressure, and ventilation (74). The rise in $[K^+]_e$ during exercise is constrained by activation of the muscle Na⁺,K⁺-pumps, which restrict arterial $[K^+]$ to ≤8 mM, thereby preventing potentially fatal cardiac arrhythmias. The cellular advantages of an acute gain in intracellular $[Na^+]$ during contractile activity are not yet clear, but two related possibilities are proposed. First, muscle K^+ loss and Na^+ gain with contraction

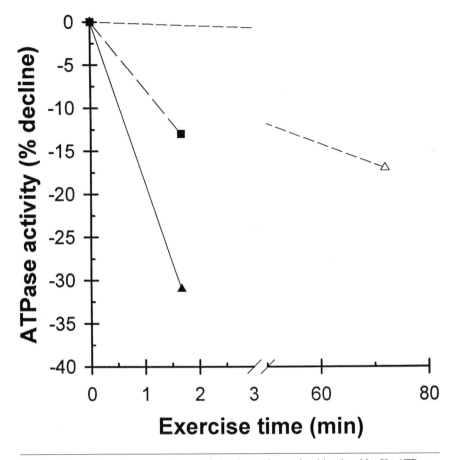

Figure 7.5.   Fatigue depresses human skeletal muscle maximal in vitro Na$^+$,K$^+$-ATPase (B) and sarcoplasmic reticulum Ca^{2+}-ATPase (Δ) activity. Data for Ca^{2+}-ATPase from quadriceps muscle sampled before and after prolonged cycling exercise to fatigue at 75% $\dot{V}O_2$ max (C, Booth et al. [6]) and 50 maximal knee extensor contractions (H, Li et al. [58]) and B, (Fraser and McKenna [23]).

contribute to muscle fatigue (2), complementing other fatigue-inducing mechanisms (e.g., impaired sarcoplasmic reticulum Ca^{2+} release) as a myoprotective mechanism. Secondly, increased intracellular [Na$^+$] probably stimulates Na$^+$,K$^+$-pump synthesis (75), ensuring increased muscle Na$^+$/K$^+$ exchange capacity and thereby allowing enhanced muscle performance under conditions of increased contractile demand.

Role of Contracting Muscle Mass and Relative Exercise Intensity

Since contracting muscles release K$^+$ and inactive muscles remove circulating K$^+$ (46), the relative amount of active versus inactive muscle should impact dramatically on K$^+$ homeostasis. The role of active versus inactive muscle mass in arterial [K$^+$] regulation was investigated recently using incremental one-arm versus two-legged cycling exercise (79). Arterial [K$^+$] during submaximal work rates was greater during one-arm cycling, whereas the peak [K$^+$] was higher during leg cycling exercise. It was shown that arterial [K$^+$] was dependent on the relative exercise intensity rather than on the amount of active versus inactive muscle mass and that this probably was due to lesser Na$^+$,K$^+$-pump activation during small muscle mass exercise. Thus, our findings are compatible with the recent report of Hallén (30), based on different experiments, that the rate of K$^+$ efflux from contracting leg muscles was greater during exercise with a small muscle mass.

Effects of Altered Physical Activity on Skeletal Muscle Na$^+$,K$^+$-Pumps

Finally, chronic alterations in physical activity clearly affect skeletal muscle Na$^+$,K$^+$-pumps and consequent ion regulation during exercise, reviewed in detail elsewhere (64, 65).

Up-Regulation of Na$^+$,K$^+$-Pumps With Training

The Na$^+$,K$^+$-pump concentration (^{3}H-ouabain binding site concentration) is up-regulated in mammalian skeletal muscle with increased physical activity. In humans, the vastus lateralis muscle Na$^+$,K$^+$-pump concentration was increased by 14–16% after sprint (67) and endurance training (18, 27, 61) (figure 7.6). It is interesting that similar gains in pump concentration occurred in these studies, despite marked differences in training modes, duration (6 d to 5 mo), and pretraining relative fitness levels (48 to 73 ml. kg^{-1}.min^{-1}, figure 7.6). Larger increases in pump concentration have been shown in animal muscles. In rat and guinea pig hindlimb muscles the pump concentration was increased by 22–43% after endurance training (51, 56). Chronic electrical stimulation in rabbit muscle increased pump concentration by 41% after 4 days and by 86% after 10 days (26). Training also induced a membrane hyperpolarization of 11 mV in canine muscle (52). However, the importance of this up-regulation in Na$^+$,K$^+$-pump concentration to K$^+$ regulation and muscle function has been largely unexplored.

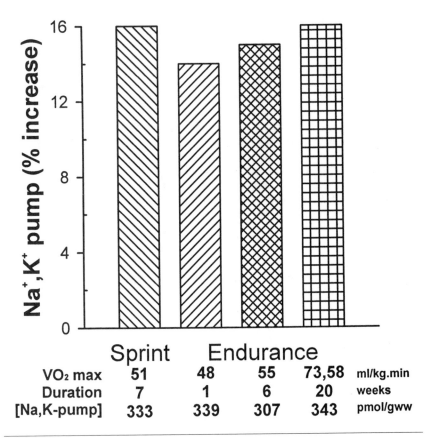

Figure 7.6.   Training increases human vastus lateralis muscle Na⁺,K⁺-pump concentration. Percent increases with training are shown for sprint and endurance training. Data from Evertsen et al. (18), Green et al. (27), Madsen et al. (61), and McKenna et al. (67).

Down-Regulation of Na⁺,K⁺-Pumps With Inactivity

Immobilization or reduced activity diminishes the Na⁺,K⁺-pump concentration in muscles of the guinea pig, rat, and sheep (41, 51, 56). These changes also were reversible with training with an ~83–93% range in pump concentration between immobilized and trained muscle (51, 56). It is likely that individuals with chronic inactivity also would show down-regulation of Na⁺,K⁺-pumps in skeletal muscle, but the direct effects of inactivity per se on Na⁺,K⁺-pumps in humans and the functional consequences remain to be clarified. Of interest are the findings that patients with chronic heart failure have low muscle Na⁺,K⁺-pump concentration and show an excessive rise in plasma [K⁺] with exercise (3, 71). Heart- and/or lung-transplant recipients also showed an excessive rise in arterial [K⁺] during exercise (29), but this did not appear to be due to reduced maximal muscle Na⁺,K⁺-pump activity (93).

Functional Effects of Altered Na⁺,K⁺-Pump Concentration in Humans

Very few studies have attempted to ascertain the functional importance of altered skeletal muscle Na⁺,K⁺-pump concentration in exercising humans. The contrasting effects of increased Na⁺,K⁺-pump concentration with endurance training, and decreased Na⁺,K⁺-pump concentration with digoxin therapy, on muscle K⁺ loss during exercise are shown in figure 7.7.

Reduced Pump Concentration.   Schmidt et al. (80) reported marked effects of functionally reduced Na⁺,K⁺-pump concentration after digoxin therapy on K⁺ regulation during exercise in congestive heart failure patients. After digoxin therapy, which blocked ~9% pumps, femoral venous [K⁺] was elevated (0.1–0.3 mM)

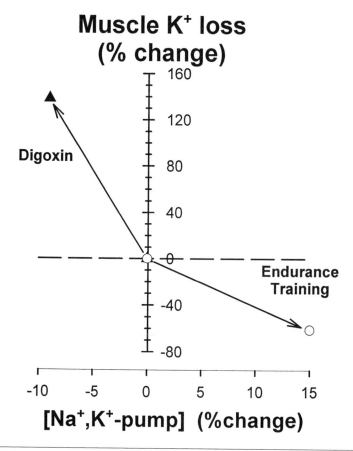

Figure 7.7.   Effects of altered Na⁺,K⁺-pump concentration in human muscles: K⁺ loss during exercise. Percent change in muscle K⁺ loss shown for digoxin therapy (Schmidt et al. [80]) and for endurance training (Kiens and Saltin [49]), where the change in pump concentration is from studies in Figure 7.6.

during exercise and recovery; the a-v [K$^+$] difference across the exercising leg was increased by 50–100% during exercise and reduced by up to 75% in recovery; and K$^+$ release from contracting leg muscles was increased by up to 138% (80). Unfortunately, the patient population used precluded rigorous performance testing, and therefore the performance effects were not investigated.

Increased Pump Concentration.   The logical benefits of Na$^+$,K$^+$-pump upregulation after training would be to reduce muscle K$^+$ loss and Na$^+$ gain, thereby minimizing fatigue and enhancing muscle performance.

Muscle K$^+$ Loss and Na$^+$ Gain.   Endurance training reduced the arteriovenous K$^+$ difference and muscle blood flow, resulting in a markedly reduced muscle K$^+$ loss (61%) during exercise at a single absolute work rate (49). Sprint training did not reduce the plasma [K$^+$]$_{a-v}$ difference (corrected for plasma water shifts) across the leg during maximal sprint cycling exercise (66). However, the a-v [K$^+$] difference in whole blood (corrected for water shifts) indicated net muscle uptake of K$^+$ during the final seconds of exercise, which was greater after training (66). Together with lower [K$^+$]$_a$ this strongly suggested reduced muscle K$^+$ loss after sprint training (66). A greater Na$^+$ uptake into active muscle was also found after sprint training (66). Estimation of the magnitude of pump activation from arteriovenous whole blood [K$^+$] differences indicate that the pump activity had increased from ~16% of the maximal theoretical pump activity before training to ~35% after sprint training.

Muscle Performance.   Several studies have investigated changes in performance and circulating [K$^+$] that accompany the ~15% increase in [Na$^+$,K$^+$-pump] in skeletal muscle after training in healthy subjects (18, 61, 67). No significant relationships were found between muscle Na$^+$,K$^+$-pump concentration and maximal repeated sprint performance (67), or high-intensity endurance exercise performance (61). In contrast, a recent study with elite cross-country skiers demonstrated significant relationships between Na$^+$,K$^+$-pump concentration and each of $\dot{V}O_2$max, 20 min treadmill run time and the rank of subjects cross-country ski performance (18). However, they concluded that the Na$^+$,K$^+$-pump was still a poor predictor of endurance performance (18).

Circulating [K$^+$] and Na$^+$,K$^+$-Pump Activation.   After endurance training, plasma [K$^+$] during incremental exercise was reduced, despite an unchanged concentration of pumps in the trained leg muscles (50). No relationship was found between the rise in [K$^+$] normalized for work output and the increase in muscle Na$^+$,K$^+$-pump concentration after sprint training (67). Sprint training resulted in reduced plasma [K$^+$] during sprint exercise, but unchanged [K$^+$] during incremental exercise (36). Together these suggest that the activation of Na$^+$,K$^+$-pumps during exercise is probably a more important determinant of systemic [K$^+$] than the concentration of pumps. However, the interaction between training, Na$^+$,K$^+$-pumps, and cation regulation in skeletal muscle has not been extensively studied.

Conclusions

The role of the Na+,K+-pumps in contracting muscle is complex, with the various pump isoforms most likely having different roles under different conditions. The pumps are activated rapidly during contraction by a variety of mechanisms, of which membrane excitation, increased intracellular [Na+], and elevated catecholamine concentrations are most important. However, despite these, the skeletal muscle Na+,K+-pumps do not appear to be maximally activated during exercise, at least in human muscles. In addition, the maximal activity of the pump in fact may decline with exercise. Consequently, during fatiguing contractions, intracellular [Na+] increases twofold to threefold, intracellular [K+] declines by 10-20%, and extracellular [K+] is doubled. It seems likely that these ionic changes are regulated phenomena and not simply a failure of homeostasis. The concentration of Na+,K+-pumps in muscle is responsive to chronic alterations in physical activity, further suggesting an important functional role. Finally, it is evident that integrated studies in humans are required to more fully understand the interactions between each of contractile activity; Na+,K+-pump concentration, isoforms, and activation in skeletal muscle; alterations in muscle [K+] and [Na+]; and muscle performance.

Acknowledgements

This paper is dedicated to the memory of John Sutton, a great and prolific scientist, but more importantly a vital, loving, and generous man.

References

1. Aledo, J. C., and Hundal, H.S. Sedimentation and immunological analyses of GLUT4 and α_2-Na,K-ATPase subunit-containing vesicles from rat skeletal muscle: evidence for segregation. *FEBS Letters* 376: 211-215; 1995.
2. Balog, E. M., and Fitts, R.H. Effects of fatiguing stimulation on intracellular Na+ and K+ in frog skeletal muscle. *J. Appl. Physiol.* 81: 679-685; 1996.
3. Barlow, C.W., Qayyum, M.S., Davey, P.P., Conway, J., Paterson, D.J., and Robbins, P.A. Effect of training on exercise-induced hyperkalaemia in chronic heart failure. Relation with ventilation and catecholamines. *Circ.* 89: 1144-1152; 1994.
4. Benders, A.G.M., van Kuppevelt, T.H.M.S.M., Oosterhof, A., Wevers, R.A., and Veerkamp, J.H. Adenosine triphosphatases during maturation of cultured human skeletal muscle cells and in adult human muscle. *Biochim. Biophys. Acta* 1112: 89-98; 1992.

5. Bigland-Ritchie, B., Bellemare, F., and Woods, J.J. Excitation frequencies and sites of fatigue. In *Human Muscle Power*. Ed. N.L. Jones, N. McCartney, and A.J. McComas, 197-211. Illinois: Human Kinetics Publishers; 1986.

6. Booth, J., McKenna, M.J., Ruell, P.A., Gwinn, T.H., Davis, G.M., Thompson, M.W., Harmer, A.R., Hunter, S.K., and Sutton, J.R. Impaired calcium pump function does not slow relaxation in human skeletal muscle after prolonged exercise. *J. Appl. Physiol.* 83: 511-521, 1997.

7. Castellino, P., Simonson, D.C., and DeFronzo, R.A. Adrenergic modulation of potassium metabolism during exercise in normal and diabetic humans. *Am. J. Physiol. (Endocrinol. Metab.)* 252: E68-E76; 1987.

8. Chin, E.R., and Green, H.J. Na$^+$-K$^+$ATPase concentration in different adult rat skeletal muscles is related to oxidative potential. *Can. J. Physiol. Pharmacol.* 71: 615-618; 1993.

9. Clausen, T. The Na$^+$,K$^+$ pump in skeletal muscle: quantification, regulation and functional significance. *Acta Physiol. Scand.* 156: 227-236; 1996.

10. Clausen, T., Andersen, S.L.V., and Flatman, J.A. Na$^+$-K$^+$ pump stimulation elicits recovery of contractility in K$^+$-paralysed rat muscle. *J. Physiol.* 472:521-536; 1993.

11. Clausen, T., and Everts, M.E. Regulation of the Na,K-pump in skeletal muscle. *Kidney International* 35:1-13; 1989.

12. Clausen, T., and Everts, M.E. K$^+$-induced inhibition of contractile force in rat skeletal muscle:role of active Na$^+$-K$^+$ transport. *Am. J. Physiol. (Cell Physiol.)* 261: C799-C807; 1991.

13. Clausen, T., Everts, M.E., and Kjeldsen, K. Quantification of the maximum capacity for active sodium-potassium transport in rat skeletal muscle. *J. Physiol.* 388:163-181; 1987.

14. Clausen, T., Hansen, O., Kjeldsen, K., and Nørgaard, A. Effect of age, potassium depletion and denervation on specific displaceable [^{3}H]ouabain binding in rat skeletal muscle in vivo. *J. Physiol.* 333: 367-381; 1982.

15. Dorup, I., Skajaa, K., and Clausen, T. A simple and rapid method for the determination of the concentrations of magnesium, sodium, potassium and sodium, potassium pumps in human skeletal muscle. *Clinical Science* 74: 241-248; 1988.

16. Everts, M.E., and Clausen, T. Excitation-induced activation of the Na$^+$-K$^+$ pump in rat skeletal muscle. *Am. J. Physiol. (Cell Physiol.)* 266: C925-C934; 1994.

17. Everts, M.E., Retterstol, K., and Clausen, T. Effects of adrenaline on excitation-induced stimulation of the sodium-potassium pump in rat skeletal muscle. *Acta Physiol. Scand.* 134:189-198; 1988.

18. Evertsen, F., Medbo, J.I., Jebens, E., and Nicolaysen, K. Hard training for 5 mo increases Na$^+$-K$^+$ pump concentration in skeletal muscle of cross-country skiers. *Am. J. Physiol. (Cell Physiol.)* 272:R1417-1424; 1997.

19. Ewart, H.S., and Klip, A. Hormonal regulation of the Na$^+$-K$^+$-ATPase: mechanisms underlying rapid and sustained changes in pump activity. *Am. J. Physiol. (Cell Physiol.)* 269: C295-C311; 1995.

20. Fambrough, D.M., Lemas, M.V., Hamrick, M., Emerick, M., Renaud, K.J., Inman, E.M., Hwang, B., and Takeyasu, K. Analysis of subunit assembly of the Na-K-ATPase. *Am. J. Physiol. (Cell Physiol.)* 266: C579-C589; 1994.

21. Fenn, W.O., and Cobb, D.M. Electrolyte changes in muscle during activity. *Am. J. Physiol.* 115:345-356; 1936.

22. Fitts, R.H. Cellular mechanisms of muscle fatigue. *Physiol. Rev.* 74: 49-94; 1994.

23. Fraser, S.F., and McKenna, M.J. Fatigue depresses human skeletal muscle Na+,K+-ATPase activity. *Proc. Aust. Physiol. Pharmacol. Soc.* 27: 47P (Abstract); 1996.

24. Fraser, S.F., McKenna, M.J., Li, J.L., Wang, X.N., and Carey, M.F. Endurance and strength training do not prevent fatigue-induced depression in Na+,K+-ATPase activity in human skeletal muscle. *Proc. 10th International Biochemistry of Exercise* 22 (Abstract); 1997.

25. Geering, K. The functional role of the β-subunit in the maturation and intracellular transport of Na,K-ATPase. *FEBS Letters* 285:189-193; 1991.

26. Green, H.J., Ball-Burnett, M., Chin, E.R., Dux, L., and Pette, D. Time-dependent increases in Na+,K+-ATPase content of low-frequency-stimulated rabbit muscle. *FEBS Letters* 310:129-131; 1992.

27. Green, H.J., Chin, E.R., Ball-Burnett, M., and Ranney, D. Increases in human skeletal muscle Na+-K+-ATPase concentration with short term training. *Am. J. Physiol. (Cell Physiol.)* 264:C1538-C1541; 1993.

28. Gullestad, L., Hallén, J., and Sejersted, O.M. K+ balance of the quadriceps muscle during dynamic exercise with and without β-adrenoceptor blockade. *J. Appl. Physiol.* 78: 513-523; 1995.

29. Hall, M.J., Snell, G.I., Side, E.A., Esmore, D.S., Walters, E.H., and Williams, T.J. Exercise, potassium, and muscle deconditioning post-thoracic organ transplantation. *J. Appl. Physiol.* 77: 2784-2790; 1994.

30. Hallén, J. K+ balance in humans during exercise. *Acta Physiol. Scand.* 156:279-286; 1996.

31. Hallén, J., Gullestad, L., and Sejersted, O.M. K+ shifts of skeletal muscle during stepwise bicycle exercise with and without β-adrenoceptor blockade. *J. Physiol.* 477:149-159; 1994.

32. Hallén, J., Saltin, B., and Sejersted, O.M. K+ balance during exercise and role of β-adrenergic stimulation. *Am. J. Physiol. (Reg. Integrative Comp. Physiol.)* 270: R1347-R1354; 1996.

33. Hallén, J., and Sejersted, O.M. Intravasal use of pliable K+-selective electrodes in the femoral vein of humans during exercise. *J. Appl. Physiol.* 75: 2318-2325; 1993.

34. Hansen, O., and Clausen, T. Quantitative determination of Na+-K+-ATPase and other sarcolemmal components in muscle cells. *Am. J. Physiol. (Cell Physiol.)* 254: C1-C7; 1988.

35. Hansen, O., and Clausen, T. Studies on sarcolemma components may be misleading due to inadequate recovery. *FEBS Letters* 384:203; 1996.

36. Harmer, A.R., McKenna, M.J., Sutton, J.R., Eager, D.M., Mackay, A., and Thompson, M.W. Plasma potassium concentration regulation is selectively enhanced during intense exercise after sprint training. *Proc. Aust. Physiol. Pharmacol. Soc.* 25:10P (Abstract); 1994.

37. Huang, W.H. and Askari, A. Regulation of (Na++K+)-ATPase by inorganic phosphate: pH dependence and physiological implications. *Biochemical and Biophysical Research Communications* 123: 438-443; 1984.

38. Hundal, H.S., Marette, A., Mitsumoto, Y., Ramlal, T., Blostein, R., and Klip, A. Insulin induces translocation of the α_2 and β_1 subunits of the Na$^+$K$^+$-ATPase from intracellular compartments to the plasma membrane in mammalian skeletal muscle. *J. Biol. Chem.* 267: 5040-5043; 1992.

39. Hundal, H.S., Marette, A., Ramlal, T., Liu, Z., and Klip, A. Expression of β subunit isoforms of the Na$^+$,K$^+$-ATPase is muscle type-specific. *FEBS Letters* 328: 253-258; 1993.

40. Hundal, H.S., Maxwell, D.L., Ahmed, A., Darakhshan, F., Mitsumoto, Y., and Klip, A. Subcellular distribution and immunocytochemical localisation of Na,K-ATPase subunit isoforms in human skeletal muscle. *Molecular Membrane Biology* 11: 255-262; 1994.

41. Jebens, E., Steen, H., Fjeld, T.O., Bye, E., and Sejersted, O.M. Changes in Na$^+$,K$^+$-adenosinetriphosphatase, citrate synthase and K$^+$ in sheep skeletal muscle during immobilization and remobilization. *Eur. J. Appl. Physiol.* 71:386-395; 1995.

42. Jewell, E.A., and Lingrel, J.B. Comparison of the substrate dependence properties of the rat Na,K-ATPase α_1, α_2 and α_3 isoforms expressed in HeLa cells. *J. Biol. Chem.* 266:16,925-16,930; 1991.

43. Jones, D.A. High- and low-frequency fatigue revisited. *Acta Physiol. Scand.* 156:265-270; 1996.

44. Jorgenson, P.L. Mechanism of the Na$^+$,K$^+$ pump protein structure and conformations of the pure (Na$^+$ + K$^+$)-ATPase. *Biochim. Biophys. Acta* 694: 27-68; 1982.

45. Juel, C. Potassium and sodium shifts during in vitro isometric muscle contraction, and the time course of the ion-gradient recovery. *Pflugers Arch.* 406: 458-463; 1986.

46. Juel, C., Bangsbo, J., Graham, T., and Saltin, B. Lactate and potassium fluxes from human skeletal muscle during and after intense, dynamic, knee extensor exercise. *Acta Physiol. Scand.* 140: 147-159; 1991.

47. Katz, A., Sahlin, K., and Juhlin-Dannfelt, A. Effect of β adrenoceptor blockade on H$^+$ and K$^+$ flux in exercising humans. *J. Appl. Physiol.* 59:336-341; 1985.

48. Keys, A. Exchanges between blood plasma and tissue fluid in man. *Science* 85: 317-318; 1937.

49. Kiens, B., and Saltin, B. Endurance training of man decreases muscle potassium loss during exercise. *Acta Physiol. Scand.* 126:20A; 1986.

50. Kjeldsen, K., Nørgaard, A., and Hau, C. Exercise-induced hyperkalaemia can be reduced in human subjects by moderate training without change in skeletal muscle Na,K-ATPase concentration. *Eur. J. Clin. Invest.* 20:642-647; 1990.

51. Kjeldsen, K., Richter, E.A., Galbo, H., Lortie, G., and Clausen, T. Training increases the concentration of [^{3}H]ouabain-binding sites in rat skeletal muscle. *Biochim. Biophys. Acta* 860:708-712; 1986.

52. Knochel, J.P., Blachley, J.D., Johnson, J.H., and Carter, N.W. Muscle cell electrical hyperpolarisation and reduced exercise hyperkalaemia in physically conditioned dogs. *J. Clin. Invest.* 75: 740-745; 1985.

53. Kristiansen, S., Hargreaves, M., and Richter, E.A. Exercise-induced increase in glucose transport, GLUT-4, and VAMP-2 in plasma membrane from human muscle. *Am. J. Physiol. (Endocrinol. Metab.)* 270: E197-E201; 1996.
54. Lavoie, L., Roy, D., Ramlal, T., Dombrowski, L., Martin-Vasallo, P., Marette, A., Carpenter, J.L., and Klip, A. Insulin-induced translocation of the Na⁺-K⁺-ATPase subunits to the plasma membrane is muscle fibre type specific. *Am. J. Physiol. (Cell Physiol.)* 270: C1421-C1429; 1996.
55. Lederer, W.J., Niggli, E., and Hadley, R.W. Sodium-calcium exchange in excitable cells: fuzzy space. *Science* 248:283; 1990.
56. Leivseth, G., Clausen, T., Everts, M.E., and Bjordal, E. Effects of reduced joint mobility and training on Na,K-ATPase and Ca-ATPase in skeletal muscle. *Muscle & Nerve* 15:843-849; 1992.
57. Levenson, R. Isoforms of the Na,K-ATPase: family members in search of function. *Rev. Physiol. Biochem. Pharm.*1: 45; 1994.
58. Li, J.L., McKenna, M.J., Fraser, S.F., Wang, X.N., and Carey, M.F. Fatigue depresses sarcoplasmic reticulum release and uptake in human skeletal muscle. *Proc. Aust. Physiol. Pharmacol. Soc.* 26:139P (Abstract); 1995.
59. Lindinger, M.I., Heigenhauser, G.J.F., McKelvie, R.S., and Jones, N.L. Blood ion regulation during repeated maximal exercise and recovery in humans. *Am. J. Physiol.* 262: R126-136; 1992.
60. Lingrel, J.B. Na,K-ATPase: isoform structure, function and expression. *Journal of Bioenergetics and Biomembranes* 24: 263-270; 1992.
61. Madsen, K., Franch, J., and Clausen, T. Effects of intensified endurance training on the concentration of Na,K-ATPase and Ca-ATPase in human skeletal muscle. *Acta Physiol. Scand.* 150: 251-258; 1994.
62. Marette, A., Krischer, J., Lavoie, L., Ackerley, C., Carpenter, J.L., and Klip, A. Insulin increases the Na⁺-K⁺-ATPase α_2-subunit in the surface membrane of rat skeletal muscle: morphological evidence. *Am. J. Physiol. (Cell Physiol.)* 265: C1716-C1722; 1993.
63. McKenna, M.J. The roles of ionic processes in muscular fatigue during intense exercise. *Sports Medicine* 13: 134-145; 1992.
64. McKenna, M.J. Effects of training on potassium homeostasis during exercise. *J. Mol. Cell. Cardiol.* 27: 941-949; 1995.
65. McKenna, M.J., Harmer, A.R., Fraser, S.F., and Li, J.L. Effects of training on potassium, calcium and hydrogen ion regulation in skeletal muscle and blood during exercise. *Acta Physiol. Scand.* 156: 335-346; 1996.
66. McKenna, M.J., Heigenhauser, G.J.F., McKelvie, R.S., MacDougall, J.D., and Jones, N.L. Sprint training enhances ionic regulation during intense exercise in men. *J. Physiol.* 501: 687-702; 1997.
67. McKenna, M.J., Schmidt, T.A., Hargreaves, M., Cameron, L., Skinner, S.L., and Kjeldsen, K. Sprint training increases human skeletal muscle Na⁺-K⁺-ATPase concentration and improves K⁺ regulation. *J. Appl. Physiol.* 75:173-180; 1993.
68. Medbø, J.I., and Sejersted, O.M. Plasma potassium changes with high intensity exercise. *J. Physiol.* 421:105-122; 1990.

69. Nielsen, O.B., and Overgaard, K. Ion gradients and contractility in skeletal muscle: the role of active Na⁺,K⁺ transport. *Acta Physiol. Scand.* 156: 247-256; 1996.

70. Nørgaard, A. Quantification of the Na,K-pumps in mammalian skeletal muscle. *Acta Pharmacol. et Toxicol.* 1-34; 1986.

71. Nørgaard, A., Bjerrgaard, P., Baandrup, U., Kjeldsen, K., Reske-Nielsen, E., and Bloch Thomsen, P.E. The concentration of Na,K-pump in skeletal and heart muscle in congestive heart failure. *Int. J. Cardiol.* 26:185-190; 1990.

72. Nørgaard, A., Kjeldsen, K., and Hansen, O. (Na⁺ + K⁺)-ATPase activity of crude homogenates of rat skeletal muscle as estimated from their K⁺-dependent 3-0-methylfluorescein phosphatase activity. *Biochim. Biophys. Acta* 770:203-209; 1984.

73. Omatsu-Kanbe, M., and Kitasato, H. Insulin stimulates the translocation of Na⁺/K⁺-dependent ATPase molecules from intracellular stores to the plasma membrane in frog skeletal muscle. *Biochem. J.* 272:727-733; 1990.

74. Paterson, D.J. Role of potassium in the regulation of systemic physiological function during exercise. *Acta Physiol. Scand.* 156:287-294; 1996.

75. Pressley, T.A. Ion concentration-dependent regulation of Na,K-pump abundance. *J. Memb. Biol.* 105:187-195; 1988.

76. Renaud, J.M., Gramolini, A., Light, P., and Comtis, A. Modulation of muscle contractility during fatigue and recovery by ATP-sensitive potassium channel. *Acta Physiol. Scand.* 156: 203-212; 1996.

77. Rolett, E.L., Strange, S., Sjøgaard, G., Kiens, B., and Saltin, B. β₂-Adrenergic stimulation does not prevent potassium loss from exercising quadriceps muscle. *Am. J. Physiol. (Reg. Integrative Comp. Physiol.)* 258: R1192-R1200; 1990.

78. Sahlin, K., and Broberg, S. Release of K⁺ from muscle during prolonged dynamic exercise. *Acta Physiol. Scand.* 136: 293-294; 1989.

79. Sangkabutra, T., McKenna, M.J., Schneider, C., Fraser, S.F., Sostaric, S., Skinner, S.L., and Crankshaw, D. Effects of contracting muscle mass and relative exercise intensity on arterial electrolytes, acid-base and pulmonary ventilation. *J. Appl. Physiol.* Submitted: 1997.

80. Schmidt, T.A., Bundgaard, H., Olesen, H.L., Secher, N.H., and Kjeldsen, K. Digoxin affects potassium homeostasis during exercise in patients with heart failure. *Cardiovascular Research* 29: 506-511; 1995.

81. Sejersted, O.M., and Hallén, J. Na,K homeostasis of skeletal muscle during activation. *Medicine Sport Sci.* 26:1-11; 1985.

82. Semb, S.O., and Sejersted, O.M. Fuzzy space and control of Na⁺,K⁺ pump rate in heart and skeletal muscle. *Acta Physiol. Scand.* 156: 213-226; 1996.

83. Sjøgaard, G. Muscle energy metabolism and electrolyte shifts during low-level prolonged static contraction in man. *Acta Physiol. Scand.* 134:181-187; 1988.

84. Sjøgaard, G., Adams, R.P., and Saltin, B. Water and ion shifts in skeletal muscle of humans with intense dynamic knee extension. *Am. J. Physiol. (Reg. Integrative Comp. Physiol.)* 248: R190-R196; 1985.

85. Skou, J.C. The influence of some cations on an adenosine triphosphatase from peripheral nerves. *Biochim. Biophys. Acta* 23: 394-446; 1957.

86. Sostaric, S., McKenna, M.J., Skinner, S.L., Brown, M.J., Sangkabutra, T., Selig, S.E., Medley, T., Medved, I., and McDonald, T. Effects of $NaHCO_3$ ingestion on arteriovenous K^+ differences during intense forearm exercise. *Proc. 10th International Biochemistry of Exercise* 21 (Abstract); 1997.

87. Sweadner, K.J. Isozymes of the Na^+/K^+-ATPase. *Biochim. Biophys. Acta* 988:185-220; 1989.

88. Tsakarides, T., Wong, P.P.C., Liu, Z., Rodgers, C.D., Vranic, M., and Klip, A. Exercise increases the plasma membrane content of the Na^+-K^+ pump and its mRNA in rat skeletal muscles. *J. Appl. Physiol.* 80: 699-705; 1996.

89. Venosa, R.A., and Horowicz, P. Density and apparent location of the sodium pump in frog sartorius muscle. *J. Memb. Biol.* 59:225-232; 1981.

90. Vøllestad, N.K., Hallén, J., and Sejersted, O.M. Effect of exercise intensity on potassium balance in muscle and blood of man. *J. Physiol.* 475: 359-368; 1994.

91. Vyskocil, F., Hnik, P., Vejsada, R., and Ujec, E. The measurement of K^+_e concentration changes in human muscles during volitional contractions. *Pflugers Arch.* 399: 235-237; 1983.

92. Westerblad, H., Lee, J.A., Lamb, A.J., Bolsover, S.R., and Allen, D.G. Spatial gradients of intracellular calcium in skeletal muscle during fatigue. *Pflugers Arch.* 415:734-740; 1990.

93. Williams, T.J., Fraser, S.F., McKenna, M.J., Li, J.L., Wang, X.N., Carey, M.F., Side, E.A., Snell, G.I., and Walters, E.H. Skeletal muscle sodium/potassium pump activity is normal post lung transplant. *Am. Rev. Resp. Crit. Care Med.* 153: A828; 1996.

CHAPTER 8

Excitation-Contraction Coupling and Fatigue in Skeletal Muscle

Graham D. Lamb

School of Zoology, La Trobe University, Bundoora, Victoria, Australia

Excitation-Contraction

Excitation-contraction (E-C) coupling is the term used to describe the sequence of events in a muscle fiber, starting with the spread of the action potential from the neuromuscular junction and culminating in the activation of the contractile apparatus and generation of force. In this brief review it is only possible to give details on some aspects of this complex process, particularly those implicated in muscle fatigue, and for more detailed descriptions the reader should refer to a number of substantial reviews on various aspects (3, 12, 18, 29, 45, 47, 59, 62, 64).

Structural Arrangement

An important feature of E-C coupling in vertebrate skeletal muscle is that not only can contraction be rapidly initiated but it can also be rapidly stopped at any time, which is clearly crucial for accurately controlled force generation. To understand this ability, one must first consider the structural arrangement of the key components in a muscle fiber (figure 8.1). Striated muscle has an extremely ordered arrangement, in which the basic functional unit, the sarcomere, is repeated many times in series to form a long chain about 1 μm in diameter, called the myofibril, which runs the entire length of the muscle fiber. The various myofibrils lie in parallel with each other, with the sarcomeres in each all aligned together, giving the whole fiber a very regular, banded pattern under polarized light. The ends of each sarcomere are delineated by the Z-lines, where the actin filaments of one sarcomere join with those of its neighbor. The length of the sarcomere is determined by the degree of overlap of the actin filaments and the myosin filaments, and at rest is about 2.0 to 2.6 μm. Wrapped around this contractile apparatus is the sarcoplasmic reticulum (SR), a membranous compartment where most of the Ca^{2+} is stored in a resting fiber. The whole muscle fiber is encompassed by the sarcolemma, or surface membrane, which is in continuity with the transverse-tubular (T-) system, a regular network of invaginations that runs all the way through the fiber in a plane perpendicular to the fiber axis and encircles each myofibril. The T-tubules in amphibian twitch fibers follow each Z-line across the fiber, whereas in most if not all mammalian skeletal fibers, there are two T-tubules per sarcomere, set on either side of the Z-line. For much of its length through the fiber, each T-tubule is closely apposed on two sides by the end chambers, or terminal cisternae, of adjacent sections of the SR (figure 8.1). This close apposition of three compartments is called the triad junction, and it is here that the surface excitation controls Ca^{2+} release and ultimately contraction.

Arrangement of Key Proteins

Communication at the triad junction is mediated by the interaction of a number of specialized proteins in the T-system and the SR membranes. Embedded in the T-system membrane are groups of four putatively identical particles arranged in a diamond configuration, and referred to as a "tetrad" (6). It is believed that these particles are dihydropyridine receptors (DHPRs) and are a type of voltage-dependent Ca^{2+} channel, which act as the "voltage-sensors" that detect the depolarization of the T-system and in some way control Ca^{2+} release from the adjacent SR (28, 29, 57, 63, 66). The DHPR tetrads occur in long chains, two or three tetrads wide, all along the T-system wherever it is in close contact with the terminal cisternae of the SR. In the apposing SR there is a matching array of even larger particles, which project all the way between the SR and T-system membranes. These particles originally were called "feet" (6, 18), and are now known to be the Ca^{2+}-release channels in the SR, and also are referred to as ryanodine receptors (RyRs), as the plant alkaloid ryanodine binds with high affinity, blocking the channels open (17, 45). Interestingly, the DHPR tetrads only associate with every alternate RyR (figure 8.1) (6), which has led to considerable speculation as to how the RyRs not facing tetrads are controlled (see later). There are also many other proteins in the triad junction, whose exact functions are unknown. There is considerable controversy as to whether a 95 kDa protein called triadin in some way links the DHPRs and the RyRs (7) or instead anchors the SR calcium binding protein, calsequestrin, close to the RyR (22 and see 18) (figure 8.1).

Dihydropyridine Receptor/Voltage Sensor

It had been clear for many decades that there must be some molecule in the T-system membrane that could sense depolarization of the T-system and somehow initiate Ca^{2+} release from the SR (24), although its identity was unknown. Such a voltage-sensing molecule must have a charged portion in order to detect changes in the electric field and hence should itself give a specific electrical signal as it moves into an activated position upon depolarization and returns to its rest position upon repolarization. Such a signal, called asymmetric charge movement, first was recorded in muscle fibers by Schneider and Chandler (63), and was subsequently shown to be generated by the DHPRs in the T-system

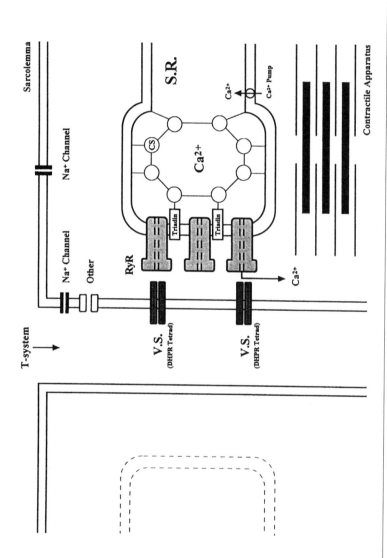

Figure 8.1. Schematic diagram of the key elements involved in the excitation-contraction coupling sequence in a vertebrate skeletal muscle fiber. An action potential travels along the sarcolemma and into the transverse-tubular (T-) system, where it activates voltage-sensor molecules (V.S.), which in turn open the ryanodine receptor (RyR)/Ca^{2+} release channels in the adjacent sarcoplasmic reticulum (S.R.). This allows Ca^{2+} to flow into the myoplasm and activate the contractile apparatus. Ca^{2+} is pumped back into the S.R., where it is predominantly bound to calsequestrin (CS).

(28, 38, 57, 66). The DHPR is presumed to function both as the voltage sensor controlling Ca^{2+} release and as a voltage-dependent Ca^{2+} channel allowing Ca^{2+} entry from the T-tubular lumen (see 29). However, it should be noted that the latter function is not critical for E-C coupling, as contraction can be triggered in vertebrate skeletal muscle even in the complete absence of extracellular Ca^{2+} (see 47). Furthermore, the skeletal muscle DHPR only opens as a Ca^{2+} channel extremely slowly (over tens of ms) and consequently will be little if at all activated during the course of a single action potential, although Ca^{2+} inflow may be quite significant during a tetanus. Each DHPR molecule actually consists of five different subunits (α_1, α_2, β, γ, δ), with the α_1 having the DHP binding site, the voltage-sensing region, the Ca^{2+} channel function, and the primary role of controlling Ca^{2+} release (65, 66, 67), although it has been recently shown that the β subunit is also critical for proper E-C coupling (4). The α_1 subunit is a 170 kDa protein in which an homologous sequence is repeated four times (called repeats I to IV) (65). Experiments with chimeras of the α_1 subunit from skeletal and cardiac DHPRs have shown that the intracellular loop between the second and third repeats of the α_1 subunit is critical for skeletal muscle-type E-C coupling in which Ca^{2+} release is controlled by the DHPR and does not depend on the inflow of extracellular Ca^{2+} (67). Thus, many investigators presume that the II-III loop of the α_1 subunit in some way physically stimulates the adjacent RyR, causing it to open and allow Ca^{2+} efflux from the SR lumen.

Ryanodine Receptor/Ca^{2+} Release Channel

The RyR/Ca^{2+} release channel is the largest channel protein known, being composed of four identical 550 kDa monomers. It has considerable homology with the other major intracellular Ca^{2+} release channel, the inositol trisphosphate ($InsP_3$) receptor, the latter being present in only relatively very small amounts in skeletal muscle (48). The homotetrameric structure of the RyR agrees well with the fourfold symmetry of the feet structures seen in three-dimensional electron microscopic image reconstructions (50, 55) and is consistent with there being four separate, though interlinked, channel pores. It is also presumed that in the case of the RyRs facing DHPR tetrads, each of the RyR monomers is linked to one of the four tetrad particles (6). Each RyR tetramer is also very tightly associated with four FK506-binding proteins (FKBPs), a ubiquitous diffusible 12 kDa protein (69). The immunosuppressant drugs, FK506 and rapamycin, bind to the FKBP, causing it to dissociate from the RyR. Such experiments have revealed that the FKBPs play a vital role in coordinating the uniform opening and closing of the four RyR monomers (1, 8, 43, 68, 69).

The activity of the RyR is modulated by many factors, such as myoplasmic (i.e., intracellular) Ca^{2+}, Mg^{2+}, ATP, inorganic phosphate (P_i), H^+, and phosphorylation (11, 45). Of critical importance is the potent inhibitory effect of myoplasmic Mg^{2+} (figure 8.2). Although Ca^{2+} can activate the RyR under some circumstances, by acting at a high affinity activation site (K_D ~1 μM), giving so-called Ca^{2+}-induced Ca^{2+} release (CICR), this effect is greatly damped in the presence of the normal level of intracellular Mg^{2+} (1 mM) (15, 33). This is due to two independent inhibitory effects of Mg^{2+}: a) competition with Ca^{2+} at its activation site, albeit with a thousandfold lower affinity, and b) binding at a low affinity inhibitory site ($K_D \sim 0.1$ mM) (39, 46). The latter site often has been called the "Ca^{2+}-inactivation" site, but it is now clear that this site is quite nonspecific for divalent metal cations (39, 46), and consequently it will be almost fully saturated with Mg^{2+} under resting conditions (1 mM). Thus, this inhibition site should be thought of functionally as a Mg^{2+}-inhibition site or a Mg^{2+}/Ca^{2+}-inhibition site (30), because even if the Ca^{2+} near the site were to reach very high levels during peak Ca^{2+} release (e.g., 100 μM), this only would cause a marginal increase in the level of saturation

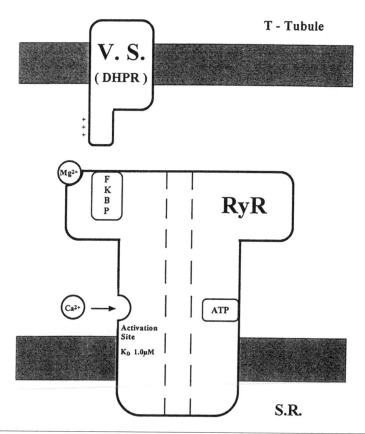

Figure 8.2. The ryanodine receptor (RyR)/Ca^{2+} release channel in skeletal muscle is subject to the excitatory and inhibitory effects of many ligands. In the absence of voltage-sensor (V.S.) stimulation, the RyR remains mostly closed and Ca^{2+}-activated Ca^{2+} release is substantially depressed, primarily due to the binding of myoplasmic Mg^{2+} at a potent, low affinity ($K_D \sim 0.1$ mM) inhibitory site. Providing that ATP is bound to a stimulatory site ($K_D \sim 1$ mM) on the RyR, voltage-sensor (V.S.) activation overcomes or removes the inhibitory effect of Mg^{2+}, possibly by a mechanism involving the FK506-binding protein (FKBP), thereby opening the channel and eliciting Ca^{2+} release.

of the site. The importance of this inhibitory site is indicated by the fact that no matter how high Ca^{2+} is raised in order to lessen the competitive effects of Mg^{2+} at the Ca^{2+}-activation site, the peak rate of Ca^{2+} release in the presence of 1mM Mg^{2+} only reaches 15–20% of the maximum possible in the absence of Mg^{2+} (15, 46). Furthermore, because ATP has a strong stimulatory effect on the RyR, when Mg^{2+} is not present on the inhibitory site, the RyR becomes appreciably activated even in the complete absence of activating Ca^{2+} (34, 46). Thus, it is clear that the RyR only remains closed in a resting fiber because of the potent inhibition occurring when Mg^{2+} is bound to this site.

E-C Coupling Sequence

Now consider the sequence of events involved in E-C coupling. The action potential generated at the neuromuscular junction spreads rapidly along the sarcolemma of the muscle fiber by the sequential activation of voltage-dependent Na^+ channels (figure 8.1). The action potential travels down the T-tubules, allowing near-synchronous activation of all parts of the fiber. The depolarization of the action potential activates the voltage-sensor/DHPR tetrads, which then activate the adjacent RyRs, perhaps by some physical interaction involving the II-III loop of the α_1 subunit. It is now clear that this communication between the voltage-sensor and the RyR does not involve either Ca^{2+} or $InsP_3$ acting as "second messenger." The principal evidence against the so-called "trigger Ca^{2+}" hypothesis is that at least a large part of voltage-sensor-activated Ca^{2+} release is entirely unaffected by heavily buffering the cytoplasm with very fast Ca^{2+} chelators (25, 52). The arguments against a role for $InsP_3$ involve the ineffectiveness of the agent in triggering Ca^{2+} release under physiological conditions, the low density of $InsP_3$ receptors, and the lack of effect of $InsP_3$ antagonists (see 54). There is now general agreement that the voltage sensors probably activate at least half of the RyRs by a relatively direct physical mechanism. It has been further proposed by some investigators that the Ca^{2+} efflux from a RyR apposed by a voltage sensor/tetrad activates the unapposed RyRs on either side by means of Ca^{2+}-induced Ca^{2+} release (CICR) (58, 62). However, it is not universally accepted that any of the RyRs normally are activated by CICR (27), and it is not clear how CICR potentially could activate even an adjacent RyR in the presence of physiological intracellular $[Mg^{2+}]$. An alternative hypothesis is that voltage sensor controlled RyRs can activate their immediate neighbors by a physical interaction, given that the RyRs are packed tightly together in an almost crystal-like arrangement.

Mg^{2+} and Voltage-Sensor Control of Ca^{2+} Release

It is important to reiterate here that, except for the possible occurrence of a brief, self-terminating surge of CICR at the commencement of activation (58), Ca^{2+} release remains under control of the voltage sensors at all times, with repolarization of the T-system causing the immediate cessation of all Ca^{2+} release. How this is achieved is not understood. We believe that the strong inhibitory effect of Mg^{2+} on the RyR (see above) is crucial in this regard (34, 35). This inhibition not only keeps the RyR channel closed when the voltage sensors are not stimulated, but it should also help the channel stop releasing Ca^{2+} when the voltage sensor activation ceases, as it reduces the tendency of the channel to undergo self-regenerating episodes of CICR. It is also clear that the voltage sensors must be able in some way to either bypass *or* remove this inhibitory effect of Mg^{2+}, because this is the physiological mechanism of triggering rapid and massive Ca^{2+} release. Now it is also important to note here that if the myoplasmic $[Mg^{2+}]$ is raised to 10 mM, T-system depolarization still activates the voltage sensors, but the voltage sensors are no longer able to activate the RyRs (34, 36). Given that the low affinity Mg^{2+}-inhibitory site on the RyR already is saturated almost fully at 1 mM Mg^{2+}, it is difficult to explain this result if the voltage sensor normally simply bypasses this inhibitory effect of Mg^{2+}. In contrast, the result is well explained by the alternative hypothesis in which the voltage sensor normally acts to remove the inhibitory effect of Mg^{2+} on the RyR by lowering the affinity of the inhibitory site for Mg^{2+}, say tenfold to twentyfold (34, 35). In the presence of physiological $[Mg^{2+}]$ (i.e., 1 mM), voltage sensor activation would cause Mg^{2+} to quickly dissociate from the inhibitory site, thereby activating the RyR channel at least partially, and then the released Ca^{2+} can activate the channel more fully by acting on the Ca^{2+}-activation site. This type of CICR is entirely under the control of the voltage sensors, because when the T-system is repolarized and the voltage sensors deactivate, the normal Mg^{2+}-affinity of the inhibitory site is restored, Mg^{2+} will rebind, and the channel will close again. However, if the $[Mg^{2+}]$ is raised to 10 mM, far above the normal resting level, voltage sensor activation would be unable to cause Mg^{2+} to dissociate from the inhibitory site and hence would not activate the channel, as observed. This hypothesis of a change in the Mg^{2+}-affinity of the RyR is still quite speculative but recently has received further support from experiments on depolarization-induced Ca^{2+} release in isolated triads (60). Furthermore, it also has been shown that a marine sponge extract, bastadin, is able to reduce the affinity of the RyR inhibitory site for Mg^{2+} by approximately tenfold (42), showing that such an affinity change is indeed possible. Interestingly, this action of bastadin on the RyR appeared to be mediated via the FKBPs (42). This appears to fit well with the recent finding that treatment of skinned fibers with FK506 or rapamycin in order to cause the dissociation of the FKBPs, results in the loss of depolarization-induced Ca^{2+} release (37). This suggests that the FKBP plays a vital role in enabling the voltage sensors to activate the RyR, and is also consistent with this involving a reduction in the affinity of the Mg^{2+}-inhibitory site.

Irrespective of whether voltage-sensor activation bypasses or removes the inhibitory effect of Mg^{2+} on the RyR, it is clear that if the $[Mg^{2+}]$ in the myoplasm rises above the normal resting level of 1 mM, the amount of Ca^{2+} released by depolarization becomes progressively reduced (34, 36). This may well be of considerable importance during fatigue (see later). It is also important to note that voltage sensor activation of Ca^{2+} release is dependent on the level of ATP in the myoplasm. It appears that ATP must be bound to a stimulatory site on the RyR ($K_D \sim 1$ mM) for the voltage sensor to be able to activate the channel (51), and this too is likely to be important in muscle fatigue (figure 8.3).

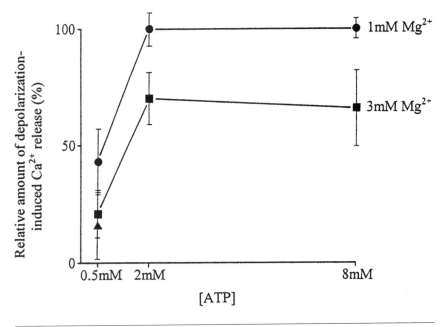

Figure 8.3.   Effect of myoplasmic [ATP] and free [Mg^{2+}] on the amount of Ca^{2+} released by depolarization in a skinned muscle fiber. Reducing the total [ATP] below 2 mM and raising the free [Mg^{2+}] above 1 mM each independently reduced depolarization-induced Ca^{2+} release. The presence of 1.5 mM AMP (triangle) did not significantly alter the amount of Ca^{2+} release at 0.5 mM ATP and 3 mM Mg^{2+}. These data suggest that depletion of ATP and a concomitant rise in free [Mg^{2+}] in the triad junction could be largely responsible for decreased Ca^{2+} release occurring in fatigue under some circumstances.

Reprinted, by permission, from Owen, Lamb, and Stephenson 1996.

Comparison With Cardiac E-C Coupling

To conclude this section, it is worth contrasting E-C coupling in cardiac and vertebrate skeletal muscle. In cardiac muscle, an action potential depolarizes the surface and T-tubular membranes, activating a voltage-dependent, DHP-sensitive Ca^{2+} channel, allowing a fast and substantial influx of extracellular Ca^{2+}. The Ca^{2+} entering the cell binds to the Ca^{2+}-activation site on a RyR/Ca^{2+} release channel, causing a self-reinforcing cycle of CICR through that and neighboring RyRs. After the prolonged release of Ca^{2+} over several hundred milliseconds, Ca^{2+} inflow and release both decline, and Ca^{2+} is both pumped back into the SR and extruded from the cell (5). In vertebrate skeletal muscle, there are a number of significant differences that are crucial for fast, accurate control of contraction, and which all relate to the direct regulation of Ca^{2+} release by the T-tubular potential. First, the skeletal muscle RyR is sensitized by a strong stimulatory effect of myoplasmic ATP but remains closed because of the potent resting inhibition produced by Mg^{2+} on the inhibitory site. Secondly, the DHPR/voltage sensor in the T-tubular membrane directly activates the adjacent RyR, without having to open as a Ca^{2+} channel. It seems that this activation mechanism must involve the lifting of resting Mg^{2+}-inhibition, rather than simple CICR. Once the inhibition has been lifted, Ca^{2+} can reinforce further Ca^{2+} release, but this can be stopped at any time by deactivating the voltage sensors. Myoplasmic Ca^{2+} then is rapidly pumped back into the SR, allowing the fiber to relax again. In skeletal muscle there are only relatively small fluxes of Ca^{2+} across the surface and T-system membranes. This may be important in a) avoiding lowering the $[Ca^{2+}]$ in the very restricted T-system space, as extracellular Ca^{2+} is critical for proper function of many ion channels (e.g., Rios and Pizarro [59]); b) preventing the $[Ca^{2+}]$ rising locally near the RyRs during activation, which still might trigger some CICR, albeit relatively weakly; and c) preventing Ca^{2+} loading of the fiber during prolonged activity, which could be deleterious and would be energetically costly, as it would have to be pumped back out of the fiber subsequently.

Aspects of Muscle Fatigue

Finally, it is worth briefly considering which sites in the E-C coupling sequence are affected in muscle fatigue and how. Muscle fatigue is not just a single state but instead is a complex phenomenon with many facets (2, 16, 64, 74). Fatigue is the reduction in force occurring after repeated activation, and it can be divided into at least three categories based on the type of stimulation and the time course of recovery.

High-Frequency Fatigue With Rapid Recovery

If a muscle is stimulated continuously at a very high frequency, there is a rapid decline in force production, and upon cessation of the stimulation there is a rapid recovery occurring in a matter of seconds. It seems that this type of muscle fatigue is caused by a progressive depolarization of the T-system, owing to an increase in the $[K^+]$ and a decrease in the $[Na^+]$ in the T-system lumen due to the repeated action potentials. This depolarization presumably results in progressively more of the voltage-dependent Na^+ channels remaining in an inactivated state, thus preventing action potential propagation down the T-system (figure 8.1) (9, 13, 26). It also is possible that some of the voltage sensors also become inactivated and hence unresponsive to depolarization, because although they require a larger and more prolonged depolarization to become inactivated, if any do move into that state, they require a much longer time to recover than do Na^+ channels (seconds compared to milliseconds) (59).

Repeated, Intermittent Stimulation Causing Metabolic Changes

If a muscle fiber is subjected to repeated, intermittent stimulation, force declines in several distinct phases, with the timing dependent on fiber type, and fully recovers again over a time course of tens of minutes after the cessation of stimulation (2). In the final, quite rapid phase of force decline, there is a reduction in amount of Ca^{2+} released by the depolarizing stimulus. When the stimulating frequency is low, it appears that the decline in Ca^{2+} release is not due to action potential failure or voltage-sensor inactivation (23, 73). With repeated stimulation there is a progressive breakdown of phosphocreatine (PCr) within the fiber, and accompanying accumulation of inorganic phosphate (P_i), H^+, and lactate, and if stimulation is severe and prolonged enough, there is a reduction in the total amount of ATP and increase in ADP, AMP, and IMP (16, 49). The initial decline in force, when normal Ca^{2+} release is still maintained, is probably due to effects of P_i and H^+ on the contractile apparatus (2). The later decline in Ca^{2+} release may have several causes.

Perhaps, the foremost of these may be direct inhibition of RyR activation by the combined effects of reduced [ATP] and raised [Mg^{2+}] in the triad junction. As mentioned above, if the [ATP] near the RyRs drops below about 2 mM, the amount of Ca^{2+} released by depolarization declines considerably (51) (figure 8.3). Furthermore, as ATP is the major Mg^{2+} buffer in the cytoplasm, this decline in [ATP] will be accompanied by a rise in free [Mg^{2+}], which itself will cause further inhibition of depolarization-induced Ca^{2+} release (34, 36, 51, 71). Even though adenosine compounds other than ATP also can stimulate the RyR, they are considerably less effective (44), and the additional presence of AMP does not help restore Ca^{2+} release at low [ATP] (51). It often is thought that low [ATP] could not be responsible for the decline in Ca^{2+} release and muscle fatigue, because the total cellular concentration only drops by 30% to 70% in extremely fatigued fibers (i.e., to ~2 to 4 mM) (16, 49), seemingly implying that there should be adequate ATP for all cellular functions. However, this is the average concentration of ATP in the fiber, and it does not indicate what happens in local regions. It is highly significant that the RyR is positioned in the triad junction, where there is a very high local density of ATPases, with many Ca^{2+} pumps in the SR terminal cisterna and Mg-ATPases and Na^+,-K^+ pumps in the T-tubular membrane, all within approximately 20 nm of the RyR. Thus, when most of the PCr has been utilized, it seems very likely that in such a region of high ATP usage the [ATP] will drop considerably below the bulk level in the myoplasm. It seems very pertinent that the ATP binding site regulating the RyR has probably the lowest affinity for ATP (K_D ~ 1 mM) of any of the important sites in the fiber and hence could act as an ideal regulatory mechanism that would reduce ATP usage if the [ATP] dropped to critically low levels, thereby preventing cellular damage that otherwise might occur if control of intracellular [Ca^{2+}] was completely lost.

It is also important to note that, contrary to common conceptions, the decline in Ca^{2+} release in the late phase of fatigue is not due to the increased [H^+]. This phenomenon still occurs even if there is no change in intracellular pH (70). Furthermore, although increased [H^+] strongly inhibits Ca^{2+} activation of the Ca^{2+} release channel (41, 61), it has little if any effect on the amount of Ca^{2+} that can be released by depolarization (32, 36).

One other factor that could contribute to reduced Ca^{2+} release is the rise in intracellular P_i. Although P_i can augment Ca^{2+} release by directly stimulating the RyR (19), it also can enter the lumen of the SR, where it can precipitate with Ca^{2+}. This would reduce the free $[Ca^{2+}]$ in the SR, as well as the total amount rapidly available for release, hence reducing the amount of Ca^{2+} released during depolarization (21, 20, 72).

Low-Frequency or Long-Lasting Fatigue

Finally, there is a third major type of muscle fatigue, in which following very prolonged stimulation the muscle fiber is unable to produce the original level of force production until a day or more later. This is often called "low-frequency" fatigue, because the response to low-frequency stimulation is substantially inhibited whereas high-frequency stimulation still can produce nearly normal force levels (14). However, it is probably more appropriately called "long-lasting" fatigue (2). The cause of this type of fatigue is not known, but the very prolonged period for recovery suggests that it is unrelated to cellular energy supplies or metabolic products, that should return to normal resting levels in a matter of hours. One possible cause is some type of cellular change resulting from elevating the intracellular $[Ca^{2+}]$ for a prolonged period. If the myoplasmic $[Ca^{2+}]$ is raised to ~10 to 20 μM for 10s, or to only 2.5 μM for 60s, it causes substantial, irreversible inhibition of depolarization-induced Ca^{2+} release in skinned muscle fibers (31). A comparable phenomenon recently has been observed in intact muscle fibers (10). The basis of this uncoupling has not been identified, but it does appear to occur in parallel with structural changes at the triad junction that are detectable with electron microscopy (31).

Another factor that might also contribute to long-lasting fatigue is the generation of oxygen-free radicals during exercise (56). Oxidation possibly could deleteriously affect the function of many proteins. We recently have found evidence that reactive disulphides, which specifically oxidize free sulfhydryl groups, abolish depolarization-induced Ca^{2+} release in skinned muscle fibers (53). This inhibitory effect was augmented by activating and inactivating the voltage sensors and could be reversed by the strong reducing agent, dithiothreitol, but less readily by the endogenous reducing agent, glutathione. It appeared that the voltage sensors were made dysfunctional by oxidizing agents, causing the loss of E-C coupling, even though oxidation actually increased the sensitivity of the RyRs (53). Thus, oxygen-free radicals generated during exercise may cause a long-lasting fatigue by acting on the voltage sensors.

Concluding Remarks

It is apparent that E-C coupling involves a complex sequence of events, and that interruption at any point can ultimately affect force production. Nevertheless, it is worth emphasizing that the coupling system actually performs its function extremely reliably, and force production can be maintained for long periods when a muscle is stimulated at physiological rates. When pushed to their limit, muscle fibers do show reduced performance, and this has given us important insight into the molecular mechanisms involved in the coupling.

Acknowledgments

I thank Professor D.G. Stephenson for his helpful comments and the National Health and Medical Research Council of Australia for support.

References

1. Ahern, G.P., Junankar, P.R., and Dulhunty, A.F. Single channel activity of the ryanodine receptor calcium release channel is modulated by FK-506. *FEBS Letters* 352:369-374; 1994.
2. Allen, D.G., Lännergren, J., and Westerblad, H. Muscle cell function during prolonged activity: cellular mechanisms of fatigue. *Experimental Physiology* 80: 497-527; 1995.
3. Ashley, C.C., Mulligan, I.P., and Lea, T.J. Ca^{2+} and activation mechanisms in skeletal muscle. *Quarterly Reviews of Biophysics* 24:1-73; 1991.
4. Bers, D.M. *Excitation-contraction coupling and cardiac contractile force.* Dordrecht, The Netherlands: Kluwer Academic; 1991.
5. Beurg, M., Sukhareva, M., Powers, P.A., Gregg, R.G., and Coronado, R. Recovery of excitation-contraction coupling and L-Type calcium current in ß1-null myotubes transfected with ß₁ dihydropyridine receptor subunit cDNA. *Biophysical Journal* 72: A377; 1997.
6. Block, B.A., Imagawa, T., Campbell, K.P., and Franzini-Armstrong, C. Structural evidence for direct interaction between the molecular components of the transverse tubule/sarcoplasmic reticulum junction in skeletal muscle. *Journal of Cell Biology* 107: 2587-2600; 1988.
7. Brandt, N.R., Caswell, A.H., Wen, S.R., and Talvenheimo, J.A. Molecular interactions of the junctional foot protein and dihydropyridine receptor in skeletal muscle triads. *Journal of Membrane Biology* 113: 237-251; 1990.
8. Brillantes, A.-M.B., Ondrias, K., Scott, A., Kobrinsky, E., Ondriasova, E., Moschella, M.C., Jayaraman, T., Landers, M., Ehrlich, B.E., and Marks, A.R. Stabilization of calcium release channel (ryanodine receptor) function by FK 506-binding protein. *Cell* 77: 513-523; 1994.
9. Cairns, S.P., and Dulhunty, A.F. High-frequency fatigue in rat skeletal muscle: role of extracellular ion concentrations. *Muscle and Nerve* 18: 890-898; 1995.
10. Chin, E.R., and Allen, D.G. The role of elevation in intracellular $[Ca^{2+}]$ in the development of low frequency in mouse single muscle fibres. *Journal of Physiology* 491: 813-824; 1996.
11. Coronado, R., Morrissette, J., Sukhareva, M., and Vaughan, D.M. Structure and function of ryanodine receptors. *American Journal of Physiology* 266: C1485-C1504; 1994.

12. Dulhunty, A.F. The voltage-activation of contraction in skeletal muscle. *Progress in Biophysics and Molecular Biology* 57:181-223; 1992.
13. Duty, S., and Allen, D.G. The distribution of intracellular calcium concentration in isolated single muscle fibres of mouse skeletal muscle during fatiguing stimulation. *Pflügers Archiv* 427: 102-109; 1994.
14. Edwards, R.H.T., Hill, D.K., Jones, D.A., and Merton, P.A. Fatigue of long duration in human skeletal muscle after exercise. *Journal of Physiology* 272: 769-778; 1977.
15. Endo, M. Calcium release from sarcoplasmic reticulum. *Current Topics in Membranes and Transport* 25: 181-230; 1985.
16. Fitts, R.H. Cellular mechanisms of muscle fatigue. *Physiological Reviews* 74: 49-94; 1994.
17. Fleischer, S., and Inui, M. Biochemistry and biophysics of excitation-contraction coupling. *Annual Review of Biophysics and Biophysical Chemistry* 18: 333-364; 1989.
18. Franzini-Armstrong, C., and Jorgensen, A.O. Structure and development of E-C coupling units in skeletal muscle. *Annual Review of Physiology* 56: 509-534; 1994.
19. Fruen, B.R., Mickelson, J.R., Shomer, N.H., Roghair, T.R., and Louis, C.F. Regulation of the sarcoplasmic reticulum ryanodine receptor by inorganic phosphate. *Journal of Biological Chemistry* 269:192-198; 1994.
20. Fryer, M.W., Owen, V.J., Lamb, G.D., and Stephenson, D.G. Effects of creatine phosphate and P_i on force development and Ca^{2+} movements in rat skinned skeletal muscle fibres. *Journal of Physiology* 482:123-140; 1995.
21. Fryer, M.W., West, J.M., and Stephenson, D.G. Phosphate transport into the sarcoplasmic reticulum of skinned fibres from rat skeletal muscle. *Journal of Muscle Research and Cell Motility* 18: 161-167; 1997.
22. Guo, W., and Campbell, K.P. Association of triadin with the ryanodine receptor and calsequestrin in the lumen of the sarcoplasmic reticulum. *Journal of Biological Chemistry* 270: 9027-9030; 1995.
23. Györke, S. Effects of repeated tetanic stimulation on excitation-contraction coupling in cut muscle fibres of the frog. *Journal of Physiology* 464: 699-710; 1993.
24. Hodgkin, A.L., and Horowicz, P. Potassium contractures in single muscle fibres. *Journal of Physiology* 153: 386-403; 1960.
25. Jacquemond, V., Csernoch, L., Klein, M.G., and Schneider, M.F. Voltage-gated and calcium-gated release during depolarization of skeletal muscle fibres. *Biophysical Journal* 60: 867-873; 1991.
26. Jones, D.A., Bigland-Ritchie, B., and Edwards, R.H.T. Excitation frequency and muscle fatigue: mechanical responses during voluntary and stimulated contractions. *Experimental Neurology* 64: 401-413; 1979.
27. Jong, D.S., Pape, P.C., Baylor, S.M., and Chandler, W.K. Calcium inactivation of calcium release in frog cut muscle fibres that contain millimolar EGTA or Fura-2. *Journal of General Physiology* 106: 337-388; 1995.
28. Lamb, G.D. Components of charge movement in rabbit muscle: the effect of tetracaine and nifedipine. *Journal of Physiology* 376: 85-100; 1986.
29. Lamb, G.D. DHP receptors and excitation-contraction coupling. *Journal of Muscle Research and Cell Motility* 13: 394-405; 1992.

30. Lamb, G.D. Ca^{2+}-inactivation, Mg^{2+}-inhibition and malignant hyperthermia. *Journal of Muscle Research and Cell Motility* 14: 554-556; 1993.

31. Lamb, G.D., Junankar, P., and Stephenson, D.G. Raised intracellular $[Ca^{2+}]$ abolishes excitation-contraction coupling in skeletal muscle fibres of rat and toad. *Journal of Physiology* 489: 349-362; 1995.

32. Lamb, G.D., Recupero, E., and Stephenson, D.G. Effect of myoplasmic pH on excitation-contraction coupling in skeletal muscle fibres of the toad. *Journal of Physiology* 448: 211-224, 1992.

33. Lamb, G.D., and Stephenson, D.G. Control of calcium release and the effect of ryanodine in skinned muscle fibres of the toad. *Journal of Physiology* 423: 519-542; 1990.

34. Lamb, G.D., and Stephenson, D.G. Effect of Mg^{2+} on the control of Ca^{2+} release in skeletal muscle fibres of the toad. *Journal of Physiology* 434: 507-528; 1991.

35. Lamb, G.D., and Stephenson, D.G. Importance of Mg^{2+} in excitation-contraction coupling in skeletal muscle. *News in Physiological Science* 7: 270-274; 1992.

36. Lamb, G.D., and Stephenson, D.G. Effect of intracellular pH and $[Mg^{2+}]$ on excitation-contraction coupling in skeletal muscle fibres of the rat. *Journal of Physiology* 489: 331-339; 1994.

37. Lamb, G.D., and Stephenson, D.G. Effects of FK506 and rapamycin on excitation-contraction coupling in skeletal muscle fibres of the rat. *Journal of Physiology* 494: 569-576; 1996.

38. Lamb, G.D., and Walsh, T. Calcium currents, charge movement and dihydropyridine binding in fast- and slow-twitch muscles of rat and rabbit. *Journal of Physiology* 393: 595-617; 1987.

39. Laver, D.R., Baynes, T.M., and Dulhunty, A.F. Magnesium-inhibition of ryanodine-receptor calcium channels: evidence for two independent mechanisms. *Journal of Membrane Biology* 156: 213-229; 1997.

40. Laver, D.R., Owen, V.J., Junankar, P.R., Taske, N.L., Dulhunty, A.F., and Lamb, G.D. Reduced inhibitory effect of Mg^{2+} on ryanodine receptor-Ca^{2+} release channels in malignant hyperthermia. *Biophysical Journal* 73: 1913-1924; 1997.

41. Ma, J., Fill, M., Knudson, M.C., Campbell, K.P., and Coronado, R. Ryanodine receptor of skeletal muscle is a gap junction-type channel. *Science* 242: 99-102; 1988.

42. Mack, M.M., Molinski, T.F., Buck, E.D., and Pessah, I.N. Novel modulators of skeletal muscle FKBP12/calcium channel complex from *Ianthella basta*. *Journal of Biological Chemistry* 269: 23,236-23,249; 1994.

43. Mayrleitner, M., Timerman, A.P., Wiederrecht, G., and Fleischer, S. The calcium release channel of the sarcoplasmic reticulum is modulated by FK-506 binding protein: effect of FKBP-12 on single channel activity of the skeletal muscle ryanodine receptor. *Cell Calcium* 15: 99-108; 1994.

44. Meissner, G. Adenine nucleotide stimulation of Ca^{2+}-induced Ca^{2+} release in sarcoplasmic reticulum. *Journal of Biological Chemistry* 259: 2365-2374; 1984.

45. Meissner, G. Ryanodine receptor/Ca^{2+} release channels and their regulation by endogenous effectors. *Annual Review of Physiology* 56: 485-508; 1994.

46. Meissner, G., Darling, E., and Eveleth, J. Kinetics of rapid Ca^{2+} release by sarcoplasmic reticulum. Effects of Ca^{2+}, Mg^{2+} and adenine nucleotides. *Biochemistry* 25: 236-244; 1986.

47. Melzer, W., Herrmann-Frank, A., and Lüttgau, H.C. The role of Ca^{2+} ions in excitation-contraction coupling of skeletal muscle fibres. *Biochimica et Biophysica Acta* 1241: 59-116; 1995.

48. Moschella, M.C., Watras, J., Jayaraman, T., and Marks, A.R. Inositol 1,4,5-trisphosphate receptor in skeletal muscle: differential expression in myofibres. *Journal of Muscle Research and Cell Motility* 16: 390-400; 1995.

49. Nagesser, A.S., Van der Laarse, W.J., and Elzinga, G. Metabolic changes with fatigue in different types of single muscle fibres of *Xenopus laevis*. *Journal of Physiology* 448: 511-523; 1992.

50. Orlova, E.V., Serysheva, I.I., Heel, M.V., Hamilton, S.L., and Chiu, W. Two structural configurations of the skeletal muscle calcium release channel. *Nature Structural Biology* 3: 547-552; 1996.

51. Owen, V.J., Lamb, G.D., and Stephenson, D.G. Effect of low [ATP] on depolarization-induced Ca^{2+} release in skeletal muscle fibres of the toad. *Journal of Physiology* 493: 309-315; 1996.

52. Pape, P.C., Jong, D.S., Chandler, W.K., and Baylor, S.M. Effect of Fura-2 on action potential-stimulated calcium release in cut twitch fibres from frog muscle. *Journal of General Physiology* 102: 295-332; 1993.

53. Posterino, G.S., and Lamb, G.D. Effect of reducing agents and oxidants on excitation-contraction coupling in skeletal muscle fibres of rats and toad. *Journal of Physiology* 496: 809-825; 1996.

54. Posterino, G.S., and Lamb, G.D. Investigation of the effect of inositol triphosphate in skinned skeletal muscle fibres with functional excitation-contraction coupling. *Journal of Muscle Research and Cell Motility* 19: 67-74; 1998.

55. Radermacher, M., Rao, V., Grassucci, R., Frank, J., Timerman, A.P., Fleischer, S., and Wagenknecht, T. Cryo-electron microscopy and three-dimensional reconstruction of the calcium release channel/ryanodine receptor from skeletal muscle. *The Journal of Cell Biology* 127: 411-423; 1994.

56. Reid, M., Haack, K.E., Franchek, K.M., Valberg, P.A., Kobzik, L., and West, M.S. Reactive oxygen in skeletal muscle I. Intracellular oxidant kinetics and fatigue in vitro. *Journal of Applied Physiology* 73: 1797-1804; 1992.

57. Rios, E., and Brum, G. Involvement of dihydropyridine receptors in excitation-contraction coupling in skeletal muscle. *Nature* 325: 717-720; 1987.

58. Rios, E., and Pizarro, G. Voltage sensors and calcium channels of excitation-contraction coupling in skeletal muscle. *News in Physiological Science* 3: 223-227; 1988.

59. Rios, E., and Pizarro, G. Voltage sensor of excitation-contraction coupling in skeletal muscle. *Physiological Reviews* 71: 849-908; 1991.

60. Ritucci, N.A., and Corbett, A.M. Effect of Mg^{2+} and ATP on depolarization-induced Ca^{2+} release in isolated triads. *American Journal of Physiology* 269: C85-C95; 1995.

61. Rousseau, E., and Pinkos, J. pH modulates conducting and gating behaviour of single channel release channels. *Pflügers Archiv* 415: 645-647; 1990.

62. Schneider, M.F. Control of calcium release in functioning muscle fibres. *Annual Review of Physiology* 56: 463-484; 1994.
63. Schneider, M.F., and Chandler, W.K. Voltage-dependent charge movement in skeletal muscle: a possible step in excitation-contraction coupling. *Nature* 242: 244-246; 1973.
64. Stephenson, D.G., Lamb, G.D., Stephenson, G.M.M., and Fryer, M.W. Mechanisms of excitation-contraction coupling relevant to skeletal muscle fatigue. In *Fatigue: Neural and Muscular Mechanisms.* Ed. S. Gandevia, R. Enoka, A. McComas, D. Stewart, and C. Thomas, 45-56. New York: Plenum; 1995.
65. Tanabe, T., Takeshima, H., Mikami, A., Flockerzi, V., Takeshima, H., Kangawa, K., Kojima, M., Matsuo, H., Hirose, T., and Numa, S. Primary structure of the receptor for calcium channel blockers from skeletal muscle. *Nature* 328: 313-318; 1987.
66. Tanabe, T., Beam, K.G., Powell, J.A., and Numa, S. Restoration of excitation-contraction coupling and slow calcium current in dysgenic muscle by dihydropyridine receptor complementary DNA. *Nature* 336: 129-134; 1988.
67. Tanabe, T., Beam, K.G., Adams, B.A., Nicodome, T., and Numa, S. Regions of the skeletal muscle dihydropyridine receptor critical for excitation-contraction coupling. *Nature* 346: 567-569; 1990.
68. Timerman, A.P., Ogunbumni, E., Freund, E., Wiederrecht, G., Marks, A.R., and Fleischer, S. The calcium release channel of sarcoplasmic reticulum is modulated by FK-506-binding protein. *Journal of Biophysical Chemistry* 268: 22,992-22,999; 1993.
69. Timerman, A.P., Wiederrecht, G., Marcy, A., and Fleischer, S. Characterization of an exchange reaction between soluble FKBP-12 and the FKBP-ryanodine receptor complex. *Journal of Biological Chemistry*, 270: 2451-2459; 1995.
70. Westerblad, H., and Allen, D.G. Changes in intracellular pH due to repetitive stimulation of single fibres from mouse skeletal muscle. *Journal of Physiology* 449: 49-71; 1992a.
71. Westerblad, H., and Allen, D.G. Myoplasmic free Mg^{2+} concentration during repetitive stimulation of single fibres from mouse skeletal muscle. *Journal of Physiology* 453: 413-434; 1992b.
72. Westerblad, H., and Allen, D.G. The effects of intracellular injections of phosphate on intracellular calcium and force in single fibres of mouse skeletal muscle. *Pflügers Archiv* 431: 964-970; 1996.
73. Westerblad, H., Duty, S., and Allen, D.G. Intracellular calcium concentration during low-frequency fatigue in isolated single fibres of mouse skeletal muscle. *Journal of Applied Physiology* 75: 382-388; 1993.
74. Westerblad, H., Lee, J.A., Lännergren, J., and Allen, D.G. Cellular mechanisms of fatigue in skeletal muscle. *American Journal of Physiology* 261: C195-C209; 1991.

CHAPTER 9

Efficiency of Energy Conversion by Muscles and Its Relation to Efficiency During Locomotion

C. J. Barclay

Department of Physiology, Monash University, Clayton, Victoria, Australia

The ability to move is a defining characteristic of animals, and in most animals skeletal muscles provide the power for movement. Muscles function by converting chemical energy, ultimately derived from the breakdown of food, into mechanical energy; they are biological energy transducers. The purpose of this article is to briefly review the current knowledge about the efficiency of energy conversion, illustrating the relative contributions of the various processes involved in muscle contraction to overall energy use by muscle, and to compare the efficiency of isolated muscle to that of whole animals during locomotion.

Efficiency of Isolated Muscles

In general terms, efficiency is the ratio of mechanical to energy used to generate that power. Precise measurements of muscle energy use only can be made using isolated muscle preparations, and experiments with these preparations are the focus of the first section of this article. The exact definition of efficiency depends on the cellular processes encompassed by the "energy used" term. At its most fundamental level, the energy used for contraction is the free energy change associated with ATP hydrolysis by actomyosin ATPase. The thermodynamic efficiency of crossbridge energy conversion (21) is defined as:

$$\epsilon_{CB} = \frac{\dot{W}}{\Delta \dot{G}}$$

where $\dot{W}$ is the power output (i.e., rate of work output), and $\Delta\dot{G}$ the rate of free energy production (the latter is the product of the rate of ATP hydrolysis and the free energy change per mole of reaction). Direct determination of ϵ_{CB} is technically difficult (e.g., 18) and a more common definition of efficiency in relation to isolated muscle is based on measurement of the enthalpy output (ΔH) from a muscle as it shortens. When a muscle shortens, enthalpy, or energy, is released as both work (W) and heat (Q). The enthalpy output that occurs during contraction is called the initial enthalpy output (ΔH_I) and reflects the net breakdown of phosphate-containing compounds, PCr and ATP. In the protocols typically used to determine efficiency, ADP formed by ATP hydrolysis is rapidly rephosphorylated at the expense of PCr, and there is little change in ATP concentration. Thus, the major process giving rise to initial enthalpy output is the net breakdown of PCr. Initial mechanical efficiency (ϵ_I) is the fraction of the initial enthalpy that appears as mechanical power and is defined as follows:

$$\epsilon_I = \frac{\dot{W}}{\Delta \dot{H}_I} = \frac{\dot{W}}{\dot{Q} + \dot{W}}$$

where $\Delta \dot{H}_I$ is the rate of enthalpy output, $\dot{W}$ is the power output, and $\dot{Q}$ is the rate of heat output. These quantities all can be readily measured with high temporal and chemical resolution. ΔH_I reflects not just the energy used by the myosin crossbridges (that perform the energy conversion) but also that used by "activation" processes. The major energy-using activation process is the ATP-powered pumping of Ca^{2+} from the sarcoplasm into the sarcoplasmic reticulum. Activation processes account for between 25% and 40% of the energy used during an isometric contraction (e.g., 3, 15). ΔH_I does not include basal metabolism.

$$\epsilon_{CB} = \frac{\dot{W}}{\Delta \dot{G}} = \frac{\dot{W}}{\Delta \dot{H}_I - \Delta \dot{H}_A} \cdot \frac{\Delta H_{PCr}}{\Delta G_{ATP}}$$

where $\Delta \dot{H}_A$ is the fraction of $\Delta \dot{H}_I$ arising from activation processes, ΔG_{ATP} is the molar free energy of ATP hydrolysis (~50 kJ mol^{-1} but depends on extent of ATP hydrolysis; 6, 18) and ΔH_{PCr} is molar enthalpy of PCr hydrolysis (~34 kJ mol^{-1}; 22). By definition, $W = \epsilon_I \cdot \Delta H_I$, so

$$\epsilon_{CB} = \epsilon_I \cdot \frac{\Delta \dot{H}_I}{\Delta \dot{H}_I - \Delta \dot{H}_A} \cdot \frac{\Delta H_{PCr}}{\Delta G_{ATP}}$$

Application of Equation 4 can be illustrated using data from frog skeletal muscle, the only muscle for which all the data required are accurately known. Reported values for ϵ_I of frog sartorius vary considerably due to variations in the rate of enthalpy output with time during a contraction (figure 9.1). Early in a tetanic contraction, enthalpy is produced at a high rate that decreases to reach a steady value after 3–4 s (12). Only when the rate of enthalpy output becomes steady does it accurately reflect the rate of PCr breakdown (5, 14, 16) and, therefore, it is only at these times that efficiency can be accurately determined using enthalpy output as an index of energy use. When efficiency is determined after 3–4 s of isometric contraction, its value is ~0.5 (e.g., 8; figure 9.1). $\Delta \dot{H}_A$ is 25% of the rate of enthalpy output during an isometric contraction ($\Delta \dot{H}_{Isom}$) and, when shortening at the velocity at which efficiency is maximal (~25 % V_{max}), $\Delta\dot{H}_I$ is ~4 $\Delta\dot{H}_{Isom}$ (13). Then:

$$\epsilon_{CB} = 0.5 \cdot \frac{4 \cdot \Delta\dot{H}_{Isom}}{4 \cdot \Delta\dot{H}_{Isom} - 0.25 \cdot \Delta\dot{H}_{Isom}} \cdot \frac{34}{50} = 0.36$$

That is, at best crossbridges can convert only ~40 % of the free energy into work. Efficiency is highly dependent on shortening velocity, so its value decreases substantially at velocities above and below 25% V_{max}. What happens to the remaining 60% of the free energy that is not converted into work? It is dissipated as heat during transitions between various crossbridge states within the crossbridge cycle. Crossbridges pass through a number of distinct biochemical states during each cycle of attachment-power generation-detachment, and the free energy changes

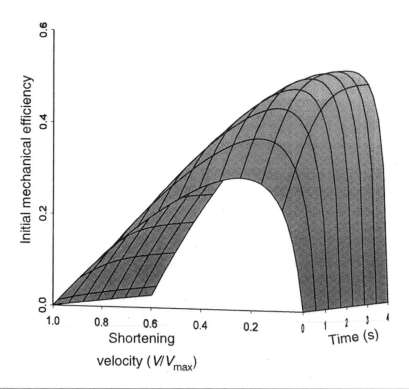

Figure 9.1. Variation in initial mechanical efficiency of frog sartorius muscle with both shortening velocity and time during a tetanus. Shortening velocity is expressed relative to the maximum velocity of shortening (V_{max}). Maximum initial mechanical efficiency increases from 0.3 at the start of a tetanus to 0.5 after ~3 s. The data were generated from empirical equations describing the time-dependence of the rate of enthalpy output from this muscle (13) and assuming that power output was independent of time.

associated with the transitions between states have been estimated on the basis of the rate and equilibrium constants for the transitions. These estimates also predict that at least 50% of the free energy is used in transitions other than those that produce work (for reviews, see 9, 17), consistent with the above estimate of ϵ_{CB}.

During a full cycle of contraction and recovery, all the initial chemical breakdown must be reversed; that is, the PCr used during the contraction must be regenerated. At moderate levels of activity, this is achieved predominantly by oxidative processes. These recovery processes also result in a net loss of free energy; that is, the free energy used by oxidative processes is greater than that gained by regenerating PCr. The thermodynamic efficiency of the recovery processes is probably between 0.6 and 0.8 (for a review, see 7). The overall efficiency of processes coupled in series is the product of the efficiencies of the individual processes (21). If we assume that the thermodynamic efficiency of the crossbridge cycle is 0.36, that of the recovery phase is 0.7 and include the cost of activation processes (~7% of the cost of crossbridge cycling when shortening at 25% V_{max}), then the overall or net efficiency (ϵ_T) is $0.36/1.07 \times 0.7 = 0.24$. Because ΔG and ΔH for the recovery processes are similar (discussed by 7), the efficiency estimated on the basis of thermodynamic efficiencies of the initial and recovery processes (i.e., based on ΔG) is very similar to that obtained on the basis of the overall enthalpy output. Thus, the net efficiency estimated above would be very similar to that determined using measurements of oxygen consumption and the enthalpy (ΔH) of this process (~20–21 J ml^{-1}):

$$\text{Overall efficiency} = \frac{\dot{W}}{\Delta \dot{G}_T} \approx \frac{\dot{W}}{20.1 \cdot \dot{V}O_2}$$

where $\dot{V}O_2$ is the rate of oxygen consumption, expressed in $\mu l \, s^{-1}$, and $\Delta \dot{G}_T$ is the rate of net free energy change of both the initial and recovery phases.

Given that muscles ultimately provide the power required for locomotion, it might be expected that the efficiency of a whole animal during locomotion would reflect the efficiency of the muscles. The value of ϵ_{CB} used in the preceding calculations is the maximum value during steady shortening at either constant velocity or against a constant load; ϵ_{CB} is lower at other velocities. During locomotion, it is unlikely that muscles shorten at the optimum velocity throughout a contraction, so overall efficiency would not be expected to be as high as the estimated value above. In addition, energy is used by other physiological systems not directly related to power output (e.g., respiratory and cardiac muscles), so it is reasonable to expect the net efficiency of an animal to be lower than 0.24 during locomotion.

Whole Animal Efficiency During Locomotion

Efficiency of animals during locomotion can be estimated using allometric equations that describe the variations in mechanical power output ($\dot{E}_{mech}$) and metabolic energy cost ($\dot{E}_{metab}$) with body size across species (10, 20). Within a species, $\dot{E}_{mech}$ and $\dot{E}_{metab}$ both vary with running velocity (for example, 1), but comparisons among species can be made by choosing a "physiologically comparable" velocity, such as the preferred speed during prolonged running or the speed corresponding to a change in gait from, say, trotting to galloping (11). The predicted variations with body mass of $\dot{E}_{mech}$ and $\dot{E}_{metab}$ (the latter excluding basal metabolism; 19) are shown in figure 9.2 for animals running at the speed corresponding to the trot-gallop transition. For a human (body mass 70 kg) this speed corresponds to ~4 m s^{-1} (~4 min km^{-1}). It is apparent (figure 9.2) that $\dot{E}_{mech}$ and $\dot{E}_{metab}$ depend on body mass in quite different ways. $\dot{E}_{mech}$ increases with body mass, whereas $\dot{E}_{metab}$ decreases with body mass. Consequently, efficiency ($\dot{E}_{mech}$ / $\dot{E}_{metab}$, with $\dot{E}_{metab}$ excluding basal metabolism) increases as size increases. That is, small animals are much less efficient than large animals when running at an equivalent speed.

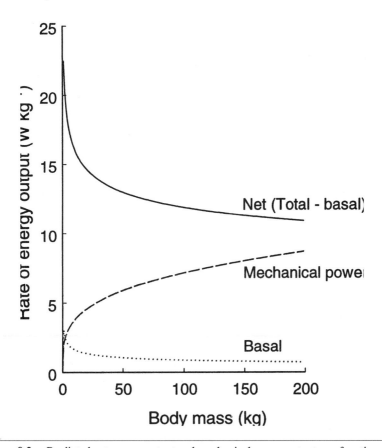

Figure 9.2. Predicted net energy output and mechanical power output as a function of body mass. Net energy output was calculated by subtracting basal metabolism (dotted line) from the total rate of energy output. Graphs are drawn from allometric equations determined using data from animals of different sizes (20, 10).

What might underlie the dependence of efficiency of the whole animal on body mass? Are muscles of small animals much less efficient than those of large animals? The answer to this question is not known, because the only accurate measurements of muscle efficiency have been made using isolated muscles from small animals (e.g., frogs, mice, and rats). However, the efficiency values for larger animals seem to be too high to be reflecting muscle efficiency. For example, even making the unlikely assumption that ϵ_{CB} in large animals is twice as great as that in small animals (for example, 0.8) and taking account of energy use by activation and recovery processes, ϵ_T would still be only 0.5 and the efficiency of the whole animal lower still. Another prediction from these equations that is inconsistent with efficiency of isolated muscles is that, for a particular species, whole animal efficiency increases monotonically with running speed. Efficiency of isolated muscles is strongly dependent on shortening velocity (figure 9.1), and it would be reasonable to expect this also to be reflected in the efficiency of the animals.

Efficiency of Running in Humans

When a 70 kg human is running at 4 m s^{-1}, both experimental data (for example, 4) and allometric equations predict that $\dot{E}_{mech}$ would be 6.8 W kg^{-1} and $\dot{E}_{metab}$ 12.8 W kg^{-1} (equivalent to $\dot{V}O_2$ of ~38 ml min^{-1} kg^{-1}; figure 9.3). Efficiency would, therefore, be ~0.52 (figure 9.3). The equations also predict that efficiency would increase with running speed. Both the magnitude of efficiency and its positive correlation with running speed have been confirmed experimentally in running humans (4). This efficiency value is much higher than the value that would be expected on the basis of muscle efficiency (< 0.25). In other words, $\dot{E}_{mech}$ is greater than expected for the magnitude of $\dot{E}_{metab}$. If it is assumed that power generation by the muscles alone would give an efficiency of 0.25, the power output corresponding to $\dot{E}_{metab}$ of 12.8 W kg^{-1} would be $0.25 \times 12.8 = 3.2$ W kg^{-1}. Therefore, 3.6 W kg^{-1}, or about half of the power output, cannot be explained on the basis of $\dot{E}_{metab}$.

The Potential of Elastic Structures to Generate Power

A popular explanation for the high efficiency of running animals compared to that expected on the basis of muscle efficiency is that elastic structures in the limbs, such as tendons, may contribute significantly to $\dot{E}_{mech}$ while not, of course, contributing to $\dot{E}_{metab}$. The force exerted on the Achilles tendon, for example, during the support phase of the stride is sufficient to stretch the tendon significantly. The energy used to stretch the tendon is then stored briefly in the tendon until the stretching force decreases during the power-developing take-off phase of the stride, when the tendon shortens and the elastic energy is released, contributing to the power output. A similar process happens with the foot; it is flattened during the support phase and recoils at lift-off. Alexander (2) has estimated that when running at ~4 m s^{-1}, the Achilles tendon and the foot can deliver ~0.8 J kg^{-1}. At 4 m s^{-1}, stride frequency is

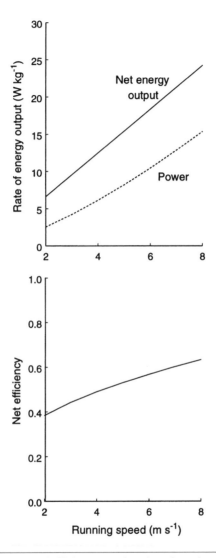

Figure 9.3.   Energy output and efficiency as a function of running velocity for a 70 kg human. Net efficiency is mechanical power output/net energy output. Figures drawn using data from Taylor et al. (20) and Heglund et al. (10).

~2.2 s^{-1}, so the Achilles/foot power output is ~1.8 W kg^{-1}. Therefore, the elastic recoil of the Achilles tendon and the foot could contribute ~25% of $\dot{E}_{mech}$ and 50% of the power that could not be explained by $\dot{E}_{metab}$. In other words, power generation by the Achilles/foot complex explains about half the power needed to account for the difference between an expected muscle efficiency of 0.25 and the observed whole human efficiency of 0.5. Presumably, the Achilles/foot complex is not the only mechanism for storing and releasing elastic energy during running, so it is reasonable to expect that at least a large fraction of the remaining "nonmetabolic" power also can be explained in these terms. To conclude, it seems entirely feasible that the high efficiency of humans (and other large animals) during running reflects not just the efficiency of the muscles but also the contribution of elastic structures to power output.

Acknowledgments

The author is supported by the National Health and Medical Research Council of Australia.

References

1. Alexander, R.M. Optimization and gaits in the locomotion of vertebrates. *Physiological Reviews* 69: 1199-1227; 1989.
2. Alexander, R.M. *Elastic Mechanisms in Animal Movement.* Cambridge: Cambridge University Press; 1988.
3. Barclay, C.J. Mechanical efficiency and fatigue of fast and slow muscles of the mouse. *Journal of Physiology* 497: 781-794; 1995.
4. Cavagna, G.A., and Kaneko, M. Mechanical work and efficiency in level walking and running. *Journal of Physiology* 268: 467-481; 1977.
5. Curtin, N.A., and Woledge, R.C. Chemical change and energy production during contraction of frog muscle: how are their time courses related? *Journal of Physiology* 288: 353-366; 1979.
6. Dawson, M.J., Gadian, D.G., and Wilkie, D.R. Muscular fatigue investigated by phosphorus nuclear magnetic resonance. *Nature* 274: 861-866; 1978.
7. Gibbs, C.L., and Barclay, C.J. Cardiac efficiency. *Cardiovascular Research* 30: 627-634; 1995.
8. Gilbert, S.H. The effect of length range on heat rate and power during shortening near in situ length in frog muscle. *Journal of Muscle Research and Cell Motility* 7: 115-121; 1986.
9. Goody, R.S., and Holmes, K.C. Cross-bridges and the mechanism of muscle contraction. *Biochimica et Biophysica Acta* 726: 13-39; 1983.
10. Heglund, N.C., Fedak, M.A., Taylor, C.R., and Cavagna, G.A. Energetics and mechanics of terrestrial locomotion. IV. Total mechanical energy changes as a function of speed and body size in birds and mammals. *Journal of Experimental Biology* 97: 57-66; 1982.
11. Heglund, N.C., Taylor, C.R., and McMahon, T.A. Scaling stride frequency and gait to animal size: mice to horses. *Science* 186: 1112-1113; 1974.
12. Hill, A.V. The effect of load on the heat of shortening. *Proceedings of the Royal Society of London B* 159: 297-318; 1964.
13. Homsher, E. Muscle enthalpy production and its relationship to actomyosin ATPase. *Annual Reviews of Physiology* 49: 673-690; 1987.

14. Homsher, E., Kean, C.J., Wallner, A., and Garibian-Sarian, V. The time course of energy balance in an isometric tetanus. *Journal of General Physiology* 73: 553-567; 1979.
15. Homsher, E., Mommaerts, W.F.H.M., Ricchiuti, N.V., and Wallner, A. Activation heat, activation metabolism and tension-related heat in frog semitendinosus muscles. *Journal of Physiology* 220: 601-625; 1972.
16. Homsher, E., Yamada, T., Wallner, A., and Tsai, J. Energy balance studies in frog skeletal muscle shortening at one-half maximal shortening velocity. *Journal of General Physiology* 84: 347-359; 1984.
17. Kodama, T. Thermodynamic analysis of muscle ATPase mechanisms. *Physiological Reviews* 65: 467-551; 1985.
18. Kushmerick, M.J., and Davies, R.E. The chemical energetics of muscle contraction. II. The chemistry, efficiency and power of maximally working sartorius muscles. *Proceedings of the Royal Society of London B* 174: 315-353; 1969.
19. Lechner, A.J. The scaling of maximal oxygen consumption and pulmonary dimensions in small mammals. *Respiration Physiology* 34: 29-44; 1978.
20. Taylor, C.R., Heglund, N.C., and Maloiy, G.M.O. Energetics and mechanics of terrestrial locomotion. I. Metabolic energy consumption as a function of speed and body size in birds and mammals. *Journal of Experimental Biology* 97: 1-21; 1982.
21. Wilkie, D.R. Thermodynamics and interpretation of biological heat measurements. *Progress in Biophysics and Biophysical Chemistry* 10: 157-183; 1960.
22. Woledge, R.C., and Reilly, P.J. Molar enthalpy change for hydrolysis of phosphorylcreatine under conditions in muscle cells. *Biophysical Journal* 54: 97-104; 1988.

CHAPTER 10

Factors Limiting ATPase Activity in Skeletal Muscle

Paavo Korge

Cardiovascular Research Laboratory, Department of Physiology,
UCLA School of Medicine, Los Angeles, CA, USA

ATPases are proteins able to bind MgATP with high affinity and subsequently change the energy level of the phosphate compound from "high" into "low" energy that allows the enzyme to perform work (force generation by cyclic interaction of myosin crossbridges with actin or ion transport against concentration gradient). For the ion transport ATPases, the energy becomes available before the phosphate compound is cleaved from the enzyme. During the catalytic cycle there is a large decrease in the K_{eq} for the hydrolysis of ATP bound to the enzyme and ion transport is coupled with the decrease in K_{eq} and not with the hydrolysis of ATP (5). When the energy deposited in the ionic gradients is used for ATP synthesis, ADP is phosphorylated on the catalytic site of the enzyme, and energy then is needed for the conversion of the phosphate compound from "low" energy into "high" energy. This occurs before ATP is released into the cytoplasm (5). Despite the fact that the detailed mechanism of energy transduction by transport ATPases is not clearly understood at present, it seems reasonable to assume that changes generated by the ATPase activity in the immediate vicinity of the catalytic site can have a profound effect on ATPase function. These changes in the ATPase reaction [products] are expected to increase during sustained contractile activity, but their possible limiting effect on the function and/or catalytic activity of various ATPases is complex and difficult to evaluate in the intact cell.

The effect of increased contractile activity on ATPase activity can be evaluated in different ways, and each approach has its limitations. First, ATPase activity and/ or function can be measured in preparations isolated from muscles at rest and after increased contractile activity. This approach has not been very informative, because it reveals changes in ATPase protein (or its lipid surrounding) that are not readily reversible and affected with isolation procedures. Generally, in these experiments ATPase surrounding media is selected to be optimal for ATPase function. Intense muscle activity, which is expected to generate highly reactive oxygen species (10, chapter 13 of this book), still can have significant impact on ATPase, via a mechanism that is independent on changes of reaction products in the vicinity of ATPase. Several ATPases, including myosin ATPase (23), sarcoplasmic reticulum (SR) Ca^{2+}-ATPase (14) and Na^+,K^+-ATPase (24), are enzymes bearing SH groups that are essential for their activity. Modification of these functionally important SH groups can selectively depress the activity of ATPase. In addition, creatine kinase (CK) is very sensitive to SH group modifications, of which two per dimer are essential for the activity of the enzyme (12). For example, CK bound to SR membranes is at least an order of magnitude more sensitive to SH group modification compared with SR Ca^{2+}-ATPase (18). This means that depression of CK activity can alter ATPase function by inhibiting local ATP regeneration in the vicinity of ATPase (see figure 10.1) before the ATPase itself is depressed due to SH group modification.

A second approach is to study the effects on ATPase activity and function of solutions mimicking intracellular milieu changes associated with intensive muscle activity and fatigue. An increase in muscle contractile activity always is accompanied by increased ATP turnover, and therefore the cell's ability to absorb the products of ATPase reaction (ADP, P_i, H^+) is an important factor regulating ATPase function. This type of regulation is likely to affect the overall reaction cycle of all ATPases involved in the energy transduction during intensive contractile activity. However, because the effects of ATPase reaction products (ligands) on ATPase function are readily reversible, those effects remain undetected when evaluated after contractile activity in the media optimal for ATPase function.

The purpose of this paper is to analyze kinetic and thermodynamic effects of ATPase reaction products on the ATPase function by mimicking changes in ATPase ligands known to be associated with high-intensity muscle activity and fatigue. Because the topic of the current meeting is connected with biomembranes, attention is focused on sarcoplasmic reticulum Ca^{2+}-ATPase.

The Effect of Changes in [ADP] on ATPase Function

Actual changes in cellular free [ADP] during contractile activity are difficult to estimate accurately. A frequently used approach to calculate [ADP] from CK reaction is based on the assumption that this reaction is in equilibrium. This assumption, while being valid for resting muscle cells, is shown to be a very rough approximation to the reality at increased workload (1). The rate of [ADP] increase in the vicinity of catalytic sites of ATPase should correspond to the rate of ATP hydrolysis. However, besides extensive binding of ADP to cellular structures, mostly to actin, the cell has several powerful mechanisms to prevent ADP buildup. Even a small increase in [ADP] can activate ADP phosphorylation via the creatine kinase (CK) reaction, because this enzyme operates as a low-threshold ADP sensor. The reported K_m of the muscle isoform of CK (MM-CK) for ADP is 10–35 μM (27). Another mechanism for ADP phosphorylation during high-intensity exercise is glycolysis. This mechanism of ATP regeneration is relatively less rapid but has higher capacity compared with the CK-catalyzed mechanism. Available experimental evidence suggests that structural proximity of ATPase and ATP-regenerating enzymes creates a microenvironment, where ADP released in ATPase reaction is immediately bound and rephosphorylated by the ATP regenerating system, so that the ATPase can reuse the regenerated ATP at its catalytic site (for review see 16, 27). Although direct determination of ADP/ATP changes in this microenvironment is impossible, the existence of functional coupling between sarcoplasmic reticulum (SR)-bound CK and Ca^{2+}-ATPase, myofibril-bound CK and myofibrillar ATPase, or sarcolemmal-bound CK and Na^+,K^+-ATPase (see 27 for review) indicates that local ATP regeneration has an impact on ATPase function. Good indirect evidence for the close structural proximity of SR-bound CK and Ca^{2+}-ATPase comes from experiments with isolated SR membrane vesicles, in which addition to the media of an ATP consuming system (apyrase or hexokinase plus glucose) quickly depressed Ca^{2+} uptake stimulated by exogenous ATP but had little effect on creatine phosphate (CP) plus ADP stimulated Ca^{2+} uptake (18). This finding, together with the fact that the apparent K_m for ATP was much lower when CK bound to the membrane was allowed to fuel the Ca^{2+} pump with ATP, can be explained by local increase in [ATP] due to the close proximity of SR-bound CK and ATP binding sites of Ca^{2+}-ATPase. In addition to being generated in Ca^{2+}-ATPase reaction, ADP is preferentially used by CK bound in the vicinity of ATPase, despite the presence of another ATP regenerating system in the medium. In these experiments CK catalyzed phosphorylation of ADP occurred after addition of CP, despite the fact that the activity of exogenous pyruvate kinase (PK) was > 20 times higher than that of endogenous CK and K_m of PK and CK for ADP are in the same range. The effect of CP was reversed fully with the CK inhibitor, dinitrofluorobenzene, indicating that membrane-bound CK was responsible for the decrease in ADP phosphorylation by exogenous PK (18). This functionally defined Ca^{2+}-ATPase microenvironment, where the concentrations of adenine nucleotides are different from that in the surrounding medium due to the high-affinity binding and close structural proximity of CK and Ca^{2+}-ATPase, is presented schematically in figure 10.1.

In experiments with SR Ca^{2+}-ATPase, enzyme function can be measured by the rate of ATP hydrolysis and coupled to that rate of ion transport. By measuring both indices one can get a coupling ratio that indicates how efficiently the enzyme transforms chemical energy into ion transport under given conditions. SR Ca^{2+}-ATPase is able to transport two Ca^{2+} ions from the cytoplasm into lumen of SR per ATP hydrolyzed (coupling ratio of two) (4). It is not known what the coupling ratio is in muscle cells at rest or whether there is any change in the transport efficiency during intense exercise and fatigue. In vitro experiments have demonstrated that Ca^{2+} transport efficiency is directly related to the increase in intravesicular free $[Ca^{2+}]$: the coupling ratio increases in parallel with an increase in Ca^{2+}-precipitating anion that is used to decrease accumulation of free Ca^{2+} inside vesicles (15). It is unknown whether or not increased Ca^{2+} cycling via Ca^{2+}-ATPase results in a $[Ca^{2+}]$ increase in the vicinity of low-affinity binding sites of Ca^{2+}-ATPase under in vivo conditions. Ca^{2+} binding proteins are located almost exclusively in the terminal part of SR (4). Ca^{2+} binding in the longitudinal part, where Ca^{2+} uptake takes place, seems to be unfavorable, because Ca^{2+} has to return to the release sites. It is not clear how fast this process occurs. Calculations based on measurements of the energy liberation in stimulated stretched muscles, where active force development was abolished (22), indicate that such a local increase in $[Ca^{2+}]$ is possible during high turnover rates of the enzyme.

In connection with the type of regulation discussed above it is important to emphasize that local ATP regeneration improves Ca^{2+}-ATPase function when Ca^{2+} transport efficiency is low (coupling ratio < 1) (15). In other words, high rates of ATP hydrolysis with the corresponding increase in Ca^{2+} cycling via the Ca^{2+}-ATPase, which is expected to take place during intense contractile activity, is likely

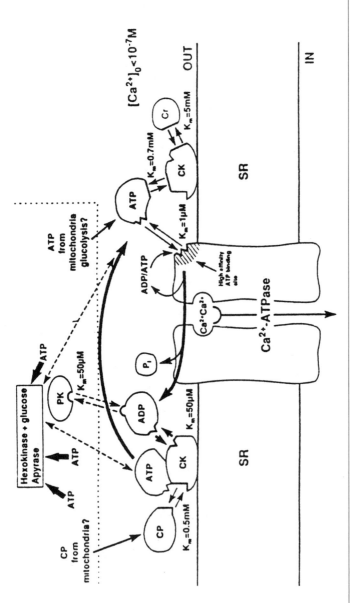

Figure 10.1. The role of membrane-bound CK in the local regeneration of ATP in the vicinity of Ca^{2+}-ATPase. ADP generated by sarcoplasmic reticulum Ca^{2+}-ATPase is bound by CK (K_m about 50 μM) and used to regenerate ATP in the presence of creatine phosphate (CP). Generated in CK reaction ATP is used preferentially by Ca^{2+}-ATPase (K_m about 1 μM). Close localization of those enzymes on SR and high-affinity binding of ADP and ATP creates a microenvironment where adenine nucleotides are not in free equilibrium with the adenine nucleotides in the surrounding medium (separated with dashed line). Under conditions where ATP turnover is high, local ATP regeneration has an important effect on Ca^{2+} transport efficiency. This general scheme is based on data from Wallimann et al. (27), Korge et al. (18), and Korge and Campbell (15).

to generate conditions where Ca^{2+} transport efficiency is low. Experimental evidence indicate that these two changes, i.e., an increase in the luminal free Ca^{2+} and accumulation of ADP generated in ATPase reaction, can stimulate Ca^{2+} efflux through the ATPase channel, producing true uncoupling of transport and catalytic activities (9, 15).

Conceivably, the importance of local ATP regeneration near catalytic sites of the ATPase increase~ when the rate of ATP hydrolysis increases. Any decrease in ATP/ADP ratio near those sites is expected to have thermodynamic consequences. In other words, the free energy available from ATP for cross-bridge cycling or active transport of ions depends on the difference in [ATP] and [ADP]+[P_i] in solution. Since maintenance of the Ca^{2+} concentration gradient by SR Ca^{2+} pump is believed to require 52 kJ/mol ATP (11), even a small drop in the free energy change may have significant effects on Ca^{2+} pump function. It is reasonable to assume that the duration of time during which the cell can maintain high rates of ATP hydrolysis will depend on the capacity of mechanisms for local ATP regeneration, provided that ATPase is not inhibited by factors other than the reaction products. When this capacity is exhausted and the ATP/ADP ratio starts to decrease, a further decrease in the free energy change of ATP hydrolysis is avoided by down-regulation of ATP consumption (13).

The Effect of Changes in [P_i] on ATPase Function

During muscle activity cellular [P_i] increases from several mM at rest to 20–30 mM after severe exercise (6, 28). This P_i increase is mainly due to CP hydrolysis. Depending on muscle fiber type [CP] is 16–32 mM at rest (19), and this amount can be almost totally depleted during intense exercise (28). The [P_i] increase has been demonstrated to decrease maximum Ca^{2+} activated tension without a significant effect on shortening velocity (3). The relationship between isometric force and ATPase activity, both measured as a function of [P_i], showed a similar reduction in soleus fibers, but in psoas (fast fiber) the reduction in ATPase activity was less than half of the reduction in force (21).

The effect of P_i on SR Ca^{2+}-ATPase function is complex because P_i has effect on ATPase cycle and at the same time improves Ca^{2+} uptake by decreasing luminal free Ca^{2+} (4). Experiments with skinned skeletal muscle fibers have provided evidence that myoplasmic P_i, depending on its concentration, can enter SR, possibly via phosphate transporter, bind Ca^{2+}, and reduce Ca^{2+} release once the calcium-phosphate solubility product is exceeded in the SR lumen (8). This proposed mechanism of P_i action also is consistent with the concentration dependent effect of P_i on the rate of Ca^{2+} uptake by isolated vesicles under conditions where P_i is the only precipitating anion (15). However, the effect of P_i on Ca^{2+} cycling through SR is apparently more complex than only binding of luminal free Ca^{2+}. First, P_i also has shown to stimulate the SR Ca^{2+} release channel (7). This effect of P_i seems to require ADP, because P_i induced Ca^{2+} efflux was reduced or abolished in the presence of active ATP regeneration by CK-CP system (25). Second, under conditions where the effect of P_i on back-inhibition is relatively small due to the presence of more potent Ca^{2+} precipitating anion (for example oxalate), P_i had dual effects on Ca^{2+} uptake. Note that 1 mM oxalate only partially decreased back inhibition. Data presented in figure 10.2 show that low $[P_i]$ significantly depressed ATP-stimulated oxalate-supported Ca^{2+} uptake, while further increases in $[P_i]$ resulted in significant reversal of this depression. Similar dual effect of P_i on Ca^{2+} uptake by SR was demonstrated in experiments with skinned papillary muscles (Zhu and Nosek [29]). The depressive effect of P_i on oxalate supported Ca^{2+} uptake was dependent on pH; at pH 6.8, 6.4, and 6.2, Ca^{2+} uptake in the presence of 1 mM oxalate was depressed maximally by 1, 3.5, and 7.5 mM P_i correspondingly (figure 10.2).

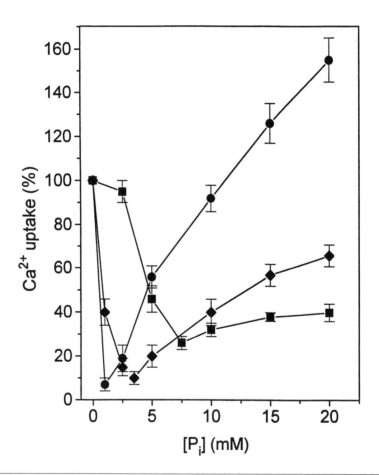

Figure 10.2.   The effect of P_i on the rate of Ca^{2+} uptake in vesicles where back inhibition was partially reduced with 1 mM oxalate. The rate of Ca^{2+} uptake in the absence of P_i was taken as 100%. 180 μg of SR vesicles, isolated from rabbit fast-twitch skeletal muscle, were incubated with 1 mM oxalate, 50 μM Ca^{2+} and indicated $[P_i]$ at pH 6.8 (●), 6.4 (◆), and 6.2 (■) for 2 min and Ca^{2+} uptake was initiated with 1 mM ATP. Data are means ± SE (n = 3-5).

The Effect of Changes in [H⁺] on ATPase Function

During high-intensity exercise the pH decrease in working muscle cells is well documented. pH_i decrease is thought to inhibit force by reducing the cross-bridge transition from low- to the high-force state (figure 10.3). In experiments with skinned or permeabilized fibers, a decrease in pH_i has been shown to decrease fiber-shortening velocity (for review see 6). Because maximal shortening velocity is considered to be proportional to myofibrillar ATPase activity, decreased pH_i is expected to also inhibit ATPase activity (6). However, myofibrillar ATPase activity showed little pH dependency between pH 6.5-8.0 in permeabilized psoas fibers, and only a slight decrease in activity associated with an increase of pH in soleus fibers (21). These results indicate that there is no simple relationship between pH effects on cross-bridge kinetics and ATP turnover.

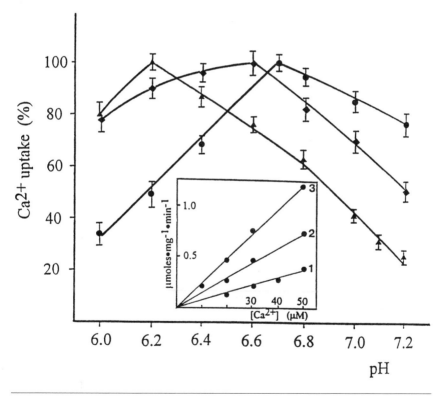

Figure 10.3.   The effect of pH on the rate of Ca^{2+} uptake by SR vesicles depends on the extent of back inhibition. SR vesicles (150 μg) were incubated with 0.75 (▲), 2.5 (◆), or 5 mM (●) oxalate and 40 μM Ca^{2+} for 2 min at the indicated pH and uptake was initiated with 1 mM ATP. Inset shows Ca^{2+} uptake in the presence of 1 mM oxalate and indicated initial extravesicular [Ca^{2+}] at pH 7.0 (1), 6.8 (2), and 6.4 (3).

It is generally believed that decreased pH inhibits SR Ca^{2+} pump function and reduces the rate of Ca^{2+} uptake by SR, which could be partially responsible for slowed relaxation associated with fatigue (6). Experiments with isolated vesicles in the presence of 5 mM oxalate support the suggestion that low pH has an inhibitory role: the rate of Ca^{2+} uptake is maximal at pH 6.7-6.8 and decreases in parallel with an increase in $[H^+]$ (figure 10.3). However, as stated above, the extent of back-inhibition of Ca^{2+} pump function by intra-SR free Ca^{2+} during intense exercise is not known. Data in figure 10.3 show that any increase in back-inhibition (decrease in oxalate) would shift the pH optimum for Ca^{2+} uptake toward lower pH. Higher rates of Ca^{2+} uptake in back-inhibited vesicles at low pH are not due to increased ATPase activity but most probably are connected with a decreased Ca^{2+} sensitivity of the low-affinity binding sites. There is also an expected decrease in Ca^{2+} binding to the high-affinity binding sites (26). However, at given extravesicular initial $[Ca^{2+}]$, the overall balance between Ca^{2+} uptake and efflux, which determines net Ca^{2+} transport, seems to be more dependent on the pH effect on the low-affinity binding sites than on the high-affinity sites (17). In support of this, several hydrophobic compounds, which are expected to decrease Ca^{2+} binding to the hydrophobic binding sites or diminish the effect initiated by that binding, stimulated Ca^{2+} transport in back-inhibited vesicles (17) by decreasing Ca^{2+} efflux and increasing the coupling ratio (2).

Taken collectively, these results suggest that SR Ca^{2+}-ATPase function is likely to be reversibly modulated by ATPase (ligand) changes during intense muscle activity. However, the effect of these ligands on Ca^{2+}-ATPase function is complex, and integration of multiple regulatory mechanisms, based on the data obtained from in vitro experiments, seems to be a complicated problem at the present time.

References

1. Aliev, M.K. and Saks, V.A. Compartmentalized energy transfer in cardiomyocytes: use of mathematical modeling for analysis of in vivo regulation of respiration. *Biophysical Journal* 73: 428-445; 1997.
2. Beeler, T.J., and Gable, K.S. Phosphate, nitrendipine and valinomycin increase the Ca^{2+}/ATP coupling ratio of skeletal muscle sarcoplasmic reticulum Ca^{2+}-ATPase. *Biochimical Biophysical Acta* 1189: 189-194; 1994.
3. Cooke, R., and Pate, E. The effect of ADP and phosphate on the contraction of muscle fibers. *Biophysical Journal* 49: 789-798; 1987.
4. de Meis, L. *The Sarcoplasmic Reticulum: Transport and Energy Transduction*. New York: J. Wiley & Sons; 1981.
5. de Meis, L. The concept of energy-rich phosphate compounds: water, transport ATPases, and entropic energy. *Archives of Biochemistry and Biophysics* 306: 287-296; 1993.
6. Fitts, R.H. Cellular mechanisms of muscle fatigue. *Physiological Reviews* 74: 49-94; 1994.
7. Fruen, B.R., Mickelson, J.R., Shomer, N.H., Roghair, T.J., and Louis, C.F. Regulation of the sarcoplasmic reticulum ryanodine receptor by inorganic phosphate. *Journal of Biological Chemistry* 269: 192-198; 1994.

8. Fryer, M.W., West, J.M., and Stephenson, D.G. Phosphate transport into the sarcoplasmic reticulum of skinned fibers from rat skeletal muscle. *Journal of Muscle Research and Cell Motility* 18: 161-168; 1997.

9. Inesi, G., and de Meis, L. Regulation of steady state filling in sarcoplasmic reticulum. Roles of back-inhibition, leakage, and slippage of the calcium pump. *Journal of Biological Chemistry* 264: 5929-5936; 1989.

10. Jenkins, R.R. Free radical chemistry: relationship to exercise. *Sports Medicine* 5: 156-170; 1988.

11. Kammermeier, H. Why do cells need phosphocreatine and phosphocreatine shuttle? *Journal of Molecular and Cellular Cardiology* 19: 115-118; 1987.

12. Kenyon, G.L., and Reed, G.L. Creatine kinase: structure-activity relationships. *Advances in Enzymology* 54: 368-426; 1983.

13. Korge, P. Factors limiting adenosine triphosphatase function during high intensity exercise. *Sports Medicine* 20: 215-225; 1995.

14. Korge, P., and Campbell, K.B. The effect of changes in iron redox state on the activity of enzymes sensitive to modification of SH groups. *Archives of Biochemistry and Biophysics* 304: 420-428; 1993.

15. Korge, P., and Campbell, K.B. Local ATP regeneration is important for sarcoplasmic reticulum Ca^{2+} pump function. *American Journal of Physiology* 267: C357-C366; 1994.

16. Korge, P., and Campbell, K.B. The importance of ATPase microenvironment in muscle fatigue: a hypothesis. *International Journal of Sports Medicine* 16: 172-179; 1995.

17. Korge, P., and Campbell, K.B. Regulation of calcium pump function in back inhibited vesicles by calcium ATPase ligands. *Cardiovascular Research* 29: 512-519; 1995.

18. Korge, P., Byrd, S.K., and Campbell, K.B. Functional coupling between sarcoplasmic reticulum bound creatine kinase and Ca^{2+}-ATPase. *European Journal of Biochemistry* 213: 973-980; 1993.

19. Kushmerick, M.J., Moerland, T.S., and Wiseman, R.W. Mammalian skeletal muscle fibers distinguished by contents of phosphocreatine, ATP, and P. *Proceedings of National Academy of Sciences USA* 89: 7521-7525; 1992.

20. Potma, E.J., van Graas, I.A., and Stienen, G.J.M. Effects of pH on myofibrillar ATPase activity in fast and slow skeletal muscle fibers of the rabbit. *Biophysical Journal* 67: 2404-2410; 1994.

21. Potma, E.J., van Graas, I.A., and Stienen, G.J.M. Influence of inorganic phosphate and pH on ATP utilization in fast and slow skeletal muscle fibers. *Biophysical Journal* 69: 2580-2589; 1995.

22. Rall, J.A. Energetics of Ca^{2+} cycling during skeletal muscle contraction. *Federation Proceedings* 41: 155-160; 1982.

23. Reisler, E., Burke, M., and Harrington, W.F. Cooperative role of two sulfhydryle groups in myosin adenosine triphosphatase. *Biochemistry* 13: 2014-2022; 1974.

24. Skou, J.C. The Na/K pump. *News in Physiological Sciences* 7: 95-99; 1992.

25. Steele, D.S., McAnish, A.M., and Smith, G.L. Effects of creatine phosphate and inorganic phosphate on the sarcoplasmic reticulum of saponin-treated rat heart. *Journal of Physiology* 483: 155-166; 1995.

26. Verjovski-Almeida, S., and de Meis, L. pH induced changes in the reactions controlled by the low- and high-affinity Ca^{2+} binding sites in sarcoplasmic reticulum. *Biochemistry* 16: 329-334; 1977.

27. Wallimann, T., Wyss, M., Brdiczka, D., Nicolay, K., and Eppenberger, H.M. Intracellular compartmentation, structure and function of creatine kinase isoenzymes in tissues with high and fluctuating energy demands: the "phosphocreatine circuit" for cellular energy homeostasis. *Biochemical Journal* 281: 21-40; 1992.

28. Westerblad, H., Lee, J.A., Lannergren, J., and Allen, D.G. Cellular mechanisms of fatigue in skeletal muscle. *American Journal of Physiology* 261: C195-C209; 1991.

29. Zhu, Y., and Nosek, T.M. Intracellular milieu changes associated with hypoxia impair sarcoplasmic reticulum Ca^{2+} transport in cardiac muscle. *American Journal of Physiology* 261: H620-H626; 1991.

CHAPTER 11

Failure of Calcium Release in Muscle Fatigue

David G. Allen, Christopher D. Balnave*, Eva R. Chin**, and Håkan Westerblad***

Department of Physiology, University of Sydney, Australia

*University Laboratory of Physiology, University of Oxford, Oxford, UK

**University of Texas Southwestern Medical Center, Dallas, TX, USA

***Department of Physiology and Pharmacology, Karolinska Institute, Stockholm, Sweden

Introduction

Muscles that are activated repeatedly at near their maximal output show a decline in performance known as muscle fatigue. This decline takes a number of forms, including reduced force production, reduced shortening velocity, and slower relaxation after contraction. As a consequence of these changes, the power production and the ability to make rapidly repeated movements decline drastically after short periods of intense muscular activity.

In this chapter we are concerned with the intracellular mechanisms of muscle fatigue. While it is accepted that fatigue has contributions from the central and peripheral nervous system in addition to muscle, it generally is agreed that changes within the muscle are the most important component under many circumstances (11, 38). This means that fatigue can be investigated conveniently in single muscle fibers, which has the advantage that many aspects of cell function (for example, the action potential, the metabolic changes, the ionic changes) can potentially be measured at the same time as the mechanical events. This approach also avoids the difficulties of attempting quantitative analysis on a muscle that is a mixture of fiber types with different functional properties. An important difference between single fibers and intact muscles is that the extracellular changes are minimized in the single fiber preparation so that this, and other possible differences, need to be borne in mind.

It long has been recognized that muscle fatigue is related to the depletion of energy sources inside muscle and/or the accumulation of metabolic waste products. Hill & Kupalov (29) showed that stimulated anoxic frog muscle became progressively weaker and, at the same time, accumulated large amounts of lactic acid; they also showed that if the lactic acid were allowed to wash away then the reduced force recovered. These observations led them to propose that lactic acid causes muscle fatigue. More recently a correlation between depletion of muscle glycogen and exhaustion has been established (10, 18). However understanding the cellular basis of these correlations between muscle fatigue and change in metabolites is much more complex and is still being explored.

There are a number of possible mechanisms whereby metabolic change might cause the decline in muscle performance. One possibility, which has a long history, is that the metabolic changes have some direct, inhibitory effect on the contractile proteins (22, 26). In the past two decades this idea has been intensively investigated using skinned muscle fibers that are exposed to solutions that simulate some of the metabolic changes that occur during fatigue. Such methods have shown that two of the metabolites that accumulate, inorganic phosphate (P_i) and protons, can have pronounced effects on the contractile proteins. Such approaches have been extensively reviewed (23, 48) and are discussed by Westerblad in the present volume. A second possibility was first proposed by Eberstein and Sandow (20) on the basis of experiments in which they showed that caffeine or high extracellular K^+ could partially overcome the reduced force of fatigued muscles. Caffeine and high K^+ bypass some stages in the normal pathway of excitation-contraction coupling and lead to increased intracellular Ca^{2+} release. Thus, Eberstein and Sandow proposed that some metabolite change occurring during fatigue caused partial failure in excitation-contraction coupling that could be overcome by such agents. This is the topic that we develop in this chapter.

Excitation-Contraction Coupling in Muscle

Excitation-contraction coupling includes all the processes involved in activating the contractile proteins. These include conduction of the action potential along the surface membrane and through the T-system, activation of the voltage sensor, transmission of a signal from the voltage sensor to the calcium release channels of the sarcoplasmic reticulum (SR) which causes them to open. Calcium then diffuses from the SR into the myoplasm. These processes result in the rapid rise of Ca^{2+} in the myoplasm. The magnitude of this rise in free intracellular calcium ($[Ca^{2+}]_i$) will depend on the flux of Ca^{2+} from the SR, the buffering power of the myoplasm, and the rate of removal of Ca^{2+} from the myoplasm that depends mainly on the SR Ca^{2+} pump. Ca^{2+} in the myoplasm binds to troponin and starts the process of crossbridge cycling.

Understanding of the processes of excitation-contraction coupling has grown rapidly in the last few years and has been the subject of many reviews (37, 42). The least understood step is the mechanism linking the voltage sensor to the SR Ca^{2+} release channel. The original hypothesis of Schneider & Chandler (43) was that a conformational change in the voltage sensor was transmitted to the SR Ca^{2+} release channel leading to channel opening. This remains a possibility, although the molecular details are not known. There is evidence that in some kinds of fatigue the SR Ca^{2+} store is loaded to its normal extent and the voltage sensor is working normally but Ca^{2+} release is reduced (see next section). This suggests that either the coupling of the voltage sensor to the SR release channel is abnormal or the release channel itself has reduced permeability when opened. For this reason the many factors that modulate the SR release channel are of great interest as possible mechanisms in fatigue. Those of possible relevance to muscle fatigue include Mg^{2+}, ATP, Ca^{2+}, SR luminal Ca^{2+}, pH, calmodulin, P_i, phosphorylation (37).

Evidence for Changes in Excitation-Contraction Coupling During Muscle Fatigue

Calcium release is the product of a number of stages of excitation-contraction coupling, and consequently, measurement of the magnitude of the rise in $[Ca^{2+}]_i$ as a result of stimulation (the Ca^{2+} transient) is a valuable way to assess possible changes in coupling. However, there are a number of difficulties in its interpretation that need to be kept in mind. As noted above, the magnitude of the Ca^{2+} transient depends on the buffering of the myoplasm and the activity of Ca^{2+} extrusion mechanisms as well as the Ca^{2+} flux from the SR. In principle, provided the properties of the SR pump and the Ca^{2+} buffer(s) are known in sufficient detail, then it is possible to calculate the Ca^{2+} release function from the measured $[Ca^{2+}]_i$ transient. The main Ca^{2+} extrusion mechanism is the SR Ca^{2+} pump and Klein et al. (1991) (31a) described a method for estimating the magnitude of the Ca^{2+} uptake by the pump based on the slope of the tail of $[Ca^{2+}]_i$ that follows a tetanus. It is known that SR Ca^{2+} pumping declines during fatigue (15, 52). Much less is known about Ca^{2+} buffering, and it seems possible that it changes during fatigue because of acidosis or the rise in resting $[Ca^{2+}]_i$. These complications render this approach difficult during fatigue, and there have been no published attempts to determine Ca^{2+} release as opposed to simply measuring the Ca^{2+} transients during muscle fatigue.

A number of studies have measured $[Ca^{2+}]_i$ in various models of fatigue that are discussed below and show that the amplitude of the Ca^{2+} transient falls during fatigue. Given the difficulties cited above of converting Ca^{2+} transients to an estimate of Ca^{2+} release, a valuable approach is to combine measurements of $[Ca^{2+}]_i$ with the use of agents, such as caffeine, which enhance Ca^{2+} release from the SR. Thus, the strongest evidence that Ca^{2+} release is failing is to show (a) that the $[Ca^{2+}]_i$ transient is reduced at the time that fatigue occurs, (b) that agents such as caffeine can increase the $[Ca^{2+}]_i$ transient, and (c) that this increase in $[Ca^{2+}]_i$ transient is associated with an appropriate increase in force production.

Using the above criteria, it now has been shown that Ca^{2+} release declines in a variety of different types of muscle fatigue, and these now will be considered briefly. Fuller accounts can be found in several recent reviews (2, 23, 48).

T-Tubular Failure

Muscles or single fibers that are continuously stimulated at or near the frequency that gives maximal force show a particularly rapid decline of force (30). Key features of this type of fatigue are that the force recovers quickly and completely after a short rest; that reduction of the stimulus frequency in the fatigued muscle often leads to a partial recovery of force, and that stimulation at lower frequencies leads to a smaller reduction of force. Jones, Bigland-Ritchie and Edwards (30) showed that the reduced force was exacerbated by reduced extracellular Na^+, and Cairns, Flatman and Clausen showed that elevated extracellular K^+ could also reduce force. These findings led to the proposal that the decline of force was caused by reductions in the amplitude or duration of the action potential, changes that were known to cause reduced Ca^{2+} release. This view was supported by measurements of the action potential that showed such changes (35).

Each action potential involves a small influx of Na^+ into the cell and a small efflux of K^+. The changes in extracellular concentration per action potential are small and quickly corrected by diffusion and the blood supply. However, within the restricted space of the T-tubules the concentration changes are larger and diffusion equilibrium is slower. Consequently, K^+ can accumulate until the membrane potential is depolarized and the Na^+ channel becomes partially inactivated, reducing the size of the action potential. The reduction in extracellular Na^+ concentration also would reduce the size of the action potential, although the fractional change in extracellular Na^+ is relatively small and seems unlikely to be functionally important. These ideas have led to the proposal that ionic changes would be greatest in the T-tubular system and that failure of the action potential to conduct through the T-system might occur in high-frequency fatigue. Evidence in support of this hypothesis has come from measurement of the distribution of Ca^{2+} across single fibers during prolonged high-frequency stimulation (47, 19). Ca^{2+} release was shown to fall more in the center of the fiber than at the edges, and this failure recovered in only a few seconds, consistent with the time required for ions to diffuse out of the T-system. Although these observations are consistent with the theory outlined above, the accumulation or deficiency of ions in the T-tubules never have been measured directly, and the proposed changes in the action potential within the T-system have not been demonstrated.

Metabolic Fatigue

In the fatigue model of repeated short tetani described above, it is well established that substantial changes in metabolites occur. Furthermore, there is a strong correlation between the rate of decline of force and the rate at which metabolic changes occur. For instance, it has been shown that the rate of decline of force depends on the duty cycle (fraction of time for which the muscle is stimulated) (9). It is known that the speed of fatigue correlates well with fiber type and metabolic profile with fibers with a higher ATP synthetic rate fatiguing more slowly than those with a lower rate (14). It also can be shown that fatigue occurs more rapidly if oxidative metabolism is prevented (29, 36). In addition, we recently have shown that fibers depleted of glycogen fatigue much more rapidly (18). All these factors suggest that the decline of force is related in some way to the metabolic status inside the fiber.

It now has been shown by a range of different methods that failure of Ca^{2+} release occurs in the final phase of the fatigue (1, 5, 50). Experiments with ion probe analysis show that the SR is still loaded with Ca^{2+} to the normal extent (27), while measurements of the charge movement associated with the voltage sensor show that it remains normal in fatigue (28). These results suggest that the critical failure in metabolic fatigue is either the coupling of the voltage sensor to the SR Ca^{2+} channel or the opening of the SR Ca^{2+} channels. The changes in metabolites that occur in this type of fatigue have been the subject of many measurements (for review see 46, 48). Approximate values at the time when force has fallen to 50% are that phosphocreatine has fallen from 35 to 2.5 mM, ATP has fallen from 6 to 4.5 mM, P_i rises from 5 to 25 mM, and ADP rises from 30 to 200 μM. Changes in pH are quite variable. In intense exercise there is usually a fall of the order of 0.5 pH units, but in long-lasting fatigue at a lower duty cycle the pH change can be much smaller or nonexistent.

If we accept that some metabolic change causes the failure of Ca^{2+} release, which metabolic change is involved and what is the mechanism of action? This question remains unresolved. One popular hypothesis is that acidosis causes the failure of Ca^{2+} release. This issue has been discussed at length elsewhere (4), and we believe the evidence against this hypothesis is overwhelming. The main reasons are that failure of Ca^{2+} release occurs in situations where there is little acidosis (51), and when acidosis does occur the failure of Ca^{2+} release appears to be very similar (Chin and Allen, in preparation). Furthermore, when muscles are made acidotic the Ca^{2+} transient becomes larger, although this may represent the known reduction in SR Ca^{2+} pumping rates in acidotic conditions (53) combined with possible changes in Ca^{2+} buffering.

Another interesting possibility is that either Mg^{2+} or ATP affects SR Ca^{2+} release. We link these possibilities because ATP has a high affinity for Mg^{2+} and normally occurs in cells in the form of MgATP. When there is net breakdown of MgATP there will be a rise in $[Mg^{2+}]_i$ and a fall in MgATP. It is well established that increasing $[Mg^{2+}]$ closes the SR Ca^{2+} release channels (45, 34), so the rising $[Mg^{2+}]_i$ during fatigue could cause or contribute to the failure of Ca^{2+} release. With this in mind we measured $[Mg^{2+}]_i$ in muscle fibers during fatigue and found that there was little change in early fatigue but a substantial rise (from 0.8 mM to 1.6 mM) during the late phase of fatigue when failure of Ca^{2+} release occurs (50). To test whether this rise in $[Mg^{2+}]_i$ could be the cause of failure of Ca^{2+} release, we injected unfatigued fibers with sufficient $MgCl_2$ to cause a higher $[Mg^{2+}]_i$ than that observed in fatigue. In these fibers we observed substantial force development despite very high levels of $[Mg^{2+}]_i$ and to reduced force to 50% required a $[Mg^{2+}]_i$ of 2.9 mM. Originally, we concluded that this suggested only a small role for elevated $[Mg^{2+}]_i$ in the failure of Ca^{2+} release. However, if we also take into account the fact that the myofibrils both have reduced sensitivity to $[Ca^{2+}]_i$ and a reduced maximum force during fatigue (53), then it may be that the effects of Mg^{2+} can explain much of the failure of Ca^{2+} release during fatigue. Further experiments in which Ca release was estimated directly in the presence of elevated $[Mg^{2+}]_i$ will be required to establish this conclusion.

It is also possible that the fall in ATP affects excitation-contraction coupling by some mechanism. As noted above, metabolic studies suggest that ATP falls from 6.0 to 4.5 mM during this type of fatigue. Calculations from the change in $[Mg^{2+}]_i$ suggest that ATP could fall as low as 1.75 mM (50). The large fall in single fibers may be because extracellular factors cause fatigue in whole muscles at an earlier stage when the ATP level is higher. How could the change in ATP affect excitation-contraction coupling? There are many possibilities that have been discussed at length elsewhere (3, 32). One possibility is the effect of ATP on the SR Ca^{2+} release channel that has been studied in bilayer experiments (44) and in skinned fibers with intact T-tubules and SR (41). Such studies have shown that ATP is required for SR channel opening and that the $K_{1/2}$ is around 0.7 mM. Even the larger fall in ATP calculated for single fibers would only have a minor effect on this basis. Another factor that reduces the likely effect of ATP is that the breakdown products in muscle (ADP and P_i) also seem able to cause some channel opening though at a lower sensitivity (37, 31). Another possibility is that the reduced ATP might cause opening of the ATP-sensitive K^+ channel, which would shorten the action potential and reduce Ca^{2+} release.

Yet another possible cause of failure of Ca^{2+} release in metabolic fatigue is that elevated myoplasmic phosphate causes precipitation of Ca^{2+} phosphate in the SR (25). This possibility was suggested from experiments on skinned fibers with raised phosphate, and it is not yet clear whether the same mechanism operates in intact fatigued muscles.

Long-Lasting Failure of Ca^{2+} Release

Humans who undertake any intense exercise without appropriate training are only too well aware that this can result in muscle pain, stiffness, and weakness, which are particularly prominent on the day following the activity. Edwards et al. (21) investigated this topic in humans and used a variety of stimulation protocols that led to muscle fatigue. They showed that the ensuing muscle weakness recovered rapidly when the muscle was tested at high frequencies of stimulation, e.g., 50–100 Hz, but if the muscle was tested at relatively low frequencies of stimulation, e.g., 10–30 Hz, then the developed force was reduced substantially, and this deficit took several days to recover. During this prolonged period of force deficit, they used muscle biopsies to show that ATP and PCr were largely recovered, and they also showed that muscle EMG was normal. They called this phenomenon low-frequency fatigue, which unfortunately can be confused with the fatigue that develops when muscles are stimulated at low frequencies.

We subsequently showed that an apparently similar phenomenon could be seen in isolated single fibers after a period of repeated tetani leading to metabolic fatigue (49). In these experiments we showed that after 10–20 min of recovery force at 100 Hz recovered to almost normal whereas the force at 20–50 Hz remained severely reduced at 60 min. It has not yet been established whether the recovery has the time course of days as observed in intact humans. An important finding in these experiments was that the tetanic $[Ca^{2+}]_i$ was reduced to about 80% of control values at all frequencies of stimulation and that the characteristic greater reduction of force at low-stimulus frequencies compared to high frequencies was a simple consequence of the shape of force-$[Ca^{2+}]_i$ relationship. These experiments also established that the Ca^{2+}-sensitivity of the contractile proteins was normal during the force deficit and that there were no radial gradients of Ca^{2+} release. We feel that a preferable name for this type of fatigue is "long-lasting failure of Ca^{2+} release." One possibility is that some aspect of fatigue leads to permanent damage to the release channel or its coupling to the voltage sensor and the time course of recovery is that of resynthesis and replacement of the damaged protein. Similar results have been obtained in *Xenopus* frog fibers (13).

In considering possible mechanisms for the putative SR release channel damage, an interesting possibility is suggested by the experiments of Lamb et al. (33). They used a skinned fiber preparation with intact T-tubular/SR coupling and showed that short periods of elevated $[Ca^{2+}]_i$ (e.g., 100 μM for 10 s) could lead to complete failure of T-tubular depolarization to trigger SR Ca^{2+} release. Following this damage they showed that the SR Ca^{2+} release channel and the surface membrane calcium channel appeared normal by antibody staining. This led us to test whether elevated $[Ca^{2+}]_i$ in the intact muscle could lead to long-lasting failure of Ca^{2+} release without the muscle necessarily having been fatigued. The idea would be that the repeated elevation of calcium associated with repeated tetani was itself enough to trigger damage to the SR Ca^{2+} release channel. Experiments were performed in which muscle fibers were given infrequent tetani so as not to cause fatigue, and the tetanic $[Ca^{2+}]_i$ was elevated with either caffeine or an inhibitor of the SR Ca^{2+} pump (tertbutyl hydroquinone). Under these conditions it was possible to produce long-lasting failure of Ca^{2+} release without fatigue (18). We subsequently showed a good correlation between the elevation of $[Ca^{2+}]_i$ during the induction period and the magnitude of the long-lasting failure of Ca^{2+} release (17).

Stretch During Contraction

Physically active humans are also well aware that exercise that involves stretching contracting muscles, sometimes called eccentric contraction, is particularly prone to cause the problems of delayed muscle pain and weakness. Thus walking down a mountain, in which the quadriceps group are stretched during contraction, is liable to cause muscle pain the next day, especially in those unaccustomed to such exercise (8).

We investigated this problem using single fibers that were stretched during a tetanus at 5 muscle lengths per s from around L_o + 10% to L_o + 60% (7) (where L_o is the length at which tetanic force is maximal). This range was chosen as it is known that long stretches on the descending limb of the tension-length relation are more likely to cause this condition in humans (40) and in isolated muscles (54). For the present purpose we intentionally used a protocol that caused little fatigue. In fibers that were stretched we found that the tetani were subsequently weaker and had reduced tetanic $[Ca^{2+}]_i$. Just as in the long-lasting failure of Ca^{2+} release caused by repeated isometric tetani, we found that the Ca^{2+} release was reduced by the same fraction at all stimulus frequencies, and as a consequence the muscle was weaker at low frequencies of stimulation than at high.

It is well known that after eccentric exercise there are ultrastructural abnormalities (24), so one issue is the extent to which the above protocol causes structural as well as excitation-contraction coupling abnormalities. We examined this issue in single fibers using rhodamine-phalloidin to stain the actin filaments that allowed us to visualize the sarcomere pattern (6). Using these fixed and stained single fibers and confocal microscopy we were able to visualize small and randomly distributed regions throughout the fiber that exhibited overstretched or misaligned sarcomeres. These were observed only with large stretches (50% L_o) during active contractions. Stretches on resting fibers had no such effect, nor did they lead to changes in Ca^{2+} release. This pattern of distribution fits well with the hypothesis proposed by Morgan (39) that suggested that overstretched sarcomeres would occur randomly throughout the fiber starting with the weakest sarcomeres. We also measured the distribution of Ca^{2+} in such fibers using conventional microscopy, but we saw only a small uniform increase in resting $[Ca^{2+}]_i$ and a small uniform reduction of tetanic $[Ca^{2+}]_i$. This suggests that the mechanism of reduced Ca^{2+} release is not limited to the regions of obvious mechanical damage but is uniform throughout the fiber.

The similarities between the changes in Ca^{2+} release following fatigue and eccentric damage suggest that they may have a similar mechanism. For instance, the coupling between the voltage sensor and the SR Ca^{2+} release channel may be damaged by a variety of interventions including the elevated Ca^{2+} associated with repeated tetani and the possible stresses associated with stretch during contraction. This concept fits well with observations in frog skeletal muscle that sometimes shows a substantial reduction of Ca^{2+} release following fatigue caused by repeated tetani (12). In this study it was found that a single stretch of a resting but fatigued muscle would sometimes greatly worsen the force response; changes in ionic strength had the same effect. Thus it seems that, particularly following fatigue, excitation-contraction coupling is unusually sensitive and can be affected by a variety of interventions.

Conclusions

A number of types of muscle fatigue have been identified in which failure of calcium release is an important component. The mechanisms involved seem to be different in each case and highlight the concept that different types of muscle activity may differentially stress different components of the system.

One frequently asked question is whether such knowledge will lead to drugs or other approaches that will minimize the effects of fatigue. While such a drug probably would be banned for use by athletes it might be of great value to patients with severe muscle or other diseases that cause fatigue to occur during essential activities. While there may be marginal gains by such approaches it seems quite likely that the failure of Ca^{2+} release is in some sense a valuable body response at least during metabolic fatigue and T-tubular failure. If muscular activity in excess of energy synthesis were to continue indefinitely then eventually the ultimate source of energy, ATP, would be depleted, and the end result of this would be muscle rigor. It seems likely that a muscle in rigor would be more incapacitating for the animal than a muscle that was simply weak. Furthermore, if some parts of a muscle approach rigor while other components are still functioning this could lead to drastic muscle damage. Thus failure of the excitation system that precedes complete metabolic depletion and protects the animal and the muscles against more serious damage may be desirable in an evolutionary sense. However, in the situations where there is long-lasting failure of Ca^{2+} release, any evolutionary advantage seems less clear. Thus if a drug or intervention was identified that minimized this type of weakness it probably would be useful. Perhaps more valuable in the long run will be to investigate how the failure of calcium release is affected by different training regimes. If it could be established that one particular component of excitation-contraction coupling was most likely to fail early in a specific type of exercise and that a particular type of training was most efficacious in minimizing that type of failure, then more appropriate training protocols might be devised.

Acknowledgments

Supported by the Australian National Health and Medical Research Council and the Swedish Medical Research Council.

References

1. Allen, D.G., Lännergren, J., and Westerblad, H. Muscle cell function during prolonged activity: cellular mechanisms of fatigue. *Experimental Physiology* 80: 497-527; 1995.
2. Allen, D.G., Lännergren, J., and Westerblad, H. The role of ATP in the regulation of intracellular Ca^{2+} release in single fibres of mouse skeletal muscle. *Journal of Physiology* 498: 587-600; 1997.
3. Allen, D.G., Lee, J.A., and Westerblad, H. Intracellular calcium and tension during fatigue in isolated single muscle fibres from *Xenopus laevis. Journal of Physiology* 415: 433-458; 1989.
4. Allen, D.G., Westerblad, H., and Lännergren, J. The role of intracellular acidosis in muscle fatigue. In *Fatigue; Neural and muscular mechanisms.* Ed. S.C. Gandevia, A.J. Enoka, G. McComas, D. Stuart, and C.K. Thomas, 57-68. New York: Plenum Press; 1995.
5. Balnave, C.D., Davey, D.F., and Allen, D.G. Distribution of sarcomere length and $[Ca^{2+}]_i$ in single fibres from mouse skeletal muscle following stretch-induced injury. *Journal of Physiology* 502: 649-659; 1997.
6. Balnave, C.D., and Allen, D.G. Intracellular calcium and force in single mouse muscle fibres following repeated contractions with stretch. *Journal of Physiology* 488: 25-36; 1995.
7. Balnave, C.D., and Thompson, M.W. Effect of training on eccentric exercise-induced muscle damage. *Journal of Applied Physiology* 75: 1545-1551; 1993.
8. Barclay, C.J., Arnold, P.D., Gibbs, C.L. Fatigue and heat production in repeated contractions of mouse skeletal muscle. *Journal of Physiology* 488: 741-752; 1995.

9. Barker, A.J., Languemare, M.C., Brandes, R., and Weiner, M.W. Intracellular tetanic calcium signals are reduced in fatigue of whole skeletal muscle. *American Journal of Physiology* 264: C577-C582; 1993.

10. Bergström, J., Hermansen, L., Hultman, E., and Saltin, B. Diet, muscle glycogen and physical performance. *Acta Physiologica Scandinavica* 71: 140-150; 1967.

11. Bigland-Ritchie, B., Furbush, F., and Woods, J.J. Fatigue of intermittent submaximal voluntary contractions: central and peripheral factors. *Journal of Applied Physiology* 61: 421-429; 1986.

12. Bruton, J.D., Lännergren, J., and Westerblad, H. Mechano-sensitive linkage in excitation-contraction coupling in frog skeletal muscle. *Journal of Physiology* 484: 737-742; 1995.

13. Bruton, J.D., Lännergren, J., and Westerblad, H. Effects of repetitive tetanic stimulation at long intervals on excitation-contraction coupling in frog skeletal muscle. *Journal of Physiology* 495: 15-22; 1996.

14. Burke, R.E., Levine, D.N., Trairis, P., and Zajac, F.E. Physiological types and histochemical profiles in muscle motor units of the cat gastrocnemius. *Journal of Physiology* 234: 723-748; 1973.

15. Byrd, S.K., Bode, A.K., and Klug, G.A. Effects of exercise of varying duration on sarcoplasmic reticulum function. *Journal of Applied Physiology* 66: 1383-1389; 1989.

16. Cairns, S.P., Flatman, J.A., and Clausen, T. Relation between extracellular [K$^+$], membrane potential and contraction in rat soleus muscle: modification by the Na$^+$-K$^+$ pump. *Pflügers Archiv European Journal of Physiology* 430: 909-915; 1995.

17. Chin, E.R., and Allen, D.G. Effects of reduced muscle glycogen concentration on force, Ca^{2+} release and contractile protein function in intact mouse skeletal muscle. *Journal of Physiology* 498: 17-29; 1997.

18. Chin, E.R., Balnave, C.D., and Allen, D.G. Role of intracellular calcium and metabolites in low-frequency fatigue in mouse skeletal muscle. *American Journal of Physiology* 272: C550-C559; 1997.

19. Duty, S., and Allen, D.G. The distribution of intracellular calcium concentration in isolated single fibres of mouse skeletal muscle during fatiguing stimulation. *Pflügers Archiv-European Journal of Physiology* 427: 102-109; 1994.

20. Eberstein, A., and Sandow, A. Fatigue mechanisms in muscle fibers. In *The Effect of Use and Disuse on the Neuromuscular Functions*, 515–526. Amsterdam: Elsevier; 1963.

21. Edwards, R.H.T., Hill, D.K., Jones, D.A., and Merton, P.A. Fatigue of long duration in human skeletal muscle after exercise. *Journal of Physiology* 272: 769-778; 1977.

22. Fabiato, A., and Fabiato, F. Effects of pH on the myofilaments and the sarcoplasmic reticulum of skinned cells from cardiac and skeletal muscles. *Journal of Physiology* 276: 233-255; 1978.

23. Fitts, R.H. Cellular mechanisms of muscle fatigue. *Physiological Reviews* 74: 49-94; 1994.

24. Fridén, J., Sjöström, M., and Ekblom, B. A morphological study of delayed muscle soreness. *Experientia* 37: 506-507; 1981.

25. Fryer, M.W., Owen, V.J., Lamb, G.D., and Stephenson, D.G. Effects of creatine phosphate and P_i on Ca^{2+} movements and tension development in rat skinned skeletal muscle fibres. *Journal of Physiology* 482: 123-140; 1995.

26. Godt, R.E., and Nosek, T.M. Changes in the intracellular milieu with fatigue or hypoxia depress contraction of skinned rabbit skeletal and cardiac muscle. *Journal of Physiology* 412: 155-180; 1989.

27. Gonzalez-Serratos, H., Somlyo, A.V., McClellan, G., Shuman, H., Borrero, L.M., and Somlyo, A.P. Composition of vacuoles and sarcoplasmic reticulum in fatigued muscle: electron probe analysis. *Proceedings of the National Academy of Sciences USA* 75: 1329-1333; 1978.

28. Györke, S. Effects of repeated tetanic stimulation on excitation-contraction coupling in cut muscle fibres of the frog. *Journal of Physiology* 464: 699-710; 1993.

29. Hill, A.V., and Kupalov, P. Anaerobic and aerobic activity in isolated muscle. *Proceedings of the Royal Society of London B* 105: 313-322; 1929.

30. Jones, D.A., Bigland-Ritchie, B., and Edwards, R.H.T. Excitation frequency and muscle fatigue: mechanical responses during voluntary and stimulated contractions. *Experimental Neurology* 64: 414-427; 1979.

31. Kermode, H., Sitsapesan, R., and Williams, A.J. ADP and inorganic phosphate activate the sheep cardiac sarcoplasmic reticulum Ca^{2+}-release channel. *Journal of Physiology* 487: 171P (Abstract); 1995.

31a. Klein, M.G., Kovacs, L., Simon, B.J., and Schneider, M.F. Decline of myoplasmic Ca^{2+}, recovery of calcium release and sarcoplasmic Ca^{2+} pump properties in frog skeletal muscle. *Journal of Physiology* 441: 639-671; 1991.

32. Korge, P., and Campbell, K.B. The importance of ATPase microenvironment in muscle fatigue: a hypothesis. *International Journal of Sports Medicine* 16: 172-179; 1995.

33. Lamb, G.D., Junankar, P.R., and Stephenson, D.G. Raised intracellular [Ca^{2+}] abolishes excitation-contraction coupling in skeletal muscle fibres of rat and toad. *Journal of Physiology* 489: 349-362; 1995.

34. Lamb, G.D., and Stephenson, D.G. Effects of Mg^{2+} on the control of Ca^{2+} release in skeletal muscle fibres of the toad. *Journal of Physiology* 434: 507-528; 1991.

35. Lännergren, J., and Westerblad, H. Force and membrane potential during and after fatiguing, continuous high-frequency stimulation of single Xenopus muscle fibres. *Acta Physiologica Scandinavica* 128: 359-368; 1986.

36. Lännergren, J., and Westerblad, H. Force decline due to fatigue and intracellular acidification in isolated fibres from mouse skeletal muscle. *Journal of Physiology* 434: 307-322; 1991.

37. Meissner, G. Ryanodine receptor/Ca^{2+} release channels and their regulation by endogenous effectors. *Annual Review of Physiology* 56: 485-508; 1994.

38. Merton, P.A. Voluntary strength and fatigue. *Journal of Physiology* 123: 553-564; 1954.

39. Morgan, D.L. New insights into the behavior of muscle during active lengthening. *Biophysical Journal* 57: 209-221; 1990.

40. Newham, D.J., Jones, D.A., Ghosh, G., and Aurora, P. Muscle fatigue and pain after eccentric contractions at long and short lengths. *Clinical Science* 74: 553-557; 1988.

41. Owen, V.J., Lamb, G.D., and Stephenson, D.G. Effect of low [ATP] on depolarization-induced Ca^{2+} release in skeletal muscle fibres of the toad. *Journal of Physiology* 493: 309-315; 1996.

42. Schneider, M.F. Control of calcium release in functioning skeletal muscle fibres. *Annual Review of Physiology* 56: 463-484; 1994.

43. Schneider, M.F., and Chandler, W.K. Voltage dependent charge movement in skeletal muscle: a possible step in excitation-contraction coupling. *Nature* 242: 244-246; 1973.

44. Smith, J.S., Coronado, R., and Meissner, G. Sarcoplasmic reticulum contains adenine nucleotide-activated calcium channels. *Nature* 316: 446-449; 1985.

45. Smith, J.S., Coronado, R., and Meissner, G. Single channel measurements of the calcium release channel from skeletal muscle sarcoplasmic reticulum. *Journal of General Physiology* 88: 573-588; 1986.

46. Vollestad, N.K., and Sejersted, O.M. Biochemical correlates of fatigue. *European Journal of Applied Physiology* 57: 336-347; 1988.

47. Westerblad, H., Lee, J.A., Lamb, A.G., Bolsover, S.R., and Allen, D.G. Spatial gradients of intracellular calcium in skeletal muscle during fatigue. *Pflügers Archiv-European Journal of Physiology* 415: 734-740; 1990.

48. Westerblad, H., Lee, J.A., Lännergren, J., and Allen, D.G. Cellular mechanisms of fatigue in skeletal muscle. *American Journal of Physiology* 261: C195-209; 1991.

49. Westerblad, H., Duty, S., and Allen, D.G. Intracellular calcium concentration during low-frequency fatigue in isolated single fibers of mouse skeletal muscle. *Journal of Applied Physiology* 75: 382-388; 1993.

50. Westerblad, H., and Allen, D.G. Myoplasmic free Mg^{2+} concentration during repetitive stimulation of single fibres from mouse skeletal muscle. *Journal of Physiology* 453: 413-434; 1992a.

51. Westerblad, H., and Allen, D.G. Changes of intracellular pH due to repetitive stimulation of single fibres from mouse skeletal muscle. *Journal of Physiology* 449: 49-71; 1992b.

52. Westerblad, H., and Allen, D.G. The contribution of $[Ca^{2+}]_i$ to the slowing of relaxation in fatigued single fibres from mouse skeletal muscle. *Journal of Physiology* 468: 729-740; 1993a.

53. Westerblad, H., and Allen, D.G. The influence of intracellular pH on contraction, relaxation and $[Ca^{2+}]_i$ in intact single fibres from mouse muscle. *Journal of Physiology* 466: 611-628; 1993b.

54. Woods, S.A., Morgan, D.L., and Proske, U. Effects of repeated eccentric contractions on structure and mechanical properties of toad sartorius muscle. *American Journal of Physiology* 265: C792-C800; 1993.

CHAPTER 12

The Role of pH and Inorganic Phosphate Ions in Skeletal Muscle Fatigue

Håkan Westerblad

Department of Physiology and Pharmacology, Karolinska Institute, Stockholm, Sweden

The performance of skeletal muscle declines during intense and/or prolonged activity. This decline is known generally as fatigue, and the underlying mechanisms have been extensively reviewed recently (2, 13). The present brief review focuses on the role of two metabolic changes occurring during intense activity: (1) acidification due to lactic acid accumulation and (2) increased concentration of inorganic phosphate ions (P_i) due to breakdown of creatine phosphate.

The present review mainly describes data obtained from our own studies, using intact, single fibers from a mouse foot muscle. The attraction of this preparation is that force is measured from a single cell during fatiguing stimulation, which makes it possible to directly correlate changes of force production to, for example, changes of the myoplasmic ion content. In most of our studies of skeletal muscle fatigue we use a stimulation protocol with repeated short contractions (tetani), and the stimulation frequency during these tetani is set so that close to maximum force is produced. Tetani are given initially every 4 s, and stimulation continues until force has declined to about 40% of the original, which usually takes less than 10 min. With this type of stimulation there is a characteristic pattern of force decline (figure 12.1): initially force falls rapidly down to about 80% of the maximal (phase 1; a to b in figure 12.1), then force production is rather stable for a variable period (phase 2; b to c), and finally there is rapid force decline until stimulation is stopped (phase 3; c to d). This pattern of force decline is not exclusive of our studies, rather it has been observed in a variety of preparations where a similar type of fatiguing stimulation has been used. The pattern first was described a century ago by Mosso (19, 3), who studied the performance of humans lifting a weight with their middle finger at 2 s intervals. It also has been observed in, for example, single motor units of cats (6) and isolated muscle fibers of frogs (20).

The force decline during phase 3 is associated with a decline of the sarcoplasmic reticulum Ca^{2+} release (27, 29). During phases 1 and 2, on the other hand, the free myoplasmic $[Ca^{2+}]$ ($[Ca^{2+}]_i$) remains high, and the force decline then would be due to impaired function of the contractile elements, the cross bridges. Acidification and accumulation of P_i may both impede cross-bridge function (15) and hence they could be responsible for the force decline during phases 1 and 2.

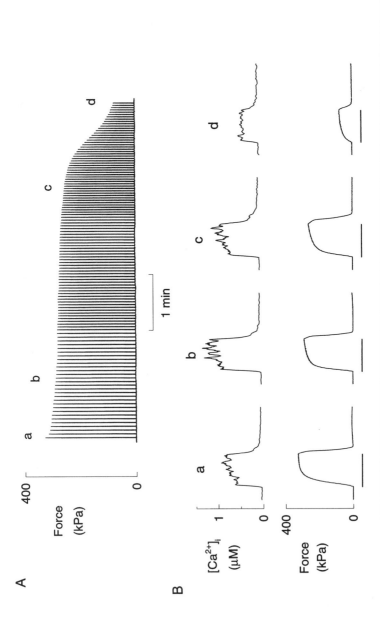

Figure 12.1.   Representative records of force and myoplasmic free [Ca²⁺] ([Ca²⁺]ᵢ) during fatigue of a single mouse muscle fiber. Panel A shows a continuous force record where each vertical line represents a 350 ms tetanus at 100 Hz. Panel B shows [Ca²⁺]ᵢ and force records from selected tetani. Periods of stimulation are indicated below records.
Adapted, by permission, from Westerblad and Allen 1993.

Besides a reduction of the force production, fatigue generally involves a reduction of the shortening velocity due to a reduced rate of cross-bridge cycling (for review, see 2). This slowing is an important factor in fatigue because most activities involve movements, and hence the power output (force × velocity) of muscles may limit the performance. The contribution of acidification and increased P_i to the reduced shortening velocity will be discussed.

Acidification

During fatigue [H^+] may increase mainly due to lactic acid accumulation, which occurs when the rate of glycolysis exceeds the mitochondrial capacity for oxidative phosphorylation or when the oxygen supply is restricted. Studies on skinned fibers have shown that acidification might reduce the isometric force and shortening velocity (8, 9, 12, 16). Furthermore, studies on human muscle fatigue often have shown a good temporal correlation between the decline of muscle pH and the reduction of force (7, 18). These findings suggest a direct coupling between acidification and impaired contractile function in fatigue. However, there are also results that indicate that decreased pH is not the cause of the force reduction. For instance, force has been found to recover more rapidly than pH in fatigued human muscle (18, 26). Furthermore, Ranatunga (25) found an increased tetanic force with acidification in rat muscle studied at 35° C, and Adams et al. (1) observed no effect of lowered pH_i on tetanic force production in cat skeletal muscles studied at 37° C.

Recent studies have further challenged the force-depressing role of H^+ in mammalian muscle fatigue by showing that the effect of H^+ very much depends on the temperature. In a study on skinned rabbit psoas fibers, Pate et al. (21) observed the expected large depressive effect of lowered pH at 10° C, but at 30° C the effect of acidification was small. We recently performed similar experiments on intact, single mouse fibers (31). In this study fibers were acidified by increasing the CO_2 content of the bath solution from 5 to 30%, which resulted in a decline of intracellular pH (pH_i) of about 0.5 pH-units. Figure 12.2 shows force records obtained at 12, 22, and 32° C under control conditions (dashed lines) and with acidification (full lines). The effect of lowered pH_i on tetanic force clearly becomes smaller as the temperature is increased. Mean data from four fibers showed that 0.5 pH-units acidification gave a force reduction of about 30% at 12° C, which is in good agreement with results from skinned fiber experiments that generally are performed at about this temperature. At 32° C, on the other hand, acidification reduced force by only 10%, which agrees with the skinned fiber results of Pate et al. (21) at 30° C. Similar results also have been obtained recently in isolated mouse muscle (32).

Acidification has been considered to be an important factor behind the reduced shortening speed in fatigue. However, in the study of Pate et al.(21) acidification was found to have little effect on the shortening speed at 30° C. Our study (31) confirmed this finding: while the maximum shortening velocity (V_0) was reduced by about 20% at 12° C, there was no significant reduction at 32° C. Thus, in mammalian muscle the cross-bridges' rate of cycling and ability to produce force are little affected by acidification at physiological temperatures.

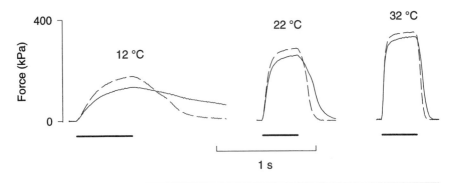

Figure 12.2. Force records from tetanic contractions of a single mouse muscle fiber under control conditions (dashed lines) and after a 0.5 pH-units acidification (full lines) produced by exposing the fiber to 30% instead of the normal 5% CO_2. The periods of stimulation are shown by the thick, horizontal lines. Observe that the effect of acidification becomes smaller as the temperature is increased.

Adapted, by permission, from Westerblad et al. 1997.

Increased Inorganic Phosphate

As discussed above, skinned fiber experiments have shown that among the metabolic changes that occur in fatigue, only acidosis and increased P_i are likely to have a large depressing effect on force production. Since acidosis does not seem to be an important cause of the depressed cross-bridge force in mammalian muscle fatigue at physiological temperature, then by exclusion the depression would be due to accumulation of P_i. This also seems to be the situation during fatigue at room temperature in single mouse muscle fibers, because these fibers fatigue without any significant acidification (28).

It is well established that increased P_i reduces the force at saturating [Ca^{2+}] in skinned fibers of skeletal muscle, presumably by reducing the number of cross-bridges in high-force producing states (17, 22). However, it must be noted that these experiments generally have been performed at low temperatures (10–15° C), and the depressive effect of P_i also may be reduced at more physiological temperatures. In fact, Dantzig et al. (11) found a smaller force-depressing effect of P_i at 20° C than at 10° C. On the other hand, the P_i analogue orthovanadate gave a similar force reduction at 5 and 25° C (23).

There are no studies on intact mammalian muscle where the effect on force production of an isolated increase of P_i has been established. In an attempt to perform such a study, Westerblad & Allen (30) injected P_i into rested, single mouse fibers. They observed no effect of P_i injection on maximum force production, but the SR Ca^{2+} release became depressed. However, fibers were allowed to recover for some minutes after the P_i injection, and it was suggested that P_i moved from the myoplasm to the SR during this period (14), which might explain both the reduced SR Ca^{2+} release and the lack of effect on cross-bridge force production.

Some recent studies have shown that a reduction of P_i results in increased cross-bridge force production: mouse soleus muscles exposed to pyruvate display a reduction of P_i and increased tetanic force (24); following a short period of repeated tetani ("warm-up"), force production is increased above the control level for about 10 min and at the same time P_i is reduced below the rested value in fast-twitch *Xenopus* fibers (4) and in mouse soleus muscles (5). Thus, if a reduction of P_i results in increased cross-bridge force production in intact muscle fibers, it seems likely that increased P_i will reduce cross-bridge force. However, the above studies of Phillips et al. (24) and Bruton et al. (5) on mammalian muscle both were performed at about room temperature (22–25° C), which means that the P_i effect on force may be markedly smaller at body temperature. Moreover, both these two studies employed soleus muscles, where the majority of fibers are slow-twitch, type I fibers, and the effect might be less marked in a muscle with mostly fast-twitch fibers. Having this in mind, we recently studied the effect of warm-up at 32° C in mouse lumbrical muscles, which mainly contain fast-twitch type IIa fibers. Figure 12.3 shows records from a representative muscle, and it can be seen that force increases above the original after the warm-up series; the force increase in five muscles was 4.8 ± 1.0% (Bruton and Westerblad, unpublished observations). Thus, reduced P_i has a significant force-increasing effect also in fast-twitch fibers studied at close to body temperature.

Skinned fiber experiments have shown that P_i has no significant effect on shortening velocity (10, 22). In accordance with this, the maximum shortening velocity was not affected under conditions where isometric force was potentiated due to a reduction of P_i in mouse soleus muscles (24) and in *Xenopus* fibers (4).

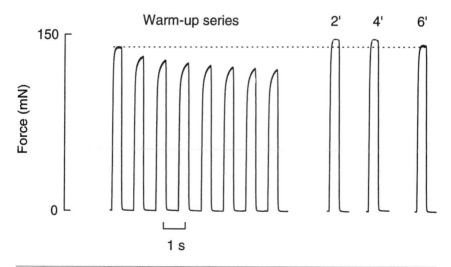

Figure 12.3.   Force is potentiated after a short series of repeated tetani (warm-up). Records from a mouse lumbrical muscle studied at 32° C. The warm-up series consisted of eight 400 ms tetani given at 1 s interval. Note that force is markedly higher than the original (indicated by dotted line) in test tetani produced 2 and 4 min after the warm-up series.

Adapted, by permission, from Westerblad et al. 1997.

Conclusions

The cross-bridge's ability to produce force and the cross-bridge cycling rate are reduced in skeletal muscle fatigue. These impairments are manifested as reduced isometric force production despite adequate Ca^{2+} activation and reduced shortening velocity. In mammalian muscle studied at physiological temperature, the reduced force production does not seem to be due to an intracellular acidification. Therefore, P_i accumulation is the most likely cause of the reduced cross-bridge force, although direct proof of this still is missing. Neither acidification nor increased P_i can explain the reduction of shortening velocity in fatigue, and the mechanism behind this slowing remains unclear.

Acknowledgements

This work was supported by grants from the Swedish Medical Research Council (Project No. 10842), the Swedish National Centre for Sports Research, and Funds at the Karolinska Institute.

References

1. Adams, G.R., Fisher, M.J., and Meyer, R.A. Hypercapnic acidosis and increased H_2PO_4 concentration do not decrease force in cat skeletal muscle. *American Journal of Physiology* 260: C805-C812; 1991.
2. Allen, D.G., Westerblad, H., and Lännergren, J. Muscle cell function during prolonged activity: cellular mechanisms of fatigue. *Experimental Physiology* 80: 497-527; 1995.
3. Asmussen, E. Muscle fatigue. *Medicine and Science in Sports* 11: 313-321; 1979.
4. Bruton, J.D., Westerblad, H., Katz, A., and Lännergren, J. "Warm up" without temperature change augments tension output in skeletal muscle. *Journal of Physiology* 493: 211-217; 1996.
5. Bruton, J.D., Wretman, C., Katz, A., and Westerblad, H. Increased tetanic force and reduced myoplasmic $[P_i]$ following a brief series of tetani in mouse soleus muscle. *American Journal of Physiology* 272: C870-C874; 1997.
6. Burke, R.E., Levine, D.N., Tsairis, P., and Zajac, F.E. Physiological types and histochemical profiles in motor units of the cat gastrocnemius. *Journal of Physiology* 234: 723-748; 1973.
7. Cady, E.B., Jones, D.A., Lynn, J., and Newham, D.J. Changes in force and intracellular metabolites during fatigue of human skeletal muscle. *Journal of Physiology* 418: 311-325; 1989.
8. Chase, P.B., and Kushmerick, M.J. Effects of pH on contraction of rabbit fast and slow skeletal muscle fibers. *Biophysical Journal* 53: 935-946; 1988.

9. Cooke, R., Franks, K., Luciani, G.B., and Pate, E. The inhibition of rabbit skeletal muscle contraction by hydrogen ion and phosphate. *Journal of Physiology* 395: 77-97; 1988.
10. Cooke, R., and Pate, E. The effects of ADP and phosphate on the contraction of muscle fibers. *Biophysical Journal* 48: 789-798; 1985.
11. Dantzig, J.A., Goldman, Y.E., Millar, N.C., Lacktis, J., and Homsher, E. Reversal of the cross-bridge force-generating transition by photogeneration of phosphate in rabbit psoas muscle fibres. *Journal of Physiology* 451: 247-278; 1992.
12. Donaldson, S.K.B., and Hermansen, L. Differential, direct effects of H^+ on Ca^{2+}-activated force of skinned fibers from the soleus, cardiac and adductor magnus muscles of rabbits. *Pflügers Archiv* 376: 55-65; 1978.
13. Fitts, R.H. Cellular mechanisms of muscle fatigue. *Physiological Reviews* 74: 49-94; 1994.
14. Fryer, M.W., West, J.M., and Stephenson, D.G. Phosphate transport into the sarcoplasmic reticulum of skinned fibres from rat skeletal muscle. *Journal of Muscle Research and Cell Motility* 18: 161-167; 1997.
15. Godt, R.E., and Nosek, T.M. Changes of intracellular milieu with fatigue or hypoxia depress contraction of skinned rabbit skeletal and cardiac muscle. *Journal of Physiology* 412: 155-180; 1989.
16. Metzger, J.M., and Moss, R.L. Greater hydrogen ion induced depression of tension and velocity in skinned single fibres of rat fast than slow muscles. *Journal of Physiology* 393: 727-742; 1987.
17. Millar, N.C., and Homsher, E. The effect of phosphate and calcium on force generation in glycerinated rabbit skeletal muscle fibers. *Journal of Biological Chemistry* 265: 20,234-20,240; 1990.
18. Miller, R.G., Boska, M.D., Moussavi, R.S., and Weiner, M.W. [31]P nuclear magnetic resonance studies of high energy phosphates and pH in human muscle fatigue. *Journal of Clinical Investigation* 81: 1190-1196; 1988.
19. Mosso, A. *Die Ermüdung*. Leipzig: Hirzel; 1892.
20. Nagesser, A.S., van der Laarse, W.J., and Elzinga, G. ATP formation and ATP hydrolysis during fatiguing, intermittent stimulation of different types of single muscle fibres from *Xenopus laevis*. *Journal of Muscle Research and Cell Motility* 14: 608-618; 1993.
21. Pate, E., Bhimani, M., Franks-Skiba, K., and Cooke, R. Reduced effect of pH on skinned rabbit psoas muscle mechanics at high temperatures: implications for fatigue. *Journal of Physiology* 486: 689-694; 1995.
22. Pate, E., and Cooke, R. Addition of phosphate to active muscle fibers probes actomyosin states within the powerstroke. *Pflügers Archiv* 414: 73-81; 1989.
23. Pate, E., Wilson, G.J., Bhimani, M., and Cooke, R. Temperature dependence of the inhibitory effects of orthovanadate on shortening velocity in fast skeletal muscle. *Biophysical Journal* 66: 1554-1562; 1994.
24. Phillips, S.K., Wiseman, R.W., Woledge, R.C., and Kushmerick, M.J. The effects of metabolic fuel on force production and resting inorganic phosphate levels in mouse skeletal muscle. *Journal of Physiology* 462: 135-146; 1993.
25. Ranatunga, K.W. Effects of acidosis on tension development in mammalian skeletal muscle. *Muscle & Nerve* 10: 439-445; 1987.

26. Sahlin, K., and Ren, J.M. Relationship of contraction capacity to metabolic changes during recovery from fatiguing contraction. *Journal of Applied Physiology* 67: 648-654; 1989.

27. Westerblad, H., and Allen, D.G. Changes in myoplasmic calcium concentration during fatigue in single mouse muscle fibers. *Journal of General Physiology* 98: 615-635; 1991.

28. Westerblad, H., and Allen, D.G. Changes of intracellular pH due to repetitive stimulation of single fibres from mouse skeletal muscle. *Journal of Physiology* 449: 49-71; 1992.

29. Westerblad, H., and Allen, D.G. The role of $[Ca^{2+}]_i$ in the slowing of relaxation in fatigued single fibres from mouse skeletal muscle. *Journal of Physiology* 46: 729-740; 1993.

30. Westerblad, H., and Allen, D.G. The effects of intracellular injections of phosphate on intracellular calcium and force in single fibres of mouse skeletal muscle. *Pflügers Archiv* 431: 964-970; 1996.

31. Westerblad, H., Bruton, J.D., and Lännergren, J. The effect of intracellular pH on contractile function of intact, single fibres of mouse declines with increasing temperature. *Journal of Physiology* 500: 193-204; 1997.

32. Wiseman, R.W., Beck, T.W., and Chase, P.B. Effect of intracellular pH on force development depends on temperature in intact skeletal muscle from mouse. *American Journal of Physiology* 271: C878-C886; 1996.

CHAPTER 13

Redox Modulation of Skeletal Muscle Contraction by Reactive Oxygen and Nitric Oxide

Michael B. Reid

Department of Medicine, Baylor College of Medicine, Houston, TX, USA

Over the past few years, a series of experimental reports have demonstrated that the contractile function of skeletal muscle is subject to redox modulation by reactive oxygen intermediates (ROI) and nitric oxide derivatives (NO). These investigations originally focused on contractile failure in fatiguing muscle, addressing the questions of whether oxidative stress plays a causal role in fatigue (it does) and whether reactive oxygen intermediates contribute to this oxidative stress (they do). In the course of this research, it became clear that ROI production was not limited to conditions of metabolic stress. Data from several laboratories showed that muscle produces ROI under resting conditions, albeit at lower rates, and that endogenous ROI influence the contractile properties of unfatigued muscle. More recently, NO production by skeletal muscle was discovered, and NO was also shown to affect contractile function. This article briefly reviews the literature in this area, describing the evidence that skeletal muscle produces ROI and NO, and that they affect the contractile function of fatiguing and unfatigued muscle; subsequent sections review emerging concepts about the intracellular mechanism(s) of redox modulation and outline a conceptual model of redox homeostasis in muscle. The goals are twofold: to provide a concise overview of the field and to highlight questions of current interest that are the likely focus of future research.

Production of ROI and NO in Muscle

ROI are an obligatory by-product of oxidative metabolism. It is estimated that 2-5% of mitochondrial oxygen consumption is shunted to superoxide anion formation by electron transfer reactions at Complex I and Complex III of the electron transport chain (20). Other potential sources of superoxide anions in skeletal muscle include membrane-associated oxidoreductases in the sarcoplasmic reticulum (24, 37) and cytosolic xanthine oxidase (39). Superoxide anions represent the starting point for a redox cascade of reactive oxygen intermediates in muscle. Superoxide anions rapidly dismute to hydrogen peroxide, either spontaneously or via enzymatic conversion by superoxide dismutase (SOD). In the presence of transition metals, superoxide anions and hydrogen peroxide can interact via Fenton-like kinetics to form hydroxyl radicals; these are among the most highly-reactive oxidants in biological systems and can be cytotoxic. Research indicates that all three of these ROI species are produced by skeletal muscle including superoxide anions (34, 52, 54), hydrogen peroxide (52), and hydroxyl radicals (23, 48).

Redox homeostasis is maintained in muscle via an array of antioxidant mechanisms. These include direct electron exchange reactions with nonspecific antioxidants, as well as enzymatic processes that selectively metabolize individual ROI species. Among nonspecific antioxidants, the most important is reduced glutathione that is present in the cytosol at near-mM concentrations. Distinct forms of SOD are present in the cytosol (CuZn-SOD) and the mitochondria (Mn-SOD) to dismute superoxide anions. Hydrogen peroxide is dehydrated to water and oxygen by catalase or can be rapidly inactivated by glutathione peroxidase in a glutathione-dependent reaction.

The ROI produced by skeletal muscle are detectable both in the cytosol (52) and in the extracellular space (23, 34, 48, 54). Passive muscle produces ROI at low rates, and endogenous antioxidants maintain free oxidants in the cytosol at low levels. However, the redox balance is shifted during periods of contractile activity; ROI production increases dramatically (23, 34, 48, 54), and cytosolic oxidant levels rise (52).

Nitric oxide synthase (NOS) generates NO enzymatically in the presence of essential co-factors; L-arginine and oxygen are substrates for this reaction, and citrulline is the by-product. Constitutive NOS expression by skeletal muscle myocytes first was reported by Kobzik and coworkers (32), who documented intracellular localization and enzymatic activity of the neuronal isoform (n-NOS) in a panel of limb and respiratory muscles. These findings subsequently were confirmed (27, 33), validating an earlier report that human muscle expresses the mRNA for n-NOS (44). Skeletal muscle myocytes also express the endothelial isoform (e-NOS) and are the first cell type in which constitutive co-expression of both n-NOS and e-NOS has been observed (33).

The two NOS isoforms distribute very differently within muscle. n-NOS is primarily localized to the sarcolemma of fast fibers (32) where it is concentrated near the motor end-plate and is associated with the dystrophin complex via interaction with the α 1-syntrophin protein (15, 16). Neurons were the first cell type in which n-NOS was identified and peripheral nerve axons that interdigitate among muscle fibers clearly express n-NOS (unpublished observations). In contrast, e-NOS expression is independent of fiber type; immunohistochemistry and cell fractionation studies indicate that e-NOS is associated with muscle mitochondria (33). Endothelial cells of the muscle microvasculature stain intensely for e-NOS (33) and represent another potential source of NO immediately adjacent to the muscle fiber sarcolemma. Recent data indicate that both n-NOS and e-NOS are inducible with exercise (11), since eight weeks of treadmill training caused a fourfold increase in n-NOS content of rat soleus and a twofold increase in e-NOS content. The inducible isoform (i-NOS) is not expressed constitutively within muscle and does not have a physiologic role in this tissue. i-NOS can be up-regulated by skeletal muscle myocytes and by adjacent nonmuscle cell types in response to an inflammatory stimulus (64) and may mediate nitrosative injury under pathologic conditions.

NO production by muscle is activity dependent. Under resting conditions, isolated skeletal muscle releases NO at the rate of approximately 1 pmol/mg/min (10, 32). NO output is markedly increased during periods of repetitive muscle contraction (10). These findings are consistent with the calcium-dependence of n-NOS and e-NOS, since the calcium transients that mediate excitation-contraction coupling would be expected to stimulate NO production by both isoforms.

Redox Modulation of Contractile Function

Oxidative Stress in Muscle Fatigue

The first evidence of free radicals in skeletal muscle came from early electron spin-resonance (ESR) studies by Commoner et al. (21) who detected free radicals in excised rabbit muscle. It was a quarter century later that Koren and coworkers (36) used ESR to demonstrate that free radical content increases in muscle during active contraction. Subsequent ESR studies confirmed the robust nature of this finding, which has been reproduced in vitro and in vivo, in amphibian muscle and mammalian muscle, in limb and respiratory skeletal muscles (14, 22, 28, 35). Recent studies have shown that ROI and NO contribute importantly to activity-dependent increases in reactive oxidant levels. We found that superoxide anion production by excised muscle was increased fourfold during repetitive, fatiguing contractions in vitro (54). Using a similar preparation, Diaz and colleagues (23) found that hydroxyl radical production was proportional to mechanical output of the contracting muscle; severely fatigued muscle produced hydroxyl radicals at rates a hundredfold greater than resting muscle. Studies by O'Neill et al. (48) demonstrated that limb muscle also produces hydroxyl radicals in situ with vascular and neural connections intact; low rates of hydroxyl radical release by passive muscle were increased more than thirtyfold during periods of repetitive contraction. NO production also increases with muscle activity but to a lesser extent; Balon and Nadler (10) found that repetitive muscle contraction caused a twofold increase in NO synthesis.

As detailed in a number of recent reviews (26, 59, 66), strenuous exercise induces biochemical indices of oxidative stress in muscle. These include increases in glutathione oxidation, lipid peroxidation, and protein oxidation. Such changes are clearly sensitive to tissue antioxidant levels. Deficiencies in nutritional antioxidants such as vitamin E promote oxidative stress and exaggerate exercise-induced changes; conversely, antioxidant supplementation blunts the effects of exercise. These findings complement measurements of increased oxidant production during exercise and document oxidative effects on muscle at the biochemical level.

The cause-effect relationship between oxidative stress and muscle fatigue has been established using exogenous antioxidants to buffer the oxidative stress that develops during exercise. Studies of isolated muscle preparations have determined that antioxidants increase force production during repetitive, fatiguing contractions (23, 30, 52, 54); this demonstrates that endogenous oxidants act directly on the myocyte to compromise function. Similar findings in perfused, neurally-stimulated muscle preparations (12, 61, 63) indicate that oxidative stress contributes to fatigue in the absence of electrical field stimulation or hyperoxia. Positive effects of antioxidants in vivo indicate that oxidative stress limits endurance during whole-body exercise (19, 46, 47) and plays a causal role in human muscle fatigue (55). Note that these results have been obtained using pharmacologic antioxidants as experimental probes. In contrast, muscle endurance generally has not been improved by administering high doses of nutritional antioxidants—vitamin E, vitamin C, beta carotene, etc.—to well-nourished individuals (26, 59).

Specific involvement of ROI in muscle fatigue has been established using ROI-selective antioxidant enzymes as mechanistic probes. Barclay and Hansel (12) documented protective effects of intravenous SOD on fatigue of canine limb muscle in situ; we found that SOD and catalase inhibited fatigue of muscle fiber bundles in vitro (52); and Supinski and coworkers (63) demonstrated that systemic administration of PEG-SOD blunts fatigue of canine diaphragm in situ. In each of these studies, the efficacy of antioxidant enzymes was comparable to that of nonspecific antioxidants, which suggests that ROI are the primary mediators of oxidant-induced fatigue. In support of this concept, Nashiwati et al. (45) has demonstrated that infusion of a ROI-generating system into exercising muscle markedly accelerates fatigue.

The effects of NO in fatiguing muscle are more enigmatic. Murrant and colleagues (41, 42, 43) have tested the effects of endothelial-derived NO and exogenous NO donors during repetitive contractions. In studies of excised rodent muscles, NO exposure was found to slow the decline of force over time. Canine limb muscle exhibited the opposite response when studied in situ; a pharmacologic NO donor significantly accelerated the fall in force during repetitive contractions. Consistent with these findings, King-VanVlack et al. (31) found that NOS blockade inhibited fatigue in the same canine limb muscle preparation. Bisnett and coworkers (13) obtained negative results after administering a NOS inhibitor to anesthetized rats prior to severe inspiratory loading; NOS blockade did not protect the diaphragm from contractile losses. Thus, to the extent that NO affects muscle fatigue, its action appears to differ in vitro and in situ. The capacity of NO to influence fatigue also depends critically on the frequency of muscle stimulation (31, 42). Further studies are required to assess the physiologic importance of NO in fatigue.

The existing data clearly establish a causal role for ROI-mediated oxidative stress in muscle fatigue. But oxidative stress is not the only mediator involved. During heavy exercise, muscle undergoes a number of metabolic changes that compromise myofilament function. These include accumulation of inorganic phosphate, phosphocreatine depletion, acidosis, and loss of calcium homeostasis as well as oxidant accumulation. Thus, while oxidative stress contributes to fatigue (antioxidants can increase sustainable force by 15–50%) it represents only one part of a complicated cascade of events. Our current task is to determine where oxidative events fit within this cascade, the mechanism(s) that accelerate reactive oxidant production, and the intracellular target(s) that are most sensitive to oxidative alteration.

ROI and NO in Unfatigued Muscle

Skeletal muscle produces ROI and NO at low rates under resting conditions, and both classes of reactive oxidants appear to modulate contractile function. Pilot studies by Regnier et al. (51) demonstrated that incubation with ROI-specific antioxidant enzymes, either SOD or catalase, would depress the twitch forces developed by single fibers from frog limb muscle. In a formal study of excised mammalian muscle, we tested the effects of SOD and catalase on isometric contractile properties (52) using SOD and catalase concentrations that had been shown to deplete cytosolic oxidant levels. Catalase exposure depressed twitch characteristics and decreased tetanic force in a dose-dependent manner; these contractile changes spontaneously reversed when the enzyme was removed. SOD had similar effects. In contrast, low concentrations of exogenous hydrogen peroxide had opposite effects; twitch characteristics and force production were increased. Later experiments documented that contractile function of unfatigued muscle was depressed similarly by other antioxidants (30, 53). These studies suggest that exogenous antioxidants can disrupt redox homeostasis in passive muscle, depleting myocytes of endogenous ROI and subjecting the cells to reductive stress. This compromises contractile function. It follows that endogenous ROI are essential for normal force development by unfatigued muscle.

The influence of endogenous NO on unfatigued muscle originally was tested using a panel of chemically distinct NOS inhibitors to block NO synthesis (32). NOS inhibition increased submaximal tetanic force, shifting the force-frequency leftward. NO donors had the opposite effect, shifting the curve rightward and reversing the effect of NOS blockade. Thus, in unfatigued skeletal muscle, endogenous NO appears to depress force production at submaximal stimulus frequencies, an effect similar to the contractile depression caused in smooth and cardiac muscle. In these other muscle types, cyclic guanosine monophosphate (cGMP) functions as a second messenger for NO (58); data from several sources support the existence of a similar NO/cGMP pathway in skeletal muscle. Two decades ago Arnold et al. (8) first observed that NO donors increase cGMP levels in muscle. We confirmed this report and went on to evaluate the functional importance of NO/cGMP signaling (1, 32). The forces developed by skeletal muscle are decreased by pharmacologic probes that amplify cGMP effects, e.g., cGMP analogues and phosphodiesterase inhibitors. Conversely, force is increased by probes that block cGMP signaling, e.g., guanylate cyclase inhibitors. The magnitude of cGMP effects on force are consistently less than those produced by NO, suggesting that NO also acts via cGMP-independent mechanisms. One possibility is that NO directly opposes the effects of ROI via antioxidant effects on target proteins as described in other cell systems (29, 65).

Target Proteins for Redox Modulation

In order to modulate contractile function, ROI and NO must alter the function of one or more redox-sensitive proteins that have been identified in skeletal muscle. The most thoroughly studied is the calcium release channel of the sarcoplasmic reticulum (SR), a large, homotetramer composed of four 565 kDa subunits. Many reports have shown that oxidation of critical sulfhydryl groups on the protein stimulate channel opening, as reviewed by Abramson and Salama (2). Recent studies of molecular mechanism (4) indicate that oxidative activation of the channel involves formation of disulfide crosslinks between adjacent subunits, causing dimerization within the protein complex. This configurational change is stimulated by exposure to physiologic oxidants, including hydrogen peroxide, and is reversible by reducing agents

(25). Such properties suggest an attractive mechanism for ROI modulation of SR calcium release under physiologic conditions. The channel also responds to NO exposure, but initial reports on this effect appear to conflict. Meszaros and coworkers (40) observed that NO donors inhibited channel opening, whereas Stoyanovsky et al. (62) found that NO donors enhanced channel opening. These observations simply may reflect the two extremes of NO action. Data from Aghdasi et al. (3) indicate that NO effects are biphasic; low levels of NO exposure appear to inhibit oxidative activation of the channel, whereas high levels stimulate channel opening and oxidative modification of the protein. NO thus may exert antioxidant effects on the channel at low concentrations and pro-oxidant effects at higher concentrations, a property well described in other biological systems (38).

Other proteins that directly influence excitation-contraction coupling also exhibit redox sensitivity. Activity of the SR calcium ATPase is depressed by oxidative modification (57), and calcium reuptake from the cytosol may be slowed by oxidative stress (56). Sulfhydryl modification has been known to influence actin-myosin interaction for the last half-century (9). Myosin heavy chains contain critical sulfhydryls, including the S1 and S2 sites, that are positioned in regions that change configuration during cross-bridge cycling (5); oxidation of these sulfhydryls inhibits myosin binding to actin (18). Evidence also suggests that sarcolemmal function is redox-sensitive. Puppi et al. (50) found that potassium contractures were sensitive to redox perturbations due, in part, to changes in the sodium-calcium exchange mechanism. Detailed experiments by Sen and coworkers (60) showed that oxidant exposure altered potassium transport in skeletal muscle cells at several sites, the most sensitive being the sodium-potassium pump.

Second messenger systems also may be activated by redox signaling. The best-characterized biological target for NO is soluble guanylate cyclase, which is present in the cytosol of skeletal muscle. NO activates this metalloenzyme by binding to a regulatory heme center, thereby stimulating cGMP synthesis (58). This is an established pathway whereby NO depresses skeletal muscle force production.

To determine the relative importance of these target proteins, investigators are using single-fiber preparations to evaluate the separate processes that compose excitation-contraction coupling. Brotto and Nosek (17) have examined hydrogen peroxide effects on skinned fibers. They found that depolarization-induced calcium release was partially inhibited and calcium-induced calcium release was abolished by hydrogen peroxide; force production by the myofilaments and SR calcium reuptake were unaffected. Posterino and Lamb (49) also studied skinned fibers. They found that the oxidizing agent 2,2'-dithiodipyridine (DTDP) abolished depolarization-induced calcium release, an effect attributed to oxidation of the voltage sensor. DTDP caused only a slight increase in calcium permeability of the SR and had no effect on force production. Studies by Andrade and coworkers (6, 7) demonstrate that intact muscle fibers respond differently. In their system, calcium activation of the myofilaments was more sensitive to hydrogen peroxide than was SR function. NO exposure simultaneously increased tetanic calcium release and blunted calcium activation of the myofilaments; the latter effect dominated, resulting in a net loss of force. These studies provide evidence that excitation-contraction coupling is sensitive to redox modulation at several sites and that these sites differ in their sensitivity to ROI and NO.

Redox Homeostasis in Muscle

The novelty of this field and its potential importance for muscle biology are exciting. However, the concepts are somewhat unfamiliar to many scientists, and the data still have many gaps to be filled. This makes it difficult to develop a cohesive picture of ROI and NO effects in muscle. One concept that has proven helpful is that of redox homeostasis, i.e., that myocytes maintain intracellular oxidant levels within a physiologic range. This general principle holds for most components of the intracellular milieu; calcium, potassium, hydrogen ion, and even water content are regulated to maintain homeostasis. Extreme fluctuations in any of these variables, either increases or decreases, will cause cellular dysfunction. Reactive oxidants appear to be regulated similarly. Resting muscle continually produces ROI and NO. Reactive oxidant production is buffered by endogenous antioxidants such that cytosolic oxidants are maintained at low levels. Under these conditions, ROI are essential to maintain the optimal redox state for effective contraction, while NO acts to oppose ROI effects and depress force. Contractile function is compromised by extreme redox perturbations, either accumulation of excess oxidants (oxidative stress) or excessive oxidant depletion (reductive stress). Oxidative stress is most likely to occur physiologically since contractile activity increases ROI production. If this oxidative challenge exceeds the buffering capacity of endogenous antioxidants, oxidative stress and contractile dysfunction ensue. Other conditions may impose more severe oxidative challenges; pharmacologic oxidants or pathologic processes can lead to oxidative injury, apoptosis, or necrotic cell death. In the other extreme, exposing passive muscle to exogenous antioxidants can deplete intracellular ROI. This markedly lowers cytosolic oxidant levels and alters redox state. Reductive stress also causes loss of contractile function and can be cytotoxic.

Conclusion

Research over the last decade has identified ROI and NO as physiologic modulators of skeletal muscle function. Both classes of reactive oxidants are produced by skeletal muscle, and both appear to influence contractile function. Continued experiments using single fibers and isolated proteins are essential to determine the intracellular mechanisms whereby redox state modulates excitation-contraction coupling and muscle fatigue.

Acknowledgment

Our research in this area has been supported by National Institutes of Health grant #HL-45721.

References

1. Abraham, R.Z., Kobzik, L., Moody, M.R., Reid, M.B., and Stamler, J.S. Cyclic GMP is a second messenger by which nitric oxide inhibits diaphragm contraction. *Comparative Biochemistry and Physiology* 119A: 177-183; 1998.
2. Abramson, J.J., and Salama, G. Critical sulfhydryls regulate calcium release from sarcoplasmic reticulum. *Journal of Bioenergetics and Biomembranes* 21: 283-294; 1989.
3. Aghdasi, B., Reid, M.B., and Hamilton, S.L. Nitric oxide protects the skeletal muscle Ca^{2+} release channel from oxidation induced activation. *Journal of Biological Chemistry* 272: 25,462-25,467; 1997.
4. Aghdasi, B., Zhang, J.Z., Wu, Y., Reid, M.B., and Hamilton, S.L. Multiple classes of sulfhydryls modulate the skeletal muscle Ca^{2+} release channel. *Journal of Biological Chemistry* 272: 3739-3748; 1997.
5. Ajtai, K., Toft, D.J., and Burghardt, T.P. Path and extent of cross-bridge rotation during muscle contraction. *Biochemistry* 33: 5382-5391; 1994.
6. Andrade, F.H., Reid., M.B., Allen, D.G., and Westerblad, H. Effect of hydrogen peroxide and dithiothreitol on contractile function of single muscle fibres from mouse. *Journal of Physiology (London)* 509: 565-575;1998.
7. Andrade, F.H., Reid, M.B., Allen, D.G., and Westerblad, H. Nitric oxide decreases myofibrillar Ca^{2+} sensitivity in single skeletal muscle fibres. *Journal of Physiology (London)* 509: 576-586;1998.
8. Arnold, W.P., Mittal, C.K., Katsuki, S., and Murad, F. Nitric oxide activates guanylate cyclase and increases guanosine 3':5'-cyclic monophosphate levels in various tissue preparations. *Proceedings of the National Academy of Science USA* 74: 3203-3207; 1977.
9. Bailey, K., and Perry, S.V. The role of sulfhydryl groups in the interaction of myosin and actin. 1947 'classical article'. *Biochimica et Biophysica Acta* 1000: 177-178; 1989.
10. Balon, T.W. and Nadler, J.L. Nitric oxide release is present from incubated skeletal muscle preparations. *Journal of Applied Physiology* 77: 2519-2521; 1994.
11. Balon, T.W. and Nadler, J.L. Evidence that nitric oxide increases glucose transport in skeletal muscle. *Journal of Applied Physiology* 82: 359-363; 1997.
12. Barclay, J.K., and Hansel, M. Free radicals may contribute to oxidative skeletal muscle fatigue. *Canadian Journal of Physiology and Pharmacology* 69: 279-284; 1991.
13. Bisnett, T., Anzueto, A., Andrade, F.H., Rodney, G.G., Napier, W.R., Levine, S.M., Maxwell, L.C., Mureeba, P., Derdak, S.D., Grisham, M.D., and Jenkinson, S. Effect of nitric oxide synthetase inhibitor on diaphragmatic function after resistive loading. *Comparative Biochemistry and Physiology* 119A: 185-190; 1998.
14. Borzone, G., Zhao, B., Merola, A.J., Berliner, L., and Clanton, T.L. Detection of free radicals by electron spin resonance in rat diaphragm after resistive loading. *Journal of Applied Physiology* 77: 812-818; 1994.

15. Brenman, J.E., Chao, D.S., Gee, S.H., McGee, A.W., Craven, S.E., Santillano, D.R., Wu, Z., Huang, F., Xia, H., Peters, M.F., Froehner, S. C., and Bredt, D.S. Interaction of nitric oxide synthase with the postsynaptic density protein PSD-95 and alpha1-syntrophin mediated by PDZ domains. *Cell* 84: 757-767; 1996.

16. Brenman, J.E., Chao, D.S., Xia, H., Aldape, K., and Bredt, D.S. Nitric oxide synthase complexed with dystrophin and absent from skeletal muscle sarcolemma in Duchenne muscular dystrophy. *Cell* 82: 743-752; 1995.

17. Brotto, M.A.P., and Nosek, T.M. Hydrogen peroxide disrupts calcium release from the sarcoplasmic reticulum of rat skeletal muscle fibers. *Journal of Applied Physiology* 81: 731-737; 1996.

18. Burke, M., Reisler, F., and Harrington, W.F. Effect of bridging the two essential thiols of myosin on its spectral and actin-binding properties. *Biochemistry* 15: 1923-1927; 1976.

19. Cazzulani, P., Cassin, M., and Ceserani, R. Increased endurance to physical exercise in mice given oral reduced glutathione (GSH). *Medical Science Research* 19: 543-544; 1991.

20. Chance, B., Sies, H., and Boveris, A. Hydroperoxide metabolism in mammalian organs. *Physiologic Reviews* 59: 527-605; 1979.

21. Commoner, B., Townsend, J., and Pake, G.E. Free radicals in biological materials. *Nature* 174: 689-691; 1954.

22. Davies, K.J.A., Quintanilha, A.T., Brooks, G.A., and Packer, L. Free radicals and tissue damage produced by exercise. *Biochemical and Biophysical Research Communication* 107: 1198-1205; 1982.

23. Diaz, P.T., She, Z.-W., Davis, W.B., and Clanton, T.L. Hydroxylation of salicylate by the in vitro diaphragm: evidence for hydroxyl radical production during fatigue. *Journal of Applied Physiology* 75: 540-545; 1993.

24. Duncan, C.J., and Rudge, M.F. Are lysosomal enzymes involved in rapid damage in vertebrate muscles? A study of the separate pathways leading to cellular damage. *Cell and Tissue Research* 253: 447-455; 1988.

25. Favero, T.G., Zable, A.C., and Abramson, J.J. Hydrogen peroxide stimulates the Ca^{2+} release channel from skeletal muscle sarcoplasmic reticulum. *Journal of Biological Chemistry* 270: 25,557-25,563; 1995.

26. Goldfarb, A.H. Antioxidants: role of supplementation to prevent exercise-induced oxidative stress. *Medicine and Science in Sports and Exercise* 25: 232-236; 1993.

27. Grozdanovic, Z., Nakos, G., Dahrmann, G., Mayer, B., and Gossrau, R. Species-independent expression of nitric oxide synthase in the sarcolemma region of visceral and somatic striated muscle fibers. *Cell and Tissue Research* 281: 493-499; 1995.

28. Jackson, M.J., Edwards, R.H.T., and Symons, M.C.R. Electron spin-resonance studies of intact mammalian skeletal muscle. *Biochimica et Biophysica Acta* 847: 185-190; 1985.

29. Kanner, J., Harel, S., and Granit, R. Nitric oxide as an antioxidant. *Archives of Biochemistry and Biophysics* 289: 130-136; 1991.

30. Khawli, F.A., and Reid, M.B. N-acetylcysteine depresses contractility and inhibits fatigue of diaphragm in vitro. *Journal of Applied Physiology* 77: 317-324; 1994.

31. King-VanVlack, C.E., Curtis, S.E., Mewburn, J.D., Cain, S.M., and Chapler, C.K. Role of endothelial factors in active hyperemic responses in contracting canine muscle. *Journal of Applied Physiology* 79: 107-112; 1995.

32. Kobzik, L., Reid, M.B., Bredt, D.S., and Stamler, J.S. Nitric oxide in skeletal muscle. *Nature* 372: 546-548; 1994.

33. Kobzik, L., Stringer, B., Balligand, J.L., Reid, M.B., and Stamler, J.S. Endothelial type nitric oxide synthase in skeletal muscle fibers: mitochondrial relationships. *Biochemical and Biophysical Research Communication* 211: 375-381; 1995.

34. Kolbeck, R.C., She, Z.-W., Callahan, L.A., and Nosek, T.M. Increased superoxide production during fatigue in the perfused rat diaphragm. *American Journal of Respiratory and Critical Care Medicine* 156: 140-145; 1997.

35. Koren, A., Sauber, C., Sentjurc, M., and Schara, M. Free radicals in tetanic activity of isolated skeletal muscle. *Comparative Biochemistry and Physiology* 74B: 633-635; 1983.

36. Koren, A., Schara, M., and Sentjurc, M. EPR measurements of free radicals during tetanic contraction of frog skeletal muscle. *Periodicum Biologorum* 82: 399-401; 1980.

37. Koshkin, V.V. Superoxide radical generation and lipid peroxidation in skeletal muscles. *Biokhimiia* 50: 1406-1410; 1985.

38. Lancaster, J.R. Nitric oxide in cells: this simple molecule plays Janus-faced roles in the body, acting as both messenger and destroyer. *American Scientist* 80: 248-259; 1992.

39. McCutchan, H.J., Schwappach, J.R., Enquist, E.G., Walden, D.L., Terada, L.S., Reiss, O.K., Leff, J.A., and Repine, J.E. Xanthine oxidase-derived H_2O_2 contributes to reperfusion injury of ischemic skeletal muscle. *American Journal of Physiology* 258: H1415-H1419; 1990.

40. Meszaros, L.G., Minarovic, I., and Zahradnikova, A. Inhibition of the skeletal muscle ryanodine receptor calcium release channel by nitric oxide. *FEBS Letters* 380: 49-52; 1996.

41. Murrant, C.L., and Barclay, J.K. Endothelial cell products alter mammalian skeletal muscle function in vitro. *Canadian Journal of Physiology and Pharmacology* 73: 736-741; 1995.

42. Murrant, C.L., Frisbee, J.C., and Barclay, J.K. The effect of nitric oxide and endothelin on skeletal muscle contractility changes when stimulation is altered. *Canadian Journal of Physiology and Pharmacology* 75: 414-422; 1997.

43. Murrant, C.L., Woodley, N.E., and Barclay, J.K. Effect of nitroprusside and endothelium-derived products on slow-twitch skeletal muscle function in vitro. *Canadian Journal of Physiology and Pharmacology* 72: 1089-1093; 1994.

44. Nakane, M., Schmidt, H.H., Pollock, J.S., Forstermann, U., and Murad, F. Cloned human brain nitric oxide synthase is highly expressed in skeletal muscle. *FEBS Letters* 316: 175-180; 1993.

45. Nashiwati, E., Dimarco, A., and Supinski, G. Effects produced by infusion of a free radical-generating solution into the diaphragm. *American Review of Respiratory Disease* 147: 60-65; 1993.

46. Novelli, G.P., Bracciotti, G., and Falsini, S. Spin-trappers and vitamin E prolong endurance to muscle fatigue in mice. *Free Radicals in Biology and Medicine* 8: 9-13; 1990.
47. Novelli, G.P., Falsini, S., and Bracciotti, G. Exogenous glutathione increases endurance to muscle effort in mice. *Pharmacology Research* 23: 149-155; 1991.
48. O'Neill, C.A., Stebbins, C.L., Bonigut, S., Halliwell, B., and Longhurst, J.C. Production of hydroxyl radicals in contracting skeletal muscle of cats. *Journal of Applied Physiology* 81: 1197-1206; 1996.
49. Posterino, G.S., and Lamb, G.D. Effects of reducing agents and oxidants on excitation-contraction coupling in skeletal muscle fibres of rat and toad. *Journal of Physiology (London)* 496: 809-825; 1996.
50. Puppi, A., Szekeres, S., and Dely, M. Correlations between the tissue redox-state and potassium-contractures. *Acta Physiologica Hungarica* 75: 253-259; 1990.
51. Regnier, M., Lorenz, R.R., and Sieck, G.C. Effects of oxygen radical scavengers on force production in single living frog skeletal muscle fibers. *FASEB Journal* 6: A1819 (Abstract); 1992.
52. Reid, M.B., Haack, K.E., Franchek, K.M., Valberg, P.A., Kobzik, L., and West, M.S. Reactive oxygen in skeletal muscle: I. Intracellular oxidant kinetics and fatigue in vitro. *Journal of Applied Physiology* 73: 1797-1804; 1992.
53. Reid, M.B. and Moody, M.R. Dimethyl sulfoxide depresses skeletal muscle contractility. *J. Appl. Physiol.* 76: 2186-2190; 1994.
54. Reid, M.B., Shoji, T., Moody, M.R., and Entman, M.L. Reactive oxygen in skeletal muscle: II. Extracellular release of free radicals. *Journal of Applied Physiology* 73: 1805-1809; 1992.
55. Reid, M.B., Stokic, D.S., Koch, S.M., Khawli, F.A., and Leis, A.A. N-acetylcysteine inhibits muscle fatigue in humans. *Journal of Clinical Investigation* 94: 2468-2474; 1994.
56. Scherer, N.M., and Deamer, D.W. Oxidation of thiols in the calcium-ATPase of sarcoplasmic reticulum microsomes. *Biochimica et Biophysica Acta* 862: 309-317; 1986.
57. Scherer, N.M., and Deamer, D.W. Oxidative stress impairs the function of sarcoplasmic reticulum by oxidation. *Archives of Biochemistry and Biophysics* 216: 589-601; 1986.
58. Schmidt, H.H.H.W., Lohmann, S.M., and Walter, U. The nitric oxide and cGMP signal transduction system: regulation and mechanism of action. *Biochimica et Biophysica Acta* 1178: 153-175; 1993.
59. Sen, C.K. Oxidants and antioxidants in exercise. *Journal of Applied Physiology* 79: 675-686; 1995.
60. Sen, C.K., Kolosova, I., Hanninen, O., and Orlov, S.N. Inward potassium transport systems in skeletal muscle derived cells are highly sensitive to oxidant exposure. *Free Radicals in Biology and Medicine* 18: 795-800; 1995.
61. Shindoh, C., Dimarco, A., Thomas, A., Manubray, P., and Supinski, G. Effect of N-acetylcysteine on diaphragm fatigue. *Journal of Applied Physiology* 68: 2107-2113; 1990.

62. Stoyanovsky, D.A., Murphy, T.D., Anno, P., Kim, Y.M., and Salama, G. Nitric oxide activates skeletal and cardiac ryanodine receptors. *Cell Calcium* 21: 19-29; 1997.
63. Supinski, G., Nethery, D., Stofan, D., and Dimarco, A. Effect of free radical scavengers on diaphragmatic fatigue. *American Journal of Respiratory and Critical Care Medicine* 155: 622-629; 1997.
64. Williams, G., Brown, T., Becker, M., Prager, M., and Giroir, B.P. Cytokine-induced expression of nitric oxide synthase in C2C12 skeletal muscle myocytes. *American Journal of Physiology* 267: R1020-R1025; 1994.
65. Wink, D.A., Hanbauer, I., Krishna, M.C., DeGraff, W., Gamson, J., and Mitchell, J.B. Nitric oxide protects against cellular damage and cytotoxicity from reactive oxygen species. *Proceedings of the National Academy of Science USA* 90: 9813-9817; 1993.
66. Witt, E.H., Reznick, A.Z., Viguie, C.A., Starke-Reed, P., and Packer, L. Exercise, oxidative damage, and effects of antioxidant manipulation. *Journal of Nutrition* 122: 766-773; 1992.

PART IV

Metabolism

CHAPTER 14

Glucose Transport and Transporters in Skeletal Muscle

Erik A. Richter, Søren Kristiansen, Jørgen F.P. Wojtaszewski, and Harinder S. Hundal

Copenhagen Muscle Research Centre, August Krogh Institute, University of Copenhagen, Denmark; and Department of Anatomy and Physiology, The Old Medical School, Smals Wynd, The University of Dundee, Scotland

Glucose is transported across the cell membrane via facilitated diffusion. A family of glucose transporter proteins named GLUT1–7 has been identified. Although the different isoforms all are capable of transporting glucose they have different characteristics and tissue distribution (table 14.1). As is apparent from table 14.1, three of the glucose transporters (GLUT1, 4, and 5) are present in adult mammalian skeletal muscle. GLUT1 is present in the sarcolemma. A large part of the GLUT1 that is found in a homogenate of skeletal muscle in fact stems from perineural sheaths from nerves located in the muscle (23). Since GLUT1 is resident in the sarcolemma independently of stimulation with insulin and/or muscle contractions, its main function is thought to be to provide basal glucose transport. GLUT4 is the most abundant and most important glucose transporter in skeletal muscle. It is unique in the sense that it is able to translocate from an intracellular storage site to the sarcolemma upon stimulation with contractions and/or insulin (15, 44). Finally, GLUT5 also is found in human skeletal muscle sarcolemma (29, 39). No intracellular storage depot of GLUT5 has been identified, and therefore GLUT5 is thought to reside permanently in the sarcolemma. As discussed below, GLUT5 is thought to function as a fructose transporter.

Translocation of Glucose Transporters

Muscle contractions and insulin are the main physiological stimulators of glucose transport in skeletal muscle. Although there is no doubt that translocation of GLUT4 to the sarcolemma and the t-tubular system is a major mechanism increasing muscle glucose transport (54), there has been some debate about whether this is the only mechanism for increasing glucose transport or whether an increase in the intrinsic activity of GLUT4 is also involved. Early

Table 14.1   The Facilitative Glucose Transporter Family

Isoform	Tissue Distribution	Function
GLUT1	Ubiquitous. High expression in brain endothelial cells and human erythrocytes	Basal glucose transport Located in plasma membrane Low K_m ($\approx$2–3 mM)
GLUT2	Liver, kidney and small intestinal epithelial cells, ß-cells	High K_m ($\approx$15–20 mM) Glucose transport in and out of hepatocytes and ß-cells
GLUT3	Neurons, placenta	Low K_m ($\approx$2–3 mM). Transport into central nervous system
GLUT4	Skeletal muscle, brown and white adipose tissue, heart	Mediates insulin and (in muscle) contraction regulated glucose transport. $K_m \approx$5 mM.
GLUT5	Small intestine, kidney skeletal muscle, human adipocytes	Fructose transporter
GLUT7	Liver	Part of the glucose-6-phosphatase complex in endoplasmic reticulum.

For extensive review of the tissue distribution and characteristics of the glucose transporter isoforms see 4, 5, 7, 20, 33, 46, 56. GLUT6 is not indicated because it encodes a pseudogene. Adapted from (54) with permission.

studies of glucose transporter translocation used subcellular fractionation techniques, and these studies showed that although insulin and/or contractions increased glucose transport many-fold in intact muscle (48, 49, 52), the plasma membrane GLUT4 content was by comparison only increased by a factor of 1.5–3 (15, 16, 44). This provided indirect evidence for the occurence of increased intrinsic activity of the glucose transporters involved. However, subcellular fractionation techniques may not give a correct picture because of the potential for cross-contamination between the different membrane fractions. Recent studies using surface labeling of GLUT4 in incubated muscle, however, show that the increase in surface labeled GLUT4 with insulin and/or contractions correlates much more with the measured increase in glucose transport, suggesting that translocation may be the major mechanism for increasing muscle glucose transport, at least during in vitro conditions (43, 60). These findings, however, do not exclude that the large changes in, for example, hormone secretion that are known to occur during exercise might modulate the intrinsic activity of the glucose transporters.

In humans, translocation studies have been few, because most techniques rely on several grams of muscle for preparation of subcellular fractions, and this is not possible to obtain from healthy humans. Recently, however, the open-muscle biopsy procedure has allowed the procurement of approximately 1 g of muscle from the vastus lateralis, which is sufficient for preparing plasma membranes and internal membrane fractions. Using this technique, it has been demonstrated that insulin causes translocation of GLUT4 from the microsomal fraction to the plasma membrane fraction (19, 22). Another technique that recently has been applied to human muscle is the preparation of sarcolemmal giant vesicles. This technique has the advantage that sufficient sarcolemmal vesicles for measurement of both glucose transport and GLUT4 transporter number can be obtained from about 300 mg wet muscle, which can be obtained with relative ease by ordinary needle biopsy of human muscle. With this technique we recently have been able to demonstrate that submaximal exercise results in a progressive increase in sarcolemmal GLUT4 and glucose transport capacity (40) (figure 14.1) paralleling the progressive increase in glucose uptake during submaximal exercise (34). Furthermore, exercise leads to increased sarcolemmal VAMP-2 protein content in human skeletal muscle (38) (figure 14.2). VAMP-2 is a so-called v-SNARE protein believed to be involved in

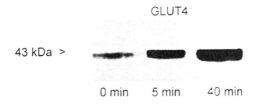

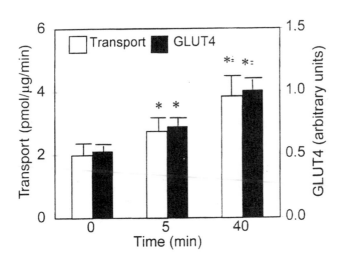

Figure 14.1.   A representative immunoblot of GLUT4 (top) and the mean GLUT4 protein content and glucose transport (bottom) in sarcolemmal giant vesicles produced from human skeletal muscle samples obtained at rest (0 min) and after 5 and 40 min of exercise. Mean values (± SE) are from nine observations. *p < 0.05 compared with values at rest. ‡ p < 0.05 compared with 5 min.

Reprinted, by permission, from Kristiansen et al. 1997.

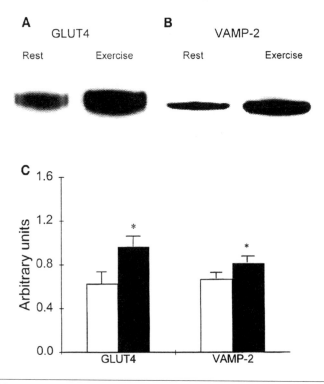

Figure 14.2. Representative immunoblots of GLUT4 (A), vesicle associated membrane protein (VAMP-2) (B), and mean concentrations of GLUT4 and VAMP-2 protein (C) expressed in arbitrary units in sarcolemmal giant vesicles produced from human muscle samples obtained at rest (open bars) and after exercise (solid bars). Mean values ($\pm$ SE) are from six experiments. *$p < 0.05$ compared with rest.

Reprinted, by permission, from Kristiansen et al. 1997.

vesicular trafficking in many tissues (6). In addition, we have demonstrated recently that the t-SNARE protein Syntaxin 4 is present in human sarcolemma (S. Kristiansen and E.A. Richter, unpublished observations). These latter findings support the hypothesis that translocation of GLUT4 to the surface membrane may involve SNARE proteins. (For review of the SNARE hypothesis, see 6.)

Signaling Involved in Increasing Muscle Glucose Uptake During Exercise

The signaling pathways initiating translocation of GLUT4 to the cell surface during exercise are still poorly understood. Early studies in amphibian muscles revealed that depolarization of the plasma membrane by itself is not responsible for the stimulation in muscle glucose transport (27). Furthermore, these studies also established that the amount of work performed per unit time during isotonic contractions (i.e., ATP utilization rate) does not affect the observed increase in glucose transport (27). However, experiments in the anesthetized rat (32) and in perfused rat hindlimbs (48, 49) have shown that different stimulation patterns result in different isometric force production and glucose transport enhancement. However, no simple correlation between force development or stimulation pattern and glucose transport was found (32).

An obvious point of interest in the study of contraction-stimulated glucose transport has been the role of intracellular calcium, because calcium is essential for the contraction process in skeletal muscle. Indeed, studies in amphibian muscle indicate that the greater the increase in intracellular calcium, brought about as a result of either potassium-induced muscle contraction (27, 58) or by NO_3-induced inhibition of sarcoplasmic reticulum calcium reuptake, the greater the rate of glucose transport (28). Moreover, the potential role of calcium as a regulator of glucose transport is not one that is just restricted to amphibian muscle, since Clausen and coworkers (10a) were able to provide the first evidence implicating calcium in the regulation of glucose transport in mammalian skeletal muscle. More recent work using the drugs W-7 or caffeine, which cause calcium release from the sarcoplasmic reticulum (63), indicate that an increase in cytosolic calcium concentration stimulates muscle glucose transport even when cytosolic calcium concentrations are too low to elicit muscle contraction (64). Thus, a rise in intracellular calcium during muscle contraction might be an important initiator for the contraction-induced increase in glucose transport.

In an attempt to further elucidate the downstream cellular events, the calcium dependent protein kinase (PKC) has been a focus of further research. PKC presumably is activated by the combination of contraction-induced increases in cytosolic calcium concentrations, by translocation to the plasma membrane and by increased production of diacylglycerol (11, 53). Decreased skeletal muscle PKC activity, induced either by prolonged incubation with phorbol esters or by using the inhibitor polymyxin B, has been reported to impair the contraction-induced increase in glucose transport (12, 24, 25, 65). Thus, activation of PKC during muscle contraction might represent a key event involved in the stimulation of glucose transport in skeletal muscle. In addition, recent experiments have indicated that muscle contractions increase the activity of the MAP kinase signaling cascade in both rat and human skeletal muscle (2, 18). However, the cell-permeable MAP-kinase kinase inhibitor PD98059 at concentrations (50 μM), which totally block the activation of the p42 and p44 MAP-kinase by contractions, did not suppress the contraction-induced increase in glucose transport in perfused rat muscles (J. Wojtaszewski, A.B. Jakobsen, and E.A. Richter, unpublished observations).

Insulin is the other major physiological stimulus increasing muscle glucose transport. Insulin and contractions both stimulate glucose transport by inducing the translocation of GLUT4 to the sarcolemma, but it is thought that the cellular and molecular events that elicit this translocation are different for the two stimuli (14, 49, 54). Still, an increased sensitivity to insulin in skeletal muscle after contractions (52) and an observed synergistic effect of contractions and submaximal insulin concentration on glucose transport in perfused rat muscle (59) suggest that the two stimuli might have a convergent course. Because activation of phosphatidylinositol 3-kinase (PI3K) is essential for the insulin stimulated glucose transport (for review, see 10) several investigators suggested that PI3K had a similar role for contraction-induced increase in glucose transport. However, muscle contractions induced as a result of electrical stimulation did not increase total PI3K activity or the subfraction of PI3K activity associated with insulin receptor substrate-1 in the anesthetized rat (17) or in perfused rat muscle (61). The PI3K inhibitor wortmannin at a concentration that totally inhibits the insulin-induced glucose transport had no effect upon the contraction-induced glucose transport in incubated rat muscles (42, 43, 62). In contrast, we recently observed that in perfused rat muscle, wortmannin at a similar concentration (1 µM) to that used in incubated muscle preparations decreased the amount of glucose taken up by the hindlimb during electrical stimulation (figure 14.3) without confounding effects on contractility, oxygen uptake, or general perfusion conditions (61). In addition, by increasing the wortmannin concentration from 1 to 10 µM it is possible to inhibit the contraction-induced glucose uptake

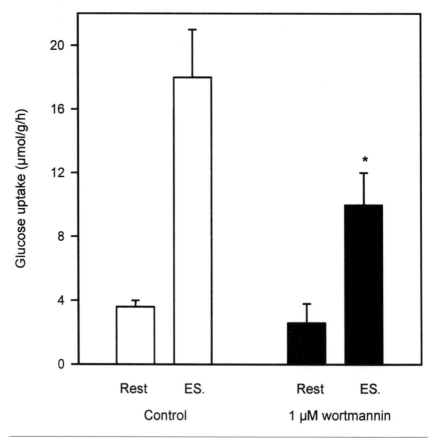

Figure 14.3.   Effect of wortmannin on contraction-stimulated glucose uptake in perfused rat hindlimb. Values are means ± SE of 10 experiments and are expressed in micromoles per gram stimulated muscle per hour. * p < 0.05 compared to control.
Adapted, by permission, from Kristiansen et al. 1997.

almost completely. However, at these high concentrations of wortmannin, force production was impaired somewhat (61). Because PI3K is not activated by electrical stimulation and because the inhibitory effect of wortmannin on contraction stimulated glucose transport was observed at concentrations that totally inhibit PI3K, our data suggest that wortmannin acts through a PI3K independent mechanism. Alternatively, another PI3K isoform, less sensitive to wortmannin, could be involved in contraction-stimulated glucose transport.

Adenosine, a well-known vascular modulator, has been suggested to mediate the observed potentiation of a submaximal insulin stimulus on glucose uptake in contracting perfused rat muscles (59). The effect seems to be mediated through the A1 adenosine receptor, but the underlying cellular interaction still is unsolved. Additional support for such a role of adenosine comes from in vivo dog myocardium studies (41) and recently from a human study in which theophylline administration reduced the whole-body glucose disposal rate only during exercise and not at rest (51). Nitric oxide is another short-lived factor produced in active muscle that might also modulate muscle glucose uptake. Data from incubated muscle suggest that sodium nitroprusside, a nitric oxide donor, increases muscle glucose uptake in a dose-dependent manner, which is additive to a submaximal insulin stimulus (3). Furthermore, inhibition of nitric oxide synthase seems to impair the stimulatory effect of muscle contractions on glucose uptake in incubated muscles (3).

Influence of Endurance Training

Endurance training has been shown to cause marked alterations in substrate utilization during exercise. In general, during exercise of the same submaximal absolute intensity, endurance training leads to greater utilization of lipids at the expense of carbohydrate (13, 26, 36, 37, 47).This decrease in carbohydrate utilization is due both to a decrease in muscle glycogenolysis (21, 36) and a decrease in muscle glucose utilization (13, 45, 47).

The mechanism behind the decrease in muscle glucose utilization during submaximal exercise in the trained individuals is not known. Theoretically, two different possibilities exist. One is that training primarily decreases glucose metabolism, which in the case of unchanged glucose transport capacity of the muscle cell membrane would lead to accumulation of glucose-6-phosphate (G-6 P) that via inhibition of hexokinase could result in accumulation of free intramuscular glucose. Decreased glucose utilization then would follow because of a decreased glucose concentration gradient from plasma to muscle cell. However, increased accumulation of glucose or G-6 P has not been reported in trained compared with untrained muscle during exercise (21, 30, 57). This suggests that decreased glucose metabolism is not the primary cause for decreased glucose utilization. The other possibility is that the primary training adaptation lessens the contraction-induced translocation of GLUT4 glucose transporters to the sarcolemma. This would result in decreased sarcolemmal glucose transport capacity during exercise in the trained muscle and would fit with the observation that there is no accumulation of glucose and G-6 P in trained muscle (21, 30, 57). We have conducted an experiment to clarify the mechanisms behind the decreased utilization rate of glucose during exercise after endurance training using the sarcolemmal giant vesicle technique.

Subjects performed knee extensions with both thighs simultaneously on two separate knee-extensor ergometers for 40 min. The subjects had endurance trained one of the thighs for three weeks prior to the experiment. The workload was the same absolute power output (77% of pretraining peak power) for each thigh. Preliminary data are given below. The full experiment is described in (55).

Training increased (p < 0.05) the total muscle GLUT4 protein content by 70% in the trained leg, whereas no change occurred in the untrained leg. At rest, glucose uptake was similar in the two thighs. During exercise, thigh glucose uptake was markedly higher compared with rest, however, the uptake during exercise was on average ≈35% lower in the trained leg than in the untrained leg (p < 0.05) (table 14.2). Sarcolemmal glucose transport capacity was similar in trained and untrained muscle at rest but increased markedly more with exercise in the untrained leg (≈150%) than in the trained leg (≈35%). Sarcolemmal GLUT4 content was similar in the two thighs at rest, but the increase with exercise was 60% larger (p < 0.05) in untrained than in trained muscle (table 14.2).

These data directly demonstrate that the mechanism involved in the smaller exercise-induced increase in glucose uptake in trained than in untrained muscle working at the same absolute submaximal workload, at least in part, is blunted exercise-induced translocation of GLUT4 protein to the sarcolemma, which in turn leads to diminished exercise-induced sarcolemmal glucose transport capacity. It is interesting that this happens in spite of markedly higher total muscle GLUT4 protein stores in trained muscle indicating that endurance training substantially decreases the fraction of the total muscle GLUT4 pool that is translocated to the sarcolemma during submaximal exercise. Furthermore, because exercise was carried out with the trained and the untrained muscles simultaneously, differences in blood-borne

Table 14.2   Thigh Glucose Uptake and GLUT4 Protein Content in Sarcolemmal Giant Vesicles at Rest and During 2-Legged Exercise With the Trained and Untrained Leg Simultaneously

	Rest	20 min	40 min
Glucose uptake, UT mmol min^{-1}	0.07 ± 0.03	0.87 ± 0.20	1.24 ± 0.22
Glucose uptake, T mmol min^{-1}	0.06 ± 0.03	0.65 ± 0.16*	0.65 ± 0.15*
Sarcolemmal GLUT4, UT arb. units	0.43 ± 0.07	ND	0.79 ± 0.07
Sarcolemmal GLUT4, T arb. units	0.44 ± 0.07	ND	0.66 ± 0.05†

Values are means ± SE of eight observations for glucose uptake and seven for GLUT4. UT: untrained. T: trained. Arb. units: Arbitrary units per ug protein compared to a rat heart standard. ND: not determined.
*p < 0.05 compared to UT.
†increase in GLUT4 is significantly (p < 0.05) smaller than in UT.
Data from Richter et al. 1998 (55).

factors cannot explain the different sarcolemmal glucose transport in trained and untrained muscle. Therefore, the adaptation to training has to reside either within the muscles themselves or be due to altered neuromuscular recruitment in the trained leg. In vitro studies are necessary to study whether the adaptation resides within the muscles themselves. Unfortunately, in vitro studies of trained and untrained rat muscle do not give a clear answer to the question of whether endurance training changes contraction-induced glucose uptake or not . Some have reported increased contraction-induced glucose transport in trained oxidative fibers but not in trained glycolytic fibers of perfused rat hindlimbs (8, 50). In contrast, chronic electrical stimulation increased total muscle GLUT4 and decreased subsequent contraction-induced glucose transport in the EDL and the red part of the tibialis anterior but increased it in the white part of the tibialis anterior muscle (31). Recently it has been described that the contraction-induced maximal rate of glucose transport is increased in trained compared with untrained rat epitrochlearis muscle but that more contractions are necessary to maximally activate glucose transport in trained compared with untrained muscle (35). Thus, it seems fair to conclude that muscles adapt to training with increased GLUT4 protein content. In some cases this is associated with an increased rate of maximally stimulated glucose transport. When stimulated submaximally, however, the glucose transport rate may be less than that in an untrained muscle.

Fructose Transport in Skeletal Muscle

GLUT5 has been identified in human skeletal muscle (29) and is known to function as a fructose transporter in the intestine (9). It also has been demonstrated that fructose can be taken up by incubated human muscle strips and can be oxidized to lactate and also serve as a substrate for glycogenesis (66). Furthermore, an increase in the arterial fructose concentration brought about by its intravenous infusion leads to a significant increase in its oxidation by exercising muscle (1). However, no details are available regarding the kinetic characteristics of fructose transport across the human sarcolemma or about the mechanism behind the exercise-induced utilization of fructose. To investigate these questions, sarcolemmal giant vesicles were produced from biopsies of human vastus lateralis at rest and after exercise (39). Fructose transport was shown to obey saturation kinetics with a transport K_m of approximately 8 mM. When studied at a hexose concentration of 5 mM, fructose transport across the sarcolemma occured at a rate about eightfold lower than that of glucose (figure 14.4). Furthermore, whereas the transport of glucose was inhibited almost entirely by cytochalasin B, fructose transport was unaffected (figure 14.4). Exercise at 100% of $\dot{V}O_2$max until exhaustion did not increase sarcolemmal fructose transport capacity or GLUT5 content. The data, therefore, indicate that human sarcolemmal vesicles contain GLUT5 and possess the capacity to take up fructose via a mechanism that displays saturation kinetics and is not sensitive to cytochalasin B. Taken together with the known substrate specificity of GLUT5 in expression systems (9), our data suggest that GLUT5 functions as a fructose transporter in human skeletal muscle.

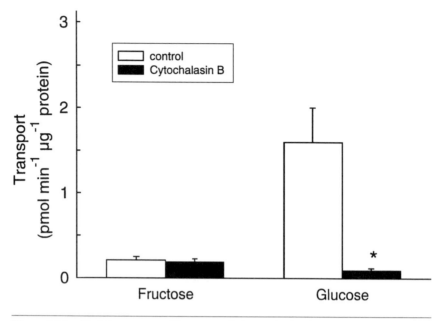

Figure 14.4.   Vesicle transport of fructose and glucose in the absence (open bars) and presence (solid bars) of 35 μM cytochalasin B. Transport was measured at a hexose concentration of 5mM. Values are means ± SE of three different preparations each measured in triplicate. *p < 0.05 vs. control.
Reprinted, by permission, from Wojtaszewski et al. 1996.

Summary

In adult human skeletal muscle, three isoforms of glucose transporter proteins have been identified. GLUT1 is present in the sarcolemma and is responsible for basal glucose transport. GLUT4 is the major glucose transporter in skeletal muscle. In the basal state GLUT4 is sequestered in one or more intracellular compartments. Upon stimulation with insulin or muscle contractions, GLUT4 is translocated to the sarcolemma and the t-tubular system. The molecular mechanism regulating this translocation is not known, but evidence is accumulating that supports that a mechanism similar to the SNARE hypothesis developed for vesicular transport of neurotransmitters in neuronal tissue may be involved in the translocation of GLUT4. The signaling leading to GLUT4 translocation in skeletal muscle during contractions is incompletely understood. Calcium release from the SR seems to be the first step, but the subsequent steps are elusive. The MAP kinase signaling cascade is activated in muscle during contractions, but this is not necessary for contraction-induced glucose transport. Endurance training is known to decrease the utilization of blood glucose during a submaximal exercise bout at a given absolute intensity. We have shown that the molecular mechanism behind the decrease in glucose utilization at least in part is blunted translocation of GLUT4 to the sarcolemma during exercise in the trained state, resulting in blunted exercise-induced sarcolemmal glucose transport.

Finally, GLUT5 has been identified in human sarcolemma that is capable of transporting fructose in a saturable manner with a K_m of ≈ 8 mM. Sarcolemmal fructose transport capacity and GLUT5 protein content is unaffected by exercise. It is proposed that GLUT5 functions as a fructose transporter in human skeletal muscle.

Acknowledgements

The authors were supported by the Danish National Research Foundation, grant #504-14, and the Wellcome Trust. Søren Kristiansen was supported by a postdoctoral training grant from the Danish Natural Science Research Council #9600433.

References

1. Ahlborg, B., and Björkman, O. Splanchnic and muscle fructose metabolism during and after exercise. *J. Appl. Physiol.* 69: 1244-1251; 1990.
2. Aronson, D., Violan, M.A., Dusfresne, S.D., Zangen, D., Fielding, R.A., and Goodyear, L.J. Exercise stimulates the mitogen-activated protein kinase pathway in human skeletal muscle. *J. Clin. Invest.* 99:1251-1257; 1997.
3. Balon, T.W., and Nadler, J.L. Evidence that nitric oxide increases glucose transport in skeletal muscle. *J. Appl. Physiol.* 82: 359-363; 1997.
4. Bell, G., Kayano, T., Buse, J., Burant, C., Takeda, J., Lin, D., Fukumoto, H., and Seino, S. Molecular biology of mammalian glucose transporters. *Diabetes Care* 13: 198-208; 1990.
5. Bell, G.I., Burant, C.F., Takeda, J., and Gould, G.W. Structure and function of mammalian facilitative sugar transporters. *J. Biol. Chem.* 268:19,161-19,164; 1993.
6. Bennett, M.K. Snares and the specificity of transport vesicle targeting. *Curr. Opin. Cell. Biol.* 7: 581-586; 1995.
7. Birnbaum, M. The insulin-sensitive glucose transporter. *Int. Rev. Cytol.* 137A: 239-297; 1992.
8. Brozinick, J.T.J., Etgen, G.J.J., Yaspelkis III, B.B., and Ivy, J.L. Effects of exercise training on muscle GLUT-4 protein content and translocation in obese Zucker rats. *Am. J. Physiol.* 265: E419-E427; 1993.
9. Burant, C.F., Takeda, J., Brot-Laroche, E., Graeme, I., Davidson, B., and Davidson, N.O. Fructose transporter in human spermatozoa and small intestine is GLUT5. *J. Biol. Chem.* 267: 14,523-14,526; 1992.
10. Cheatham, B., and Kahn, C.R. Insulin action and the insulin signaling network. *Endocrine Reviews* 16: 117-142; 1995.
10a. Clausen, T., Elbrink, J., Dahl-Hansen, A.B. The relationship between the transport of glucose and cations across cell membranes in isolated tissues:

IX. The role of cellular calcium in the activation of the glucose transport system in rat soleus muscle. *Biochem. Phys. Acta* 375: 2392-2408; 1975.

11. Cleland, P.J., Abel, K., Rattigan, S., and Clark, M. Long-term treatment of isolated rat soleus muscle with phorbol ester leads to loss of contraction-induced glucose transport. *Biochem. J.* 267: 659-663; 1990.

12. Cleland, P.J., Appleby, G., Rattigan, S., and Clark, M. Exercise-induced translocation of protein kinase C and production of diacylglycerol and phosphatidic acid in rat skeletal muscle in vivo. *J. Biol. Chem.* 264: 17,704-17,711; 1989.

13. Coggan, A., Kohrt, W., Spina, R., Bier, D., and Holloszy, J. Endurance training decreases plasma glucose turnover and oxidation during moderate-intensity exercise in men. *J. Appl. Physiol.* 68: 990-996; 1990.

14. Constable, S., Favier, R., Cartee, G., Joung, D., and Holloszy, J. Muscle glucose transport: interactions of in vitro contractions, insulin and exercise. *J. Appl. Physiol.* 64: 2329-2332; 1988.

15. Douen, A., Ramlal, T., Rastogi, S., Bilan, P., Cartee, G., Vranic, M., Holloszy, J., and Klip, A. Exercise induces recruitment of the insulin-responsive glucose transporter. *J. Biol. Chem.* 265: 13,427-13,430; 1990.

16. Goodyear, L.J., Chung, P.Y., Sherwood, D., Dufresne, S.D., and Moller, D.E. Effects of exercise and insulin on mitogen-activated protein kinase signaling pathways in rat skeletal muscle. *Am. J. Physiol.* 271: E403-E408; 1996.

17. Goodyear, L.J., Giorgino, F., Balon, T.W., Condorelli, G., and Smith, R.J. Effects of contractile activity on tyrosine phosphoproteins and PI 3-kinase activity in rat skeletal muscle. *Am. J. Physiol.* 268: E987-E995; 1995.

18. Goodyear, L.J., Hirshman, M.F., and Horton, E.S. Exercise-induced translocation of skeletal muscle glucose transporters. *Am. J. Physiol.* 261: E795-E799; 1991.

19. Goodyear, L.J., Hirshman, M.F., Napoli, R., Calles, J., Markuns, J.F., Ljungqvist, O., and Horton, E.S. Glucose ingestion causes GLUT4 translocation in human skeletal muscle. *Diabetes* 45: 1051-1056; 1996.

20. Gould, G.W., and Holman, G.D. The glucose transporter family: structure, function and tissue-specific expression. *Biochem. J.* 295: 329-341; 1993.

21. Green, H.J., Jones, S., Ball-Burnett, M., Farrance, B., and Ranney, D. Adaptations in muscle metabolism to prolonged voluntary exercise and training. *J. Appl. Physiol.* 78: 138-145; 1995.

22. Gumà, A., Zierath, J.R., Wallberg-Henriksson, H., and Klip, A. Insulin induces translocation of GLUT-4 glucose transporters in human skeletal muscle. *Am. J. Physiol.* 268: E613-E622; 1995.

23. Handberg, A., Kayser, L., Høyer, P.E., and Vinten, J. A substantial part of GLUT-1 in crude membranes from muscle originates from perineurial sheaths. *Am. J. Physiol.* 262: E721-E727; 1992.

24. Henriksen, E., Sleeper, M., Zierath, J., and Holloszy, J. Polymyxin B inhibits stimulation of glucose transport in muscle by hypoxia or contractions. *Am. J. Physiol.* 256: E662-E667; 1989.

25. Henriksen, E.J., Rodnick, K.J., and Holloszy, J.O. Activation of glucose transport in skeletal muscle by phospholipase C and phorbol ester. Evaluation of the regulatory roles of protein kinase C and calcium. *J. Biol. Chem.* 264: 21,536-21,543; 1989.

26. Henriksson, J. Training induced adaptations of skeletal muscle and metabolism during submaximal exercise. *J. Physiol. (London)* 270: 661-675; 1977.

27. Holloszy, J.O., and Narahara, H.T. Enhanced permeability to sugar associated with muscle contraction. *J. Gen. Physiol.* 50: 551-562; 1967.

28. Holloszy, J.O., and Narahara, H.T. Nitrate ions: potentiation of increased permeability to sugar associated with muscle contractions. *Science* 155: 573-575; 1967.

29. Hundal, H.S., Ahmed, A., Guma, A., Mitsumoto, Y., Marette, A., Rennie, M.J., and Klip, A. Biochemical and immunocytochemical localization of the GLUT5 glucose transporter in human skeletal muscle. *Biochem. J.* 286: 339-343; 1992.

30. Jansson, E., and Kaijser, L. Substrate utilization and enzymes in skeletal muscle of extremely endurance-trained men. *J. Appl. Physiol.* 62: 999-1005; 1987.

31. Jóhannsson, E., Jensen, J., Gundersen, K., Dahl, H.A., and Bonen, A. Effect of electrical stimulation patterns on glucose transport in rat muscles. *Am. J. Physiol.* 271: R426-R431; 1996.

32. Jóhannsson, E., McCullagh, K.J.A., Han, X.-X., Fernando, P.K., Jensen, J., Dahl, H.A., and Bonen, A. Effect of overexpressing GLUT-1 and GLUT-4 on insulin- and contraction-stimulated glucose transport in muscle. *Am. J. Physiol.* 271: E547-E555; 1996.

33. Kasanicki, M., and Pilch, P. Regulation of glucose-transporter function. *Diabetes Care* 13: 219-227; 1990.

34. Katz, A., Sahlin, K., and Broberg, S. Regulation of glucose utilization in human skeletal muscle during moderate dynamic exercise. *Am. J. Physiol.* 260: E411-E415; 1991.

35. Kawanaka, K., Tabata, I., & Higuchi, M. More tetanic contractions are required for activating glucose transport maximally in trained muscle. *J. Appl. Physiol.* 83: 429-433; 1997.

36. Kiens, B., Essen-Gustavsson, B., Christensen, N.J., and Saltin, B. Skeletal muscle substrate utilization during submaximal exercise in man: effect of endurance training. *J. Physiol. (London)* 469: 459-478; 1993.

37. Klein, S., Coyle, E.F., and Wolfe, R.R. Fat metabolism during low-intensity exercise in endurance-trained and untrained men. *Am. J. Physiol.* 267: E934-E940; 1994.

38. Kristiansen, S., Darakhshan, F., Richter, E.A., and Hundal, H.S. Fructose transport and GLUT-5 protein in human sarcolemmal vesicles. *Am. J. Physiol.* 273: E543-E548; 1997.

39. Kristiansen, S., Hargreaves, M., and Richter, E.A. Exercise-induced increase in glucose transport, GLUT4, and VAMP-2 in plasma membrane from human muscle. *Am. J. Physiol.* 270: E197-E201; 1996.

40. Kristiansen, S., Hargreaves, M., and Richter, E.A. Progressive increase in glucose transport and GLUT-4 in human sarcolemmal vesicles during moderate exercise. *Am. J. Physiol.* 272: E385-E389; 1997.

41. Law, W., McLane, M., and Raymond, R. Adenosine is required for myocardial insulin responsiveness in vivo. *Diabetes* 37: 842-845; 1988.

42. Lee, A.D., Hansen, P.A., and Holloszy, J.O. Wortmannin inhibits insulin-stimulated but not contraction-stimulated glucose transport activity in skeletal muscle. *Febs Lett* 361: 51-54; 1995.

43. Lund, S., Holman, G.D., Schmitz, O., and Pedersen, O. Contraction stimulates translocation of glucose transporter GLUT4 in skeletal muscle through a mechanism distinct from that of insulin. *Proc. Natl. Acad. Sci.* 92: 5817-5821; 1995.

44. Marette, A., Richardson, J.M., Ramlal, T., Balon, T.W., Vranic, M., Pessin, J.E., and Klip, A. Abundance, localization, and insulin-induced translocation of glucose transporters in red and white muscle. *Am. J. Physiol.* 263: C443-C452; 1992.

45. Mendenhall, L.A., Swanson, S.C., Habash, D.L., and Coggan, A.R. Ten days of exercise training reduces glucose production and utilization during moderate-intensity exercise. *Am. J. Physiol.* 266: E136-E143; 1994.

46. Mueckler, M. Facilitative glucose transporters. *Eur. J. Biochem.* 219: 713-725; 1994.

47. Phillips, S.M., Green, H.J., Tarnopolsky, M.A., Heigenhauser, G.J.F., Hill, R.E., and Grant, S.M. Effects of training duration on substrate turnover and oxidation during exercise. *J. Appl. Physiol.* 81: 2182-2191; 1996.

48. Ploug, T., Galbo, H., and Richter, E.A. Increased muscle glucose uptake during contractions: no need for insulin. *Am. J. Physiol.* 247: E726-E731; 1984.

49. Ploug, T., Galbo, H., Vinten, J., Jørgensen, M., and Richter, E.A. Kinetics of glucose transport in rat muscle: effects of insulin and contractions. *Am. J. Physiol.* 253: E12-E20; 1987.

50. Ploug, T., Stallknecht, B.M., Pedersen, O., Kahn, B.B., Ohkuwa, T., Vinten, J., and Galbo, H. Effect of endurance training on glucose transport capacity and glucose transporter expression in rat skeletal muscle. *Am. J. Physiol.* 259: E778-E786; 1990.

51. Raguso, C.A., Coggan, A.R., Sidossis, L.S., Gastaldelli, A., and Wolfe, R.R. Effect of theophylline on substrate metabolism during exercise. *Metabolism* 45: 1153-1160; 1996.

52. Richter, E.A., Cleland, P.J.F., Rattigan, S., and Clark, M.G. Contraction-associated translocation of protein kinase C in rat skeletal muscle. *Febs Lett.* 217: 232-236; 1987.

53. Richter, E.A., Garetto, L.P., Goodman, M.N., and Ruderman, N.B. Enhanced muscle glucose metabolism after exercise: modulation by local factors. *Am. J. Physiol.* 246: E476-E482; 1984.

54. Richter, E.A. Glucose utilization. In *Handbook of Physiology. Section 12: Exercise: Regulation and integration of multiple systems*. Ed. L.B. Rowell and J.T. Shepherd, 912-951. New York: Oxford University Press; 1996.

55. Richter, E.A., Jensen, P., Kiens, B., and Kristiansen, S. Sarcolemmal glucose transport and GLUT4 translocation during exercise is diminished by endurance training. *Am. J. Physiol.* 274: E89-E95, 1998.

56. Thorens, B., Charron, M., and Lodish, H. Molecular physiology of glucose transporters. *Diabetes Care* 13: 209-218; 1990.

57. Turcotte, L.P., Richter, E.A., and Kiens, B. Increased plasma FFA uptake and oxidation during prolonged exercise in trained vs. untrained humans. *Am. J. Physiol.* 262: E791-E799; 1992.

58. Valant, P., and Erlij, D. K+-stimulated sugar uptake in skeletal muscle: role of cytoplasmic Ca++. *Am. J. Physiol.* 245: C125-C132; 1983.

59. Vergauen, L., Hespel, P., and Richter, E.A. Adenosine receptors mediate synergistic stimulation of glucose uptake and transport by insulin and by contractions in rat skeletal muscle. *J. Clin. Invest.* 93: 974-981; 1994.
60. Wilson, C.M. and Cushman, S.W. Insulin stimulation of glucose transport activity in rat skeletal muscle: increase in cell surface GLUT4 assessed by photolabelling. *Biochem. J.* 299: 755-759; 1994.
61. Wojtaszewski, J.F.P., Hansen, B.F., Ursø, B., and Richter, E.A. Wortmannin inhibits both insulin- and contraction-stimulated glucose uptake and transport in rat skeletal muscle. *J. Appl. Physiol.* 81: 1501-1509; 1996.
62. Yeh, J.I., Gulve, E.A., Rameh, L., and Birnbaum, M.J. The effects of wortmannin on rat skeletal muscle. *J. Biol. Chem.* 270: 2107-2111; 1995.
63. Youn, J.H., Gulve, E.A., Henriksen, E.J., and Holloszy, J.O. Interaction between effects of W-7, insulin and hypoxia on glucose transport in skeletal muscle. *Am. J. Physiol.* 267: R888-R894; 1994.
64. Youn, J.H., Gulve, E.A., and Holloszy, J.O. Calcium stimulates glucose transport in skeletal muscle by a pathway independent of contraction. *Am. J. Physiol.* 260: C555-C561; 1991.
65. Young, J.C., Kurowski, T.G., Maurice, A.M., Nesher, R., and Ruderman, N.B. Polymyxin B inhibits contraction-stimulated glucose uptake in rat skeletal muscle. *J. Appl. Physiol.* 70: 1650-1654; 1991.
66. Zierath, J.R., Nolte, L.A., Wahlstrøm, E., Galuska, D., Shepherd, P.R., Kahn, B.B., and Walberg-Henriksson, H. Carrier-mediated fructose uptake significantly contributes to carbohydrate metabolism in human skeletal muscle. *Biochem. J.* 311: 517-521; 1995.

CHAPTER 15

Lactate Exchange and pH Regulation in Skeletal Muscle

Carsten Juel and Henriette Pilegaard

Copenhagen Muscle Research Centre, August Krogh Institute,
University of Copenhagen

Lactate Production, Accumulation, and Release

Lactate is produced in skeletal muscle when the rate of glycolysis exceeds the mitochondrial removal of pyruvate. During high-intensity exercise large amounts of lactate can be formed, and when the rate of production is higher than the rate of release to the surroundings, lactate accumulates in the muscle fibers. After an intense exercise bout or when the work intensity is reduced, lactate accumulated in the muscle cells during the exercise will either be metabolized in the muscle fibers or be released from the fiber to the interstitial space. From the interstitium lactate either can be taken up by neighboring muscle fibers or it can enter into the blood and subsequently be taken up by other tissues including resting as well as active muscle. The fraction that will be released to the blood will depend on the actual conditions, of which muscle blood flow and muscle mass involved in the exercise are important factors. For instance, it was estimated that one-third of the produced lactate was released to the blood during an exhaustive one-legged knee extensor exercise bout of 3.2 min duration, and for the entire period of exercise and recovery 82% of the lactate was released (4). The half-time for lactate release from muscle to blood during recovery from intense exercise in man falls within 4 to 10 min (21, 29, 50).

The lactate that remains inside a prior intensely contracting muscle fiber and lactate that enters a fiber can be used as precursor for oxidation or glyconeogenesis. Estimations of the relative contribution of the two pathways have given widely different results with a large fraction to glycogen synthesis in some studies (20) but only a minor contribution in others (5, 56). Nonetheless, the flux of lactate between different compartments of the body constitutes a transfer of energy-rich compounds. Therefore, the traditional view of lactate as a simple waste product seems to be too narrow, because lactate may play an important role as an energy-rich substance during sustained high-intensity exercise and during prolonged exercise with high-intensity bouts included.

Lactate/H⁺ Transport

Basic Characteristics

During short, intense exercise and in early recovery from such exercise the intracellular lactate concentration is clearly higher than the concentration in blood plasma. The existence of this gradient between muscle and blood shows that the sarcolemma and/or the capillary membrane constitute a barrier for lactate translocation.

The main part of the lactate flux across the sarcolemma is mediated by a group of membrane bound monocarboxylate transporter proteins; e.g., the lactate/H⁺ carrier. Although H⁺ and lactate are the primary substrates for this transport system other monocarboxylates also can be transported by the carrier, but of these only pyruvate and ketone bodies are of physiological relevance. The co-transport of lactate and protons is coupled in a 1:1 ratio, resulting in an electroneutral transport process. Values reported for K_m of lactate transport in skeletal muscle range from 10 to 45 mM, which are rather high compared with the lactate levels obtained during physiological conditions (for review, see 27).

The high K_m of lactate transport may explain the difficulties in detecting the activity of the lactate/H⁺ carrier in human in vivo experiments, where no saturation of lactate release is apparent when efflux data are related to the muscle to plasma lactate gradients (4). Therefore, a model system needs to be used for examination of the lactate/H⁺ carrier properties. Skeletal muscle lactate transport has been characterized in animals by use of both giant sarcolemmal vesicles obtained by collagenase treatment of muscle tissue (24, 28) as well as small vesicles produced by homogenization and ultra-centrifugation (35, 48). However, only the giant vesicle technique has been applied on human muscle biopsies, and it has been used to characterize the carrier mediated sarcolemmal lactate flux in human skeletal muscle (25). In such experiments a clear saturation of the transmembrane flux rate was observed at increasing lactate concentrations (figure 15.1). In addition, by means of specific transport inhibitors, it was found that the fraction of the total lactate transport mediated by the co-transporter was between 50 and 90% (25).

Isoforms

Until now two isoforms of the lactate/H⁺ carrier have been cloned. These isoforms have been denoted MCT1 and MCT2 (monocarboxylate transporter), and they are both present in red muscle fibers, but only in small amounts in white muscle (16, 17). The latter observation shows that further MCT isoforms must exist in skeletal muscle, because a pronounced carrier mediated lactate transport also is present in white muscle (28, 45). This suggestion is in agreement with the statement made by Jackson et al. (23) that more MCT isoforms are under way. It has been suggested that different isoforms may be involved in uptake and release of lactate (34). However, the muscle fiber type distribution of the various MCT isoforms has only partly been investigated, and based on the finding that the K_m for efflux and influx of lactate is similar (26), it appears that the MCT isoforms transport lactate symmetrically. Thus,

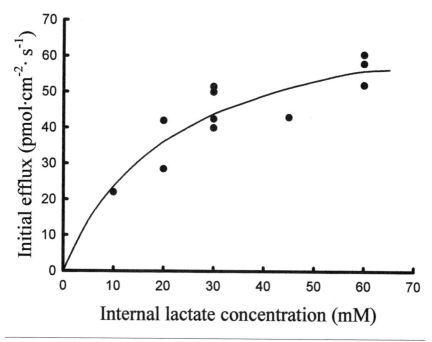

Figure 15.1.   Lactate/H⁺ transport in human skeletal muscle displays saturation kinetics. Initial transport rate was measured in giant vesicles made from muscle needle biopsies obtained from vastus lateralis. Each point represents the initial efflux rate from one experiment.

it is too early for final conclusions about the functional roles of the various MCT isoforms in skeletal muscle.

Fiber Type Dependency

The relation between sarcolemmal lactate transport capacity and muscle fiber type distribution has been investigated in rats, using the giant vesicle technique. Dividing rat hindlimb muscles into red and white muscles showed that red muscles had approximately 50% higher lactate transport capacity than white muscles (28), indicating that muscle fibers with a high oxidative potential possess a higher lactate transport capacity than more glycolytic fibers. This suggestion was confirmed by the findings that the rat soleus muscle, which is mainly composed of slow-twitch oxidative fibers (3), had approximately 15% and 32% higher lactate transport capacity than the red gastrocnemius (dominated by fast-twitch oxidative glycolytic fibers) (3) and the white gastrocnemius muscle (primarily composed of fast-twitch glycolytic fibers), respectively (45) (figure 15.2). Furthermore, a positive correlation between lactate transport capacity and percentage occurrence of slow-twitch fibers in the vastus lateralis muscle was observed in a human cross-sectional study (43).

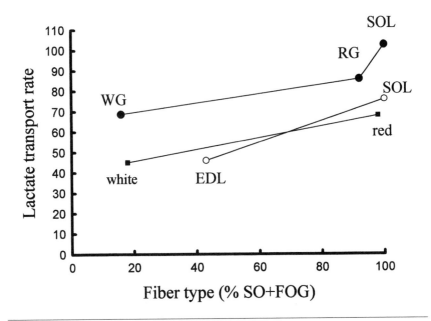

Figure 15.2.   Muscle fiber type and lactate/H⁺ transport capacity. The total lactate transport rate with 30 mM lactate was measured in giant vesicles obtained from soleus (SOL), red gastrocnemius (RG), white gastrocnemius (WG) (●) (data from Pilegaard and Juel [45]), mainly "red" or "white" muscles (■) (data from Juel et al. [23]), and in isolated SOL and extensor digitorum longus muscle (EDL) (○) (data from Bonen and McCullagh [10]). The total lactate transport is the sum of carrier mediated transport and simple diffusion, where the latter constitutes approximately 30% of the total transport. The transport rates are expressed in three different units: pmol cm⁻² s⁻¹, the rate constant k · 1000 (s⁻¹), and nmol mg⁻¹ protein min⁻¹, respectively.

The latter finding is in agreement with the existence of a similar relation in human muscle, but the correlation between lactate transport capacity and fiber type distribution was low, showing that the two parameters are not necessarily tightly coupled in human skeletal muscle. A positive correlation between the occurrence of oxidative fibers and the lactate/H⁺ transport capacity may seem surprising at first, since glycolytic fibers are specialized for recruitment during short-term, high-intensity exercise with large productions of lactate and protons. However, muscle fibers with a large lactate gradient across the sarcolemma are less dependent on the lactate transport capacity than fibers with a low concentration difference between the intra- and extracellular compartments. In addition, the distribution of lactate/H⁺ transport capacities among the various fiber types may suggest that the lactate/H⁺ carrier also is important for uptake of lactate by skeletal muscles both because of the correlation between lactate transport capacity and ability to oxidize lactate and the notion that inward directed lactate gradients generally are rather small in magnitude. It may be speculated that lactate uptake by skeletal muscle could play a special role during prolonged exercise where high-intensity intervals are included.

Adaptations

It may be speculated that changes in the ability of muscle fibers to release lactate and H⁺ could be important for the work capacity during high-intensity exercise. Moreover, the existence of a carrier mediated membrane transport system allows adaptations to occur in the efflux of lactate and H⁺ in response to alterations in the level of muscle activity. In accordance with this statement, the lactate transport capacity of rat skeletal muscle was increased by up to 78% by six to seven weeks of moderate (36, 44) and high-intensity treadmill training (44), and doubled by chronic low-frequency stimulation for seven days (34). In contrast, lactate transport of rat skeletal muscle was unaffected by six to seven weeks of swim training (44), low-intensity treadmill training (49), and training with short-lasting work bouts at a very high intensity (49) (figure 15.3).

A human cross-sectional study revealed that athletes had about 30% higher lactate transport capacity in the vastus lateralis muscle than untrained and less trained subjects (figure 15.3). The results suggested that subjects with an elevated

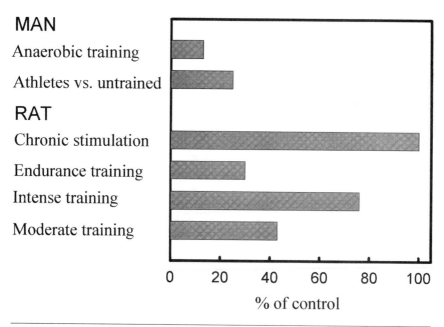

Figure 15.3.   Lactate/H⁺ transport increase and level of muscle activity. The effects of six to seven weeks of moderate (McDermott and Bonen [33]; Pilegaard et al. [44]), intense (Pilegaard et al. [44]) treadmill training, and seven days of chronic stimulation (McCullagh et al. [34]) on lactate/H⁺ transport in rats were measured with giant or small sarcolemmal vesicles. The lactate/H⁺ transport rate measured in human sarcolemmal vesicles was measured after eight weeks of anaerobic training (Pilegaard, unpublished) and in athletes compared with untrained individuals (Pilegaard et al. [43]).

lactate transport capacity also had a high maximal oxygen uptake ($\dot{V}O_2$max), which could indicate that a large volume of training is needed to obtain a high lactate transport capacity. However, a high $\dot{V}O_2$max was not always associated with a high lactate transport capacity. Moreover, the observations that the subjects with the highest transport capacities were elite 4 km track bicyclists and that the one with the highest lactate transport capacity was a bronze medalist at the Olympic Games in Barcelona in 1992 may reflect the importance of including high-intensity in the training (43). Recently, we conducted a longitudinal human training study, where eight weeks of anaerobic training with one leg resulted in 12% higher total lactate transport in the trained than in the untrained muscle (47).

Various studies have been dealing with the effect of lowering the level of muscle activity. It was demonstrated that the lactate transport rate decreased (9–15% depending on the muscle type) already after one day of denervation of rat skeletal muscle. Prolonging the denervation period induced a further decline in the lactate transport with 36–50% lower lactate transport rate after three weeks of denervation, compared with control muscles (45). In agreement with these findings lactate uptake into isolated rat soleus strips was reduced after three days of denervation (33), although the magnitude of the change was larger in the latter study (68%) than in the former (24%). Furthermore, the rate of lactate transport was reduced by 36% upon four weeks of rat tail suspension when measured at a very low lactate concentration (11). These observations show that lactate/H^+ transport in rat skeletal muscle is affected by the level of muscle activity. In addition, the recent finding that the carrier mediated lactate transport capacity was approximately 25% lower in the thigh muscle of spinal cord-injured individuals than in the muscle of untrained, normally physically active subjects suggests that a similar conclusion can be drawn for humans (47).

The mechanisms underlying differences and changes in lactate transport capacity are not understood fully yet. Improved knowledge about this issue is expected when antibodies directed toward the various isoforms of the lactate carrier are available. However, Michaelis-Menten parameters may give some information about the mechanism behind differences in transport capacities, because the maximal transport rate (V_{max}) can be assumed to reflect the number of transporters involved (or the turnover number of the transport proteins), and the K_m of the transport can be taken as an indication of the affinity of the carrier for the substrate.

The distinct lactate transport capacity in red and white rat skeletal muscles appeared to be associated with different V_{max} values, suggesting that more transporter proteins are involved in lactate transport in red muscles than in the white (28). The data describing the effects of changes in muscle activity are conflicting but may suggest that more than one mechanism is involved. Thus, rat studies have indicated both a decreased K_m (35, 44) and an increased V_{max} (44) after training, and chronic stimulation induced improvements in lactate transport appeared to be mediated mainly through an elevated V_{max} and to a minor extent by lowering of the K_m (34). Furthermore, it was reported that K_m of the lactate transport seemed to be higher in rat muscles subjected to four weeks of unweighting compared with control muscles (11).

Muscle Fatigue

For a long time, lactic acid accumulation has been suggested to be involved in muscle fatigue. The lactate ion per se generally has not been regarded as having negative effects on muscle functions, although exercise-induced muscle fiber swelling has been shown to be related to muscle lactate accumulation (9), but it recently has been shown that elevated muscle lactate concentrations may impair muscle contractions (13).

Traditionally, it has been suggested that the accumulation of free H^+ in muscle is the possible cause of muscle fatigue associated with a large lactate production. Many studies with isolated muscle preparations have demonstrated that lowering of muscle pH can impair force production by interfering at various steps in the excitation-contraction coupling sequence (for review, see 14, 61) or by inhibiting formation of ATP (53). However, despite the vast number of studies showing inhibitory effects of reduced muscle pH, several experiments have shown that muscle pH cannot be the sole factor involved in muscle fatigue (4, 29, 55, 61). Nonetheless, elevated concentrations of H^+ and maybe also of lactate appear to be involved in muscle fatigue at least during certain conditions, and it might be expected that the ability of muscle fibers to delay an accumulation of H^+ and lactate as well as to remove accumulated lactate and H^+ can be important during and initially after high-intensity exercise.

Regulation of Muscle pH

The lactate associated H^+ production will be the major contributor to the formation of H^+ during high-intensity exercise, and the magnitude of the concomittant decrease in intracellular pH will be determined by the ability of the muscle fibers to remove and sequester the protons. This regulation of muscle pH can be accomplished by three different mechanisms: 1) membrane transport systems, 2) physiochemical buffering, and 3) metabolic processes.

Membrane Transport Systems in pH Regulation

Resting skeletal muscle fibers are continuously exposed to an acid load resulting from the passive influx of H^+ driven by the electrochemical gradient. The observation that internal pH is kept at a level much higher than predicted from a passive distribution of H^+ according to the Nernst equation shows that acid extrusion mechanisms operate even in resting muscle fibers as it is the case in most other cell types. A pH regulating membrane transport system can be defined as a system that removes H^+, and at least three different types are involved in pH regulation in skeletal muscle: lactate/H^+ co-transport, Na^+/H^+ exchange, and HCO_3-dependent transport systems (figure 15.4).

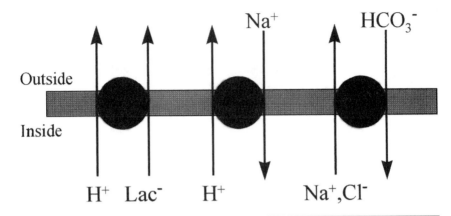

Figure 15.4.   The three membrane transport systems involved in skeletal muscle pH regulation: lactate/H^+ co-transport, Na^+/H^+ exchange, and Na^+ dependent Cl^-/HCO_3^- exchange.

Lactate/H^+ Transport in pH Regulation.   The lactate/H^+ co-transport is driven mainly by the lactate gradient across the sarcolemma and, therefore, it cannot mediate an uphill release of H^+ against the electrochemical gradient in resting muscles. However, a large outward directed lactate gradient can be created during intense muscle activity with pronounced intracellular accumulation of lactate, and during such conditions the lactate/H^+ transporter will contribute the most to the release of H^+.

During 3 min exhaustive one-legged knee extensor exercise two-thirds of the measured H^+ release was found to be coupled to lactate (4, 29) showing that the lactate/H^+ transporter operates as the dominant pH regulating mechanism during such type of exercise. During less intense exercise with minor muscle lactate accumulation only a small fraction of total H^+ release will be coupled to lactate (6).

Interestingly, if different parts of the body are working at different intensities the orientations of the lactate and H^+ fluxes can be opposite in muscles that are working at a high and low intensity, respectively. Thus, thigh muscles working at a moderate intensity switched from net release to net uptake of lactate when the blood lactate concentration was elevated by adding intense arm exercise to the leg exercise, even though there was a net release of H^+ from the muscles (6). Therefore, other mechanisms for H^+ release also must be present.

Na^+/H^+ Exchange and HCO_3^- Dependent Systems in pH Regulation.   Generally, the Na^+/H^+ exchanger has been considered to be the most important pH regulation system of the body, and this assumption is true for resting skeletal muscle, where the efflux of H^+ predominantly is mediated by this system (2). Although the Na^+/H^+ exchanger is already active at resting pH, a further activation occurs when internal pH is lowered. However, the need for H^+ removal during intense exercise exceeds the maximal capacity of the Na^+/H^+ exchanger, which is therefore insufficient during such conditions.

The capacity of the Na^+/H^+ exchanger has been found to be enhanced in rats by high-intensity treadmill training, but not with endurance training (Juel, unpublished data). The Na^+/H^+ exchange system has been reported to be up-regulated in both hypertensive rats (54) and humans (12). Interestingly, it was shown by use of nuclear magnetic resonance that hypertensive subjects had a faster muscle pH recovery after exercise than normal subjects (12), demonstrating the functional importance of the Na^+/H^+ exchanger for pH regulation during recovery from exercise.

Membrane transport systems involving fluxes of bicarbonate also are assumed to be involved in muscle pH homeostasis both at rest and during exercise (18). The capacity of the bicarbonate dependent systems in fast-twitch skeletal muscles of rats has been found to be similar to the capacity of the Na^+/H^+ exchanger (18). Examinations of basal transport characteristics as well as responsiveness to changes in level of muscle activity remain to be conducted for human skeletal muscle Na^+/H^+ exchange and bicarbonate dependent systems. Furthermore, detailed in vivo investigations of the role of bicarbonate fluxes for H^+ extrusion during muscle activity are complicated by the simultaneous fluxes of CO_2 and the pH changes in muscle and blood, and such studies have not been performed in humans.

Physio-Chemical Buffer Capacity

Buffering involves sequestering of H^+ so that the number of free H^+ is reduced and the presence of a buffer increases the amount of acid that can be added to obtain a given change in pH. Thus, in general, buffering can be considered as the ability to minimize changes in pH (58).

The buffer value was introduced by Van Slyke (58) as a measure of buffer action, and it was defined as the differential ratio dB/dpH (where B = base), which is the slope of the titration curve at any point. The letter β was chosen to denote the ratio. Moreover, as a numerical measure of the buffer value the amount of acid or base that must be added to produce a pH change of one unit corresponding to a tenfold change in H^+ concentration was proposed (58). To simplify the unit of the buffer value it was suggested that it was abbreviated and just called "a slyke" (62). The efficiency of a buffer substance is dependent on the concentration of the buffer and the H^+ sequestering capability of the buffer in the pH range of interest. The latter is a function of the dissociation constant (K_a) of the buffer and the maximal buffer effect is obtained when the H^+ concentration equals K_a corresponding to pH = pK_a (58).

Components of Physio-Chemical Buffering.   The components of physio-chemical buffering in skeletal muscle will include bicarbonate, free phosphate, and a major contribution from dipeptides and proteins (1, 7, 22, 40). The histidine containing amino acid compounds is important in the buffering process in skeletal muscle, because only the imidazole group of histidine has a pKa value in the physiological pH range (1, 7, 22, 62). Special attention has been given to the buffer action of the dipeptide carnosine, which is found in rather high concentrations in skeletal muscle of many species (7), including humans (41).

Metabolic Processes

The creatine kinase reaction involves H^+ with an uptake of H^+ during creatine phosphate (CP) utilization. The high concentrations of CP in skeletal muscle and the almost complete depletion of muscle CP during high-intensity exercise, therefore, will make creatine phosphate hydrolysis the major metabolic buffering process during intense exercise (see 22).

Determination of Buffer Capacity: Methodological Comments

Determination of the buffer value of homogenized muscle samples can be accomplished by titration with strong acid or base as first described by Furusawa and Kerridge (15), but it can also be done by means of changes in PCO_2 (19). The concentration of added acid and base has to be high so that the change in volume of the sample can be neglected, and Van Slyke (58) recommended that the increase should be kept below 50% of the original volume. Generally the amount of H^+ (or OH^-) required to cause a certain change in pH is used as the buffer value, although the slope of the tangent of the titration curve will be the precise measure of the buffer value at a given pH. Thus, the former approach gives an average buffer value over a chosen pH range.

The buffer value obtained by the homogenate technique often is denoted the in vitro buffer capacity, and it is a measure of the nonbicarbonate physio-chemical buffering of the sample without the exchange of substances across the sarcolemma. Therefore, if the purpose is to examine the overall ability of a muscle to regulate pH, the H^+ removal by membrane transport systems and metabolic processes must be considered and added to the in vitro determined buffer value.

In an attempt to get a more functional measure, studies have used changes in PCO_2 on in situ and in vivo skeletal muscle or with a muscle suspended in a small chamber (see 19). Moreover, it was suggested that an in vivo buffer capacity could be determined in humans by means of an in vivo titration of a muscle during intense exercise. This in vivo buffer capacity was defined as the ratio between the change in muscle lactate and the change in muscle pH (ΔlactateΔpH) during an isometric contraction to fatigue. This procedure has been repeated in other studies using both isometric and dynamic exercise of various duration (39, 52, 57). Originally, it was assumed that the circulation to the muscle was occluded during the contraction and that the sarcolemmal passage of lactate and H^+ could be disregarded (51). However, that may not always be the case (51). It then might be argued that the in vivo buffer capacity is a functional measure that takes into account all the various H^+ removing mechanisms. However, the following discussion shows that several concerns should be considered when the in vivo buffer capacity is used, and comparisons and interpretations of training data can be critical. First, the duration and intensity of the exercise bout will influence the result, which made McKenna et al. (37) propose the use of the ratio ΔH^+/work as a more physiological measure of training effects on muscle H^+ regulation instead of the in vivo buffer capacity. Second, whether the exercise will occlude the circulation to the working muscle needs to be taken into account. Third, the removal of H^+ mediated by the various membrane transport systems will affect the in vivo buffer capacity differently. Thus, operation of the Na^+/H^+ exchanger will increase the in vivo buffer capacity, because it removes H^+ without lactate being involved. In contrast, the lactate/H^+ transporter will reduce the in vivo buffer capacity, although it removes lactate and H^+ in a 1:1 manner, because of the logarithmic scale in the denominator of the expression. This notion may explain why the difference between in vitro and in vivo determined buffer capacities can be rather small (30) although in vitro measurements include neither bicarbonate nor metabolic and membrane processes. Furthermore, training induced increases in the capacities of the Na^+/H^+ exchanger and the lactate/H^+ transporter will affect the in vivo buffer capacity in opposite directions and may even cancel each other. The above discussed points have to be considered when in vitro and in vivo buffer capacities are examined and interpreted; a statement that is in agreement with the conclusion made by Mannion et al. (30) that it is important to take into account the method used to determine the buffer capacity and that problems may exist in comparing inter-individual and training data.

Buffer Capacity and Fiber Type

Both in vitro (1, 42, 59) and in vivo buffer capacities (57) have been found to be related to the fiber type distribution in rat and fish muscle with higher values in muscles dominated by glycolytic fibers than in more oxidative muscles. The content of histidine-related compounds was also higher in white than in red fish muscle (1). In contrast, a significant correlation between buffer capacity and fiber type distribution has generally not been observed for human muscle (32, 38, 51).

Buffer Capacity and Training

Several studies have investigated the importance of the level of physical activity for the buffer capacity of skeletal muscle applying both in vitro and in vivo determinations. However, the results are not consistent. Cross-sectional human studies reported that 800 m runners and rowers had almost 50% higher in vitro buffer capacity of deproteinized muscle samples than marathon runners and untrained subjects (41). Moreover, athletes competing in ball games with high anaerobic contribution had 18% higher in vivo buffer capacity than untrained subjects (51), whereas the in vivo buffer capacity was similar in endurance trained and untrained subjects (52). The in vitro buffer capacity was found to be enhanced by approximately 16% both after sprint training (8) and when cyclists replaced a portion of the habitual training with high-intensity interval training (60), whereas no effect was detected after isokinetic (31) or sprint training in other studies (37, 39). Furthermore, high-intensity training of rats and sprint training of humans (52) improved the in vivo buffer capacity by 11–24% and approximately 35%, respectively, but the in vivo buffer capacity was unaffected by sprint training in other human studies (37, 39). Thus, the data are conflicting, and it is very important to define the methods used if buffer capacity and training are considered.

Acknowledgement

The study was supported by the Danish National Research Foundation.

References

1. Abe, H., Dobson, G., Hoeger, U., and Parkhouse, S. Role of histidine-related compounds to intracellular buffering in fish skeletal muscle. *American Journal of Physiology* 249: R449-R454; 1985.
2. Aickin, C.C., and Thomas, R.C. An investigation of the ionic mechanism of intracellular pH regulation in mouse soleus muscle fibers. *Journal of Physiology* 273: 295-316; 1977.
3. Armstrong, R.B., and Phelps, P.O. Muscle fiber type composition of the rat hindlimb. *American Journal of Anatomy* 171: 259-272; 1984.
4. Bangsbo J., Gollnick, P.D., Graham, T.E., Juel, C., Kiens, B., Mizuno, M., and Saltin, B. Anaerobic energy production and O_2 deficit-debt relationship during exhaustive exercise in humans. *Journal of Physiology* 422: 539-559; 1991.
5. Bangsbo, J., Johansen, L., Graham, T., and Saltin, B. Lactate and H^+ efflux from human skeletal muscles during intense, dynamic exercise. *Journal of Physiology* 462: 115-133; 1993.
6. Bangsbo, J., Juel, C., Hellsten, Y., and Saltin, B. Dissociation between lactate and proton exchange in muscle during intense exercise in man. *Journal of Physiology* 504 (2): 489-499; 1997.
7. Bate Smith, E.C. The buffering of muscle in rigor; protein, phosphate and carnosine. *Journal of Physiology* 92: 336-343; 1938.
8. Bell, G.J. and Wenger, H.A. The effect of one-legged sprint training on intramuscular pH and nonbicarbonate buffering capacity. *European Journal of Applied Physiology* 58: 158-164; 1988.
9. Bergstrom, J., Guarnier, G., and Hultman, E. Carbohydrate metabolism and electrolyte changes in human muscle tissue during heavy work. *Journal of Applied Physiology* 1: 122-125; 1971.
10. Bonen, A., and McCullagh, K.J.A. Effects of exercise on lactate transport into mouse skeletal muscles. *Canadian Journal of Applied Physiology* 19: 275-285; 1994.
11. Dubouchaud, H., Grainer, P., Mercier, J., Le Peuch, C., and Prefaut, C. Lactate uptake by skeletal muscle sarcolemma vesicles decreased after 4 weeks of hindlimb unweighting in rats. *Journal of Applied Physiology* 80: 416-421; 1996.
12. Dudley, C.R.K., Taylor, D.J., Ng, L.L., Kemp, G.J., Ratcliffe, P.J., Radda, G.K., and Ledingham, G. Evidence for abnormal Na^+/H^+ antiport activity detected by phosphorus nuclear magnetic resonance spectroscopy in exercising skeletal muscle of patients with essential hypertension. *Clinical Science* 79: 491-497; 1990.

13. Favero, T.G., Zable, A.C., Colter, D., and Abramson, J.J. Lactate inhibits Ca²⁺-activated Ca²⁺-channel activity from skeletal muscle sarcoplasmic reticulum. *Journal of Applied Physiology* 82 (2): 447-452; 1997.

14. Fitts, R.H. Cellular mechanisms of muscle fatigue. *Physiological Reviews* 74: 49-94; 1994.

15. Furusawa, K., and Kerridge, P.M.T. The hydrogen ion concentration of the muscles of the cat. *Journal of Physiology* 63: 33-41; 1927.

16. Garcia, C.K., Brown, M.S., Pathak, R.K., and Goldstein, J.L. cDNA cloning of MCT2, a second monocarboxylate transporter expressed in different cells than MCT1. *Journal of Biological Chemistry* 270: 1843-1849; 1995.

17. Garcia, C.K., Goldstein, J.L., Pathak, R.K., Anderson, R.G.W., and Brown, M.S. Molecular characterization of a membrane transporter for lactate, pyruvate and other monocarboxylates: implications for the Cori cycles. *Cell* 76: 865-873; 1994.

18. Grossie, J.C., Collins, C., and Julian, M. Bicarbonate and fast-twitch muscle: evidence for a major role in pH regulation. *Journal of Membrane Biology* 105: 265-272; 1988.

19. Heisler, N., and Piiper, J. The buffer value of rat diaphragm muscle tissue determined by PCO₂ equilibration of homogenates. *Respiratory Physiology* 12: 169-178; 1971.

20. Hermansen, L., and Osnes, J.B. Blood and muscle pH after maximal exercise in man. *Journal of Applied Physiology* 32: 304-308; 1972.

21. Hermansen, L., and Vaage, O. Lactate disappearance and glycogen synthesis in human muscle after maximal exercise. *American Journal of Physiology* 133(5): E422-E429; 1977.

22. Hultman, E., and Sahlin, K. Acid-base balance during exercise. *Exercise and Sports Science Review* 7: 41-128; 1980.

23. Jackson, V.N., Price, N.T., and Halestrap, A.P. cDNA cloning of MCT1, a monocarboxylate transporter from rat skeletal muscle. *Biophysica Acta* 1238: 193-196; 1995.

24. Juel, C. Muscle lactate transport studied in sarcolemmal giant vesicles. *Biochimica et Biophysica Acta* 1065: 15-20; 1991.

25. Juel, C., Bangsbo, J., Graham, T., and Saltin, B. Lactate and potassium fluxes from human skeletal muscle during and after intense, dynamic, knee extensor exercise. *Acta Physiologica Scandinavica* 140: 147-159; 1990.

26. Juel, C., Honig, A., and Pilegaard, H. Muscle lactate transport in sarcolemmal giant vesicles: dependence on fiber type and age. *Acta Physiologica Scandinavica*. 143: 361-365; 1991.

27. Juel, C., Kristiansen, S., Pilegaard, H., Wojtaszewski, J., and Richter, E.A. Kinetics of lactate transport in sarcolemmal giant vesicles obtained from human skeletal muscle. *Journal of Applied Physiology* 76: 1031-1036; 1994.

28. Juel, C. Lactate-proton cotransport in skeletal muscle. *Physiological Reviews* 77(2): 321-358; 1997.

29. Juel, C. Symmetry and pH dependency of the lactate/proton carrier in skeletal muscle studied with rat sarcolemmal giant vesicles. *Biochem Biophys Acta* 1283: 106-110; 1996.

30. Mannion, A.F., Jakeman, P.M., and Willan, P.L.T. Determination of human skeletal muscle buffer value by homogenate technique: methods of measurement. *Journal of Applied Physiology* 75(3): 1412-1418; 1993.

31. Mannion, A.F., Jakeman, P.M., and Willan, P.L.T. Effects of isokinetic training of the knee extensors on high-intensity exercise performance and skeletal muscle buffering. *European Journal of Physiology* 68: 356-361; 1994.

32. Mannion, A.F., Jakeman, P.M., and Willan, P.L.T. Skeletal muscle buffer value, fiber type distribution and high intensity exercise performance in man. *Experimental Physiology* 80: 89-101; 1995.

33. McCullagh, K.J.A., and Bonen, A. Reduced lactate transport in denervated rat skeletal muscle. *American Journal of Physiology* 268: R884-R888; 1995.

34. McCullagh, K.J.A., Juel, C., O'Brian, M., and Bonen, A. Chronic muscle stimulation increases lactate transport in rat skeletal muscle. *Molecular and Cellular Biochemistry* 156: 51-57; 1996.

35. McDermott, J.C., and Bonen, A. Lactate transport by skeletal muscle sarcolemmal vesicles. *Molecular and Cellular Biochemistry* 122: 113-121; 1993.

36. McDermott, J.C., and Bonen, A. Endurance training increases skeletal muscle lactate transport. *Acta Physiologica Scandinavica* 147: 323-327; 1993.

37. McKenna, M.J., Harmer, A.R., Fraser, S.F., and Li, J.L. Effects of training on potassium, calcium and hydrogen ion regulation in skeletal muscle and blood during exercise. *Acta Physiologica Scandinavica* 156: 335-346; 1996.

38. Mizuno, M., Juel, C., Bro-Rasmussen, T., Mygind, E., Schibye, B., Rasmussen, B., and Saltin, B. Limb skeletal muscle adaptation in athletes after training at altitude. *Journal of Applied Physiology* 68: 496-502; 1990.

39. Nevill, M.E., Boobis, L.H., Brooks, S., and Williams, C. Effect of training on muscle metabolism during treadmill sprinting. *Journal of Applied Physiology* 67: 2376-2382; 1989.

40. Parkhouse, W.S., and McKenzie, D.C. Possible contribution of skeletal muscle buffers to enhanced anaerobic performance: a brief review. *Medicine and Science in Sports Exercise* 16(4): 328-338; 1984.

41. Parkhouse, W.S., McKenzie, D.C., Hochacka, P.W., and Ovalle, W.K. Buffering capacity of deproteinized human vastus lateralis muscle. *Journal of Applied Physiology* 58(1): 14-17; 1985.

42. Pilegaard, H., and Asp, S. Effect of prior eccentric contractions on lactate/H^+ transport in rat skeletal muscle. *American Journal of Physiology* 274: E554-E559; 1998.

43. Pilegaard, H., Bangsbo, J., Richter, E.A., and Juel, C. Lactate transport studied in sarcolemmal giant vesicles from human muscle biopsies: relation to training status. *Journal of Applied Physiology* 77(4): 1858-1862; 1994.

44. Pilegaard, H., Juel, C., and Wibrand, F. Lactate transport studied in sarcolemmal giant vesicles from rats: effect of training. *American Journal of Physiology* 264: E156-E160; 1993.

45. Pilegaard, H., and Juel, C. Lactate transport studied in sarcolemmal giant vesicles from rat skeletal muscles: effect of denervation. *American Journal of Physiology* 269: E679-E682; 1995.

46. Pilegaard, H., Mohr, T., Kjaer, M., and Juel, C. Lactate/H⁺ transport in skeletal muscle from spinal-cord-injured patients. *Scandinavian Journal of Medicine and Science in Sport* 8: 98-101; 1998.

47. Pilegaard, H., Domino, K., Noland, T., Hellsten, Y., Saltin, C.J.B., and Bangsbo, J. Effect of high intensity training on lactate/H⁺ transport in human skeletal muscle; 1998 (submitted).

48. Roth, D.A., and Brooks, G.A. Lactate transport is mediated by a membrane-bound carrier in rat skeletal muscle sarcolemmal vesicles. *Archives of Biochemistry and Biophysics* 279: 377-385; 1990.

49. Roth, D.A., and Brooks, G.A. Training does not affect zero-trans lactate transport across mixed skeletal muscle sarcolemmal vesicles. *Journal of Applied Physiology* 75: 1559-1565; 1993.

50. Sahlin, K., Harris, R.C., Nylind, B., and Hultman, E. Lactate content and pH in muscle samples obtained after dynamic exercise. *Pflügers Archiv* 367: 143-149; 1976.

51. Sahlin, K., and Henriksson, J. Buffer capacity and lactate accumulation in skeletal muscle of trained and untrained men. *Acta Physiologica Scandinavica* 122: 331-339; 1984.

52. Sharp, R.L., Costill, D.L., Fink, W.J., and King, D.S. Effects of eight weeks of bicycle ergometer sprint training on human muscle buffer capacity. *International Journal of Sports Medicine* 7: 13-17; 1986.

53. Spriet, L.L. Phosphofructokinase activity and acidosis during short-term tetanic contractions. *Canadian Journal of Physiology and Pharmacology* 69: 298-304; 1991.

54. Sjøgaard, G. Exercise induced muscle fatigue: the significance of potassium. *Acta Physiologica Scandinavica* 140: 5-43; 1990.

55. Syme, P.D., Arnolda, L., Green, Y., Aronson, J.K., Graham-Smith, D.G., and Radda, G.K. Evidence for increased *in vivo* Na⁺-H⁺ antiporter activity and an altered skeletal muscle contractile response in the spontaneously hypertensive rat. *Journal of Hypertension* 8: 1027-1036; 1990.

56. Stevenson, R.W., Mitchell, D.R., Hendrick, G.K., Rainey, R., Cherrington, A.D., and Frizzell, R. Lactate as substrate for glycogen resynthesis after exercise. *Journal of Applied Physiology* 62: 2239-2240; 1987.

57. Troup, J.P., Metzger, J.M., and Fitts, R.H. Effect of high-intensity exercise training on functional capacity of limb skeletal muscle. *Journal of Applied Physiology* 60: 1743-1751; 1986.

58. Van Slyke. On the measurement of buffer values and on the relationship of buffer value to the dissociation constant of the buffer and the concentration and reaction of the buffer solution. *Journal of Biological Chemistry* 52: 525-570; 1922.

59. Weston, A.R., Wilson, G.R., Noakes, T.D., and Myburgh, K.H. Skeletal muscle buffering capacity is higher in the superficial vastus than in the soleus of spontaneously running rats. *Acta Physiologica Scandinavica* 157: 211-216; 1996.

60. Weston, A.L., Myburgh, K.H., Lindsay, F.H., Dennis, S.C., Noakes, T.D., and Hawley, J.A. Skeletal muscle buffering capacity and endurance performance after high-intensity interval training by well-trained cyclists. *European Journal of Applied Physiology* 75: 7-13; 1997.

61. Westerblad, H., Lee, J.A., Lannergren, J., and Allen, D.G. Cellular mechanisms of fatigue in skeletal muscle. *American Journal of Physiology* 261: C195-C209; 1991.

62. Woodbury, J.W. Regulation of pH. In *Physiology and Biophysics*. Ed T.C. Ruch and H.D. Patton, 899-934. Philadelphia: Saunders; 1965.

CHAPTER 16

Fatty Acid Binding Proteins and Muscle Lipid Metabolism

Lorraine P. Turcotte

Metabolic Regulation Lab, Department of Exercise Science,
University of Southern California, Los Angeles, CA, USA

Introduction

It is well accepted that long-chain free fatty acids (FFA) are an important energy source for skeletal muscle. Their contribution to total oxidative metabolism is dependent on a variety of factors, including exercise intensity and duration, as well as dietary and training status. Central to their many cellular functions is the process by which FFA are taken up by muscle cells. This process involves a complex sequence of steps, namely: dissociation from albumin, transport across the endothelial and plasma membranes, binding to cytoplasmic fatty acid binding proteins, and intramuscular metabolism. Although the molecular mechanisms by which long-chain FFA travel from the vasculature to the cytoplasm of muscle cells are not understood completely, recent studies have shed some light on some of the steps that comprise this cellular uptake mechanism.

Transmembrane Transport of FFA

Due to their poor solubility in the aqueous media that prevail in most biological systems, 99.9% of FFA circulate in plasma bound to albumin, a soluble protein that binds FFA with high affinity and allows total plasma FFA concentration to reach values as high as 2 mmol/L. While there may exist as many as eight binding sites for FFA on each albumin molecule, only a few of these sites have a high affinity for FFA (27, 32). Thus, when total plasma FFA concentration increases, these sites fill up, and the concentration of FFA not bound to albumin increases exponentially. Despite this, only a very small fraction of FFA (~0.1%) is actually dissolved in the plasmatic water, and this unbound fraction is assumed to be in equilibrium with the albumin-bound fraction (26). Following dissociation from albumin, FFA must cross the endothelium lining of the microvascular compartment, the interstitial space and the plasma membranes of muscle cells before being bound to cytoplasmic fatty acid binding proteins or possibly directly to metabolic enzymes (51). It has been suggested that transmembrane transport of FFA is driven by the difference in FFA concentration between blood plasma and the cytoplasm of muscle cells (50, 51). In line with this notion, it recently has been shown that the intracellular concentration of FFA in resting dog muscle is extremely low (30.7 ± 6.7 nmol/g wet wt), compared with the arterial plasma concentration of FFA (511 ± 108 nmol/ml) (50). With a ratio of arterial plasma concentration to intramuscular concentration of ~17, a steep FFA gradient could be established between the vascular and cytoplasmic compartments in muscles. This finding supports the notion that a gradient of FFA could be one of the driving forces for the net uptake of FFA by muscle (50).

Transendothelial Transport of FFA

Transendothelial transport of FFA from blood plasma to the interstitial space is the first step in the transmembrane transport of FFA. Little is known about the molecular mechanisms of action of this transport step (51). From studies performed in cardiac muscle, it has been suggested that diffusion of FFA alone or bound to albumin through the clefts of the endothelium would be too slow to contribute significantly to total transendothelial transport because of limitations in protein trafficking through the clefts and in surface area of diffusion (6, 51). Rather, it has been postulated that membranous albumin binding proteins located on the luminal side of endothelial cells mediate transendothelial transport of FFA by facilitating the release of FFA from plasma albumin and allowing the diffusion of FFA across the cytoplasm of endothelial cells, either as unbound FFA or bound to cytoplasmic fatty acid binding proteins (3, 29, 51). In the interstitial space, FFA are believed to be carried to the muscle plasma membranes bound to albumin (51) (figure 16.1).

Transsarcolemmal Transport of FFA

According to the conventional theory of cellular uptake for protein-bound ligand, only unbound ligand participates in the uptake process (11), and because of their lipid nature, FFA flux across plasma membranes has long been considered a passive diffusional process (23, 45). This notion was reinforced by the existence of a linear relationship between total plasma FFA concentration and FFA utilization in humans and dogs at rest and during exercise (17, 20, 22). Arguing against the hypothesis of simple diffusion of FFA across plasma membranes are several structural and physiological considerations. One argument is that the unbound FFA concentration is far too low to explain the high cellular influx rates observed in vivo (26). Therefore, accelerated dissociation of the FFA: albumin complex at the plasma membranes would have to occur, and this seems unlikely when considering the high affinity of albumin for FFA ($10^8 M^{-1}$) (26, 37). Another consideration

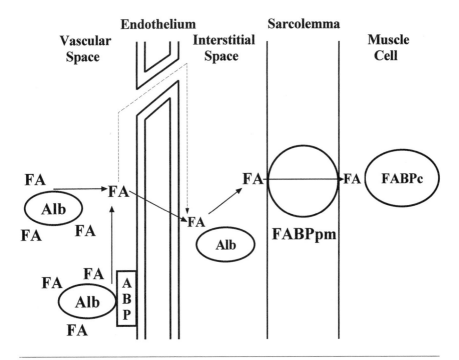

Figure 16.1.   Schematic representation of transendothelial and transsarcolemmal fatty acid transport in skeletal muscle. ABP, albumin binding protein; Alb, albumin; FA, fatty acids; FABP$_C$, cytosolic fatty acid binding protein; FABP$_{PM}$, plasma membrane fatty acid binding protein. Adapted, with permission, from Rowell and Shepherd 1996.

relates to the tight structure of the phospholipid bilayer of plasma membranes, whose polar groups facing the extra- and intracellular spaces may hinder FFA permeation (26). Moreover, at physiologic pH, FFA exist in plasma as anions and thus have to be taken up against an unfavorable electrical gradient due to the negative charge on the cytosolic side of the plasma membrane (26, 37). These theoretical considerations raise the possibility that transport mechanisms more efficient than simple diffusion exist for FFA to account for the high rates of FFA uptake that have been measured (26, 37). Thus, FFA uptake would consist of two components: one a linear and the other a saturable function of the unbound FFA concentration. Within the range of unbound FFA concentration normally encountered in vivo, most of the uptake of FFA would occur via the saturable carrier-mediated pathway, whereas in hyperlipidemic conditions a higher percentage of the total flux might occur via the diffusional pathway (26, 37). In recent years the search for a plasmalemmal FFA transport system has intensified and has led to the identification of several putative FFA transporters in a variety of mammalian tissues (2, 15, 28, 34, 40). Identification of a membranous FFA transport system would be of considerable significance since this pathway might represent a site for the metabolic and hormonal control of long-chain FFA metabolism.

Binding and Uptake Kinetics. Binding of a ligand to a specific membrane protein would represent the first step of a carrier-mediated transport system. To determine whether high affinity binding sites for FFA are located in the plasma membranes of cells, FFA binding studies were performed with plasma membranes isolated from different tissues (12, 13, 33, 34, 36). Membrane binding of [^{14}C] long-chain FFA can be assessed by a vacuum filtration assay, and the difference between binding to native and heat-denatured plasma membrane proteins represents the specific binding rate. In all cell types studied to date, specific binding of [^{14}C] long-chain FFA to plasma membrane proteins has been shown to be a saturable function of the unbound FFA concentration with a range in maximal binding capacities of 1–10 nmol/mg protein and in dissociation constant values of 10–30 nmol/L (33, 34, 36). Specific binding of long-chain FFA also was found to be inhibitable by prior treatment of plasma membrane proteins with phloretin and trypsin, reversible by the addition of excess cold oleate and time-, pH-, and temperature-dependent (12, 13, 33, 34, 36). In skeletal muscle we have shown that specific [^{14}C] palmitate binding to isolated plasma membrane proteins is also a saturable function of the unbound palmitate concentration at palmitate:albumin molar ratios of 0.25:1 to 5:1 ($V_{max} = 4.8 \pm 0.3$ nmol/mg protein/15 min, $K_d = 22.1 \pm 4.4$ nmol/L) (49). In line with data collected in other tissues (13, 33, 34, 36), the optimum temperature for specific binding was found to be 30°C, and phloretin treatment of plasma membrane proteins resulted in a 27–35% decrease in specific [^{14}C] palmitate binding. Thus, in all tissues studied to date, the demonstration of saturable, phloretin inhibitable specific binding agrees with the notion that high affinity binding sites for FFA exist in the plasma membranes of cells.

To strengthen the above conclusion, it is critical to show that, independent of binding and intracellular metabolism, transsarcolemmal transport of FFA reveals criteria of a carrier-mediated transport mechanism. In isolated cell systems the initial rate of FFA uptake can be used as a measure of FFA transport, because the initial rate of uptake has been shown to be maximal and linear for the first 15–30 seconds of uptake and to be independent of glucose availability (1, 35). Furthermore, at all total FFA concentrations studied, the initial rate of FFA uptake was less than the rate of spontaneous dissociation of the incubated FFA-albumin complex by a factor of more than 100 (37). These characteristics indicate that the initial rate of FFA uptake is not limited by intracellular metabolism or by the FFA-albumin dissociation rates (1, 35, 37). Thus, under similar experimental conditions, the determination of the initial rate of FFA uptake as a function of increasing unbound FFA concentration can distinguish diffusion that shows a linear relationship from transport that is a saturable process. In all cell systems studied to date, the initial rate of FFA uptake was found to saturate with an increase in unbound FFA concentration (1, 30, 31, 33, 36). Comparable K_m and V_{max} values were measured independently, substantiating the significance of this finding, which is considered to be a necessary characteristic of a carrier-mediated transport system (26). The initial rate of FFA uptake also has been shown to exhibit other characteristics of a carrier-mediated transport system, namely specificity, inhibition, countertransport, and transstimulation (26, 31). In some isolated cell systems, pretreatment with ouabain, an inhibitor of the Na$^+$/K$^+$-ATPase, resulted in a significant inhibition of the initial rate of FFA uptake, suggesting that the transport mechanism may be linked to the activity of the enzyme (26).

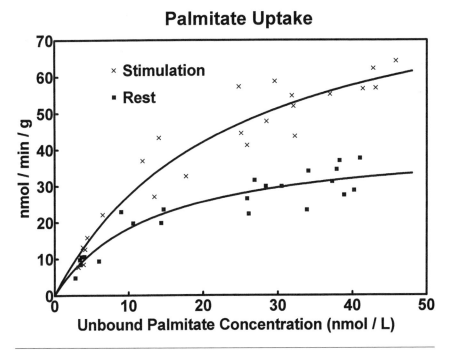

Figure 16.2.   Palmitate uptake in perfused rat hindquarters at rest and during electrical stimulation as a function of unbound perfusate plasma palmitate concentration. The data points were fitted to a rectangular hyperbolic function and analyzed by the program PRISM (GraphPad, San Diego, CA).

Evidence also is accumulating to suggest that similar saturation-type kinetics exist for the rate of FFA uptake in skeletal muscle. In untrained human subjects performing two or three hours of one-legged dynamic knee extension exercise, FFA uptake into the exercising skeletal muscles, determined either as net uptake or with tracer techniques, was found to saturate with an increase in plasma unbound FFA concentration (24, 42). Similarly, in isolated perfused rat skeletal muscle, palmitate uptake displayed saturation kinetics when plotted against the unbound perfusate palmitate concentration (41) (figure 16.2). The maximal velocity of the uptake of palmitate was shown to increase with acute muscle contractions induced by electrical stimulation and decrease with low carbohydrate availability, suggesting that physiological stimuli may alter membrane transport of FFA (43, 44). However, the measurement of the rate of FFA uptake in a hindlimb perfusion system could be affected by intracellular metabolism and cannot be used as definite evidence for the existence of a carrier-mediated transport system in skeletal muscle.

Recently, Bonen et al. (10) developed a transport assay in giant sarcolemmal vesicles that measures initial rates of FFA influx independently of intramuscular FFA metabolism and esterification (10, 14). Using giant sarcolemmal vesicles prepared from red and white skeletal muscle, palmitate transport was shown to be a saturable function of the unbound palmitate concentration, was inhibited by phloretin and, in line with the established fiber-type differences in metabolic capacities for lipid metabolism (5), was found to be 35–50% higher in red muscle vesicles than in white muscle vesicles (10, 14). These data agree with the data described above and indicate that transsarcolemmal transport of FFA in skeletal muscle may occur via a carrier-mediated transport mechanism similar to that described for adipocytes, hepatocytes, and cardiac myocytes.

Fatty Acid Transporters.   Multiple plasma membrane proteins have been identified as putative transporters/receptors of FFA in mammalian cells (table 16.1). Although isolated by different procedures, these proteins have been shown to be distinct from the 12–15 kDa cytosolic fatty acid binding proteins found in most tissues (2, 15, 16, 28, 34, 40). Among these, the best characterized are the plasma membrane fatty acid binding protein ($FABP_{PM}$), fatty acid transporter (FAT), and fatty acid transport protein (FATP) (2, 28, 34). Although it has become clear that all three proteins play important roles in transsarcolemmal transport of FFA in some tissues, the exact nature of the proteins and their interrelationships remains a subject of investigation.

Fatty Acid Binding Protein.   The plasma membrane fatty acid binding protein ($FABP_{PM}$) first was isolated from liver tissue using oleate-agarose affinity column chromatography of solubilized plasma membrane proteins (26, 34). The purified protein yielded a single band on SDS-PAGE with a molecular mass of 40–43 kDa, and its affinity for long-chain FFA was confirmed by co-chromatography with tracer amounts of [^{14}C] palmitate (26, 34). Other 40–43 kDa proteins with slightly different isoelectric points and amino acid sequences have been isolated and purified from other tissues including heart, gut, adipose tissue, and sheep placenta (12, 13, 30, 31, 33), and we have isolated a similar protein from skeletal muscle (Turcotte and Tucker, unpublished data). Evidence of a role for $FABP_{PM}$ in transsarcolemmal transport of FFA includes 1) the selective inhibition of long-chain FFA binding, uptake, and transport by antibodies to $FABP_{PM}$, 2) the stimulation/inhibition of $FABP_{PM}$ protein expression following chronic adaptation to different physiological conditions, and 3) the parallel, progressive

Table 16.1   Putative Fatty Acid Transporters

	Size (kDa)	Method of Identification	Location	References
$FABP_{PM}$	40	Affinity Chromatography	AT, H, L, SM	8, 25, 30, 31, 34, 46, 47, 48
FAT	88	Labeled sulfo-N-succinimidyl derivative	AT, SM	2, 8, 19, 51
FATP	63	cDNA Library	AT, H, SM	8, 28

AT, adipose tissue; $FABP_{PM}$, plasma membrane fatty acid binding protein; FAT, fatty acid translocase; FATP, fatty acid transporter protein; H, heart; L, liver; SM, skeletal muscle.

expression of $FABP_{PM}$ on the cell surface and of saturable FFA uptake during differentiation of 3T3-L1 fibroblasts to an adipocyte morphology and in adipocytes transfected with the cDNA for mitochondrial aspartate aminotransferase (mAspAT).

The effects of monoclonal and polyclonal antibodies raised against $FABP_{PM}$ on the binding, transport, and uptake rates of FFA have been examined in different tissues (30, 31, 33, 36). In isolated cell systems, pretreatment with antibodies to $FABP_{PM}$ as opposed to pretreatment with the pre-immune serum was accompanied by a significant inhibition of FFA binding to plasma membrane proteins and of the initial rate of FFA uptake (30, 31, 33, 36). In skeletal muscle, prior treatment of plasma membrane proteins with monoclonal and polyclonal antibodies raised against the rat hepatic $FABP_{PM}$ were shown to decrease specific binding of palmitate to plasma membrane proteins by ~60% (49) and palmitate transport by giant sarcolemmal vesicles by 65% (49). In contrast, transport of palmitate at 4°C and of glucose, another actively transported substrate, were not affected by pre-incubation of the giant sarcolemmal vesicles with antibodies to $FABP_{PM}$. These data indicate that the role of $FABP_{PM}$ in transsarcolemmal transport is specific to long-chain FFA and not representative of a more generalized transport mechanism.

To test the hypothesis that $FABP_{PM}$ content increases under conditions of increased FFA utilization, $FABP_{PM}$ content was measured in red and white skeletal muscles of fasted, exercised, and endurance-trained rats (46, 47, 48) (figure 16.3). Immunoblotting of the enriched plasma membrane fractions with a polyclonal antibody to $FABP_{PM}$ showed that, compared to white muscle, $FABP_{PM}$ content of red muscle was $83 \pm 18\%$ higher (47, 48). When compared with their respective control group, $FABP_{PM}$ content of red muscle was 61% and 56% higher in fasted and trained animals, respectively, but was not changed by prolonged running (46, 47, 48). No significant differences in $FABP_{PM}$ content of white muscle were observed among groups. A significant correlation was found ($r = 0.78$, $p < 0.05$) between $FABP_{PM}$ content and muscle oxidative capacity as assessed by succinate dehydrogenase activity (47, 48). Similarly, in humans, three weeks of intense one-legged endurance training increased the content of $FABP_{PM}$ by 49%, whereas no change was observed in the muscle of the contralateral untrained leg (25). These results support the hypothesis that $FABP_{PM}$ content in skeletal muscle is related to oxidative potential and can be increased during metabolic conditions known to be associated with a chronic increase in FFA utilization. Together, these data suggest that cellular expression of $FABP_{PM}$ may play a role in the regulation of FFA metabolism in skeletal muscle.

The 3T3-L1 cell line undergoes differentiation to an adipocyte morphology in a sequence that is controlled by the induction of enzymes involved in lipid metabolism (53). Studies have shown that while preconfluent fibroblasts are characterized by a slow linear rate of FFA uptake and very low levels of $FABP_{PM}$, differentiating adipocytes show a gradual increase in the expression of $FABP_{PM}$ that is associated with the development of a saturable component for FFA uptake (53). In differentiated 3T3-L1 adipocytes, the increase in the maximal velocity of FFA uptake was associated with a dramatic increase in $FABP_{PM}$ levels, indicating that $FABP_{PM}$ expression and the kinetics of FFA uptake are subject to parallel regulation (53).

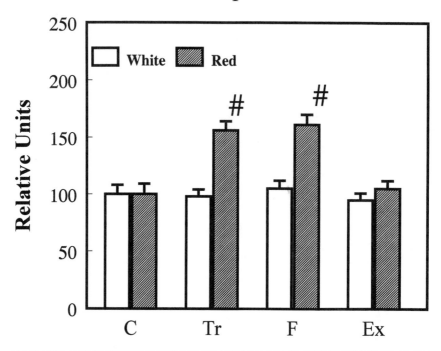

Figure 16.3. FABPpm protein content in red and white skeletal muscle plasma membranes of rats after prolonged exercise (Ex), 8 weeks of endurance training (Tr) and 48 hours of fasting (F). $FABP_{PM}$ protein content was measured by immunoblotting with a polyclonal antibody to the rat hepatic $FABP_{PM}$ and quantitated by scanning densitometry. Each value represents the mean ± SE of percentage value with respect to the control condition of independent immunoblots of separate plasma membrane preparations (n = 6-8 for each experimental conditions). # $p < 0.05$ compared with control group.

It has also been reported that $FABP_{PM}$ is related to mitochondrial aspartate aminotransferase (mAspAT) (7, 8, 38, 39). Although originally disputed, comparisons between $FABP_{PM}$ and mAspAT have revealed that the two proteins have similar or identical molecular masses, N-terminal amino acid sequence, amino acid composition, peptides from tryptic digests, aspartate aminotransferase activity, isoelectric point, mobility on SDS-PAGE, and behavior in multiple chromatographic systems (7, 8, 21, 38, 39). More recent transfection studies in 3T3-L1 fibroblasts provide strong evidence for a role of mAspAT in transsarcolemmal FFA transport (21). In these studies, 3T3-L1 fibroblasts were transfected with a plasmid containing a full-length mAspAT cDNA under the control of a Zn^{2+}-inducible metallothionein promoter (pMAAT2), and the rate of FFA uptake was measured (21). When compared with control cells, cells transfected with pMAAT2 demonstrated a significant increase in the rate of FFA uptake, and this was correlated with an increase in the content of $FABP_{PM}$ (21) (figure 16.4). Antisera to rat $FABP_{PM}$ selectively inhibited

Oleate Uptake

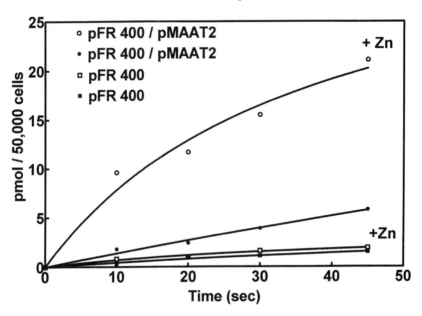

Figure 16.4.   Effect of culture in the presence (○, □) or absence (●, ■) of zinc on [³H]oleate uptake by 3T3 cells transfected with plasmid pFR400 alone (□, ■) or with pFR400 and pMAAT2 (○, ●). Uptake was measured in the presence of oleate (250 μM) and bovine serum albumin (500 μM) resulting in an unbound oleate concentration of 43 nM. Values are mean ± SE.

Reprinted, by permission, from Isola et al. 1995.

the rate of FFA uptake by 61% in transfected cells (21). In isolated adipocytes of lean and obese Zucker rats, the initial rate of FFA uptake also was found to be closely correlated with mRNA levels of mAspAT (9). Together, these data support the dual hypotheses that mAspAT and FABP$_{PM}$ are analogous proteins and that they mediate saturable long-chain FFA uptake (21).

Fatty Acid Translocase.   The fatty acid translocase (FAT) first was identified in rat adipocytes by labeling plasma membrane proteins with nonpermeable sulfo-*N*-succinimidyl derivatives of fatty acids (18, 19). The initial rate of FFA uptake was inhibited by 14- and 18-carbon FFA derivatives but not by a 3-carbon derivative (18, 19). The identified protein was found to have a molecular weight of 88 kDa (2). A cDNA for the adipocyte membrane protein was isolated by screening with a synthetic oligonucleotide derived from the amino terminal sequence of the protein (2), and the revealed sequence of FAT was found to be 85% homologous with that of CD36, a membrane protein expressed in lactating mammary epithelium and human platelets (2). Consistent with this, polyclonal antibodies against CD36 reacted with adipocyte plasma membranes and detected a single band at 88 kDa (2). In a pattern similar to that observed for albumin, CD36 has been shown to bind various long-chain FFA, suggesting that the protein has a high affinity for FFA and could compete effectively for FFA bound to albumin (4). Tissue distribution of the FAT protein generally documents strong expression in tissues with high metabolic requirements for long-chain FFA and with high rates of lipid synthesis. FAT mRNA has been shown to be abundant in heart, adipose tissue, intestines and spleen, low in testes, and not present in kidney or liver (2). While skeletal muscle does not completely adhere to this description for FAT protein distribution, FAT is expressed in skeletal muscles and presumably would contribute to the uptake of long-chain FFA (2). In line with the different oxidative capacities already established for various fiber types (5), the levels of FAT protein and mRNA, respectively determined by Western blotting with a monoclonal antibody to CD36 and by Northern blotting with a cDNA for FAT, were found to be 5–6 higher in red than in white muscles (10, 52). Further evidence of a role of FAT in the transsarcolemmal transport of FFA includes the induction of FAT mRNA levels during the differentiation process in cultured adipose cell lines and by the treatment of preadipocytes with dexamethasone, two conditions that were also associated with an increase in the initial rate of FFA uptake (2). The exact role of FAT/CD36 in the regulation of FFA metabolism in skeletal muscle is not known, and its interrelationship with FABPpm remains to be determined.

Fatty Acid Transport Protein.   Using an expression cloning strategy and a cDNA library from 3T3-L1 adipocytes, Schaffer and Lodish (28) have identified a cDNA that augments the initial rate of FFA uptake upon expression of the protein in cultured cells. This cDNA was shown to encode the sequence for a 63 kDa plasma membrane protein that is highly expressed in mouse heart, adipose tissue, and skeletal muscle (28). As shown for the FAT protein, transcripts for FATP were found to be present in red and white muscles, with red muscle containing more FATP mRNA than white muscle (10). In line with the data collected for $FABP_{PM}$/mAspAT, FATP mRNA levels have been shown to be higher in adipocytes of obese Zucker rats when compared with their lean counterparts (9). In skeletal muscle cells that simultaneously express $FABP_{PM}$/mAspAT, FAT/CD36, and FATP it remains to be established whether the three proteins represent separate parallel pathways of transsarcolemmal transport of FFA or work together as components of a single transport system (21).

Other FFA Transporters.   Two other research teams have identified plasma membrane proteins that could play a role in transsarcolemmal transport of FFA (15, 40). Using isolation procedures similar to the ones described by others (26, 36), a 56 kDa fatty acid binding protein has been isolated and purified from rat renal basolateral membrane (15). Partial characterization of this protein has shown that optimal binding characteristics of the protein for long-chain FFA are similar to the ones described for $FABP_{PM}$ by others (15, 33, 36). However, it remains to be determined whether this protein is specific to renal tissue. Using photoaffinity labeling with 11-*m*-diazirinophenoxy[11-^{3}H] undecanoate, it was shown that differentiation of 3T3-L1 fibroblasts to adipocyte morphology was accompanied by a twofold increase in the intensity of labeling of an apparently high affinity fatty acid binding protein located in the plasma membranes of these cells (40). This protein was estimated to have a molecular weight of 22 kDa and to be different from the other putative FFA transporters (40). The expression of the 56 kDa and 22 kDa proteins has not been studied extensively in other tissues, and their putative roles in transsarcolemmal transport of FFA in skeletal muscle have not been defined at this time.

Conclusion

In conclusion, evidence is accumulating to show that a significant part of FFA uptake by skeletal muscle is achieved by a carrier-mediated system and that the plasma membrane fatty acid binding protein ($FABP_{PM}$) is a putative transporter protein that may be involved in the process. The rates of binding of FFA to plasma membrane proteins and of uptake of FFA by skeletal muscle cells saturate with an increase in the concentration of unbound FFA concentration and are inhibitable in the presence of antibodies to $FABP_{PM}$. $FABP_{PM}$ protein expression in skeletal muscle correlates with the oxidative capacity of the muscle and can be modified under different physiological conditions associated with high rates of FFA metabolism. Recent evidence collected in transfected cells suggests that $FABP_{PM}$ may be analogous to mitochondrial aspartate aminotransferase. Other putative plasma membrane FFA transporter proteins have been identified, namely FAT/CD36 and FATP, but their role in the regulation of FFA uptake in skeletal muscle is not known and their relationship to $FABP_{PM}$ remains to be determined.

Acknowledgements

The author wishes to thank Michelle Z. Tucker and Jason R. Swenberger for their expert technical assistance. The author was supported by a Zumberge Research and Innovation Fund from the University of Southern California.

References

1. Abumrad, N.A., Perkins, R.C., Park, J.H., and Park, C.R. Mechanism of long chain fatty acid permeation in the isolated adipocyte. *J. Biol. Chem.* 256: 9183-9191; 1981.
2. Abumrad, N.A., El-Maghrabi, M.R., Amri, E.-Z. Lopez, E., and Grimaldi, P.A. Cloning of a rat adipocyte membrane protein implicated in binding or transport of long-chain fatty acids that is induced during preadipocyte differentiation. *J. Biol. Chem.* 268: 17,665-17,668; 1993.
3. Antohe, F., Dobrila, L., Heltianu, C., Simionescu, N., and Simionescu, M. Albumin-binding proteins function in the receptor-mediated binding and transcytosis of albumin across cultured endothelial cells. *Eur. J. Cell Biol.* 60: 268-275; 1993.

4. Baillie, A.G.S., Coburn, C.T., and Abumrad, N.A. Reversible binding of long-chain fatty acids to purified FAT, the adipose CD36 homolog. *J. Membr. Biol.* 153: 75-81; 1996.
5. Baldwin, K.M., Klinkerfuss, G.H., Terjung, R.L., Mole, P.A., and Holloszy, J.O. Respiratory capacity of white, red, and intermediate muscle: adaptive response to exercise. *Am. J. Physiol.* 222: 373-378; 1972.
6. Bassingthwaighte, J.B., Noodleman, L., van der Vusse, G.J., and Glatz, J.F.C. Modeling of palmitate transport in the heart. *Mol. Cell. Biochem.* 88: 51-59; 1989.
7. Berk, P.D., Wada, H., Horio, Y., Potter, B.J., Sorrentino, D., Zhou, S.-L., Isola, L.M., Stump, D., Kiang, C.-L., and Thung, S. Plasma membrane fatty acid-binding protein and mitochondrial glutamic-oxaloacetic transaminase of rat liver are related. *Proc. Natl. Acad. Sci. USA.* 87: 3484-3488; 1990.
8. Berk, P.D., Bradbury, M., Zhou, S.-L., Stump, D., and Han, N.I. Characterization of membrane transport processes: lessons from the study of BSP, bilirubin, and fatty acid uptake. *Seminars in Liver Disease* 16(2): 107-120; 1996.
9. Berk, P.D., Zhou, S.-L., Kiang, C.-L., Stump, D., Bradbury, M., and Isola, L.M. Uptake of long chain free fatty acids is selectively up-regulated in adipocytes of Zucker rats with genetic obesity and non-insulin-dependent Diabetes Mellitus. *J. Biol. Chem.* 272: 8830-8835; 1997.
10. Bonen, A., Dyck, D.J., Kiens, B., Kristiansen, S., Luiken, J.J.F.P., Liu, S., van der Vusse, G.J., Glatz, J.F.C., Ibrahimi, A., and Abumrad, N.A. Palmitate transport, FATP mRNA, FAT/CD36 mRNA and protein in rat muscles. *10th International Conference Biochemistry of Exercise* (Poster Abstract, p.62); 1997.
11. Brauer, R.W., and Pessotti, R.L. The removal of bromosulphthalein from blood plasma by the liver of the rat. *J. Pharmacol. Exp. Ther.* 97: 358-370; 1949.
12. Campbell, F.M., Gordon, M.J., and Dutta-Roy, A.K. Plasma membrane fatty acid-binding protein ($FABP_{PM}$) of the sheep placenta. *Biochim. Biophys. Acta.* 1214: 187-192; 1994.
13. Campbell, F.M., Taffesse, S., Gordon, M.J., and Dutta-Roy, A.K. Plasma membrane fatty acid-binding protein in human placenta: identification and characterization. *Biochem. Biophys. Res. Comm.* 209: 1011-1017; 1995.
14. Dyck, D.J., Lui, S., Kiens, B., Richter, E.A., Gorski, J., Glatz, J., Spriet, L., van der Vusse, G., and Bonen, A. Effects of muscle activity on palmitate transport in skeletal muscle giant sarcolemmal vesicles. *The Physiologist* 39: A13; 1996.
15. Fujii, S., Kawaguchi, H., and Yasuda, H. Isolation and partial characterization of an amphiphilic 56-kDa fatty acid binding protein from rat renal basolateral membrane. *J. Biochem.* 101: 679-684; 1987.
16. Glatz, J.F.C., and van der Vusse, G.J. Intracellular transport of lipids. *Mol. Cell. Biochem.* 88: 37-44; 1989.
17. Hagenfeldt, L. Turnover of individual free fatty acids in man. *Fed. Proc.* 34: 2246-2249; 1975.
18. Harmon, C.M., Luce, P., Beth, A.H., and Abumrad, N.A. Labeling of adipocyte membranes by sulfo-*N*-succinimidyl derivatives of long-chain fatty acids: Inhibition of fatty acid transport. *J. Membr. Biol.* 121: 261-268; 1991.

19. Harmon, C.M., and Abumrad, N.A. Binding of sulfosuccinimidyl fatty acids to adipocyte membrane proteins: isolation and amino-terminal sequence of an 88-kD protein implicated in transport of long-chain fatty acids. *J. Membr. Biol.* 133: 43-49; 1993.

20. Havel, R.J., Pernow, B., and Jones, N.L. Uptake and release of free fatty acids and other metabolites in the legs of exercising men. *J. Appl. Physiol.* 23: 90-99; 1967.

21. Isola, L.M., Zhou, S.-L., Kiang, C.-L., Stump, D.D., Bradbury, M.W., and Berk, P.D. 3T3 fibroblasts transfected with a cDNA for mitochondrial aspartate aminotransferase express plasma membrane fatty acid binding protein and saturable fatty acid uptake. *Proc. Natl. Acad. Sci. USA.* 92: 9866-9870; 1995.

22. Issekutz, B., Jr., Bortz, W.M., Miller, H.I., and Paul, P. Turnover rate of plasma FFA in humans and in dogs. *Metabolism* 16: 1001-1009; 1967.

23. Kamp, F., Zakim, D., Zhang, F., Noy, N., and Hamilton, J.A. Fatty acid flip-flop in phospholipid bilayers is extremely fast. *Biochem.* 34: 11,928-11,937; 1995.

24. Kiens, B., Essen-Gustavsson, B., Christensen, N.J., and Saltin, B. Skeletal muscle substrate utilization during maximal exercise in man: effect of endurance training. *J. Physiol. (London)* 469: 459-478; 1993.

25. Kiens, B., Kristiansen, S., Jensen, P., Richter, E.A., and Turcotte, L.P. Membrane associated fatty acid binding protein (FABP$_{PM}$) in human skeletal muscle is increased by endurance training. *Biochem. Biophys. Res. Comm.* 231: 463-465; 1997.

26. Potter, B.J., Sorrentino, D., and Berk, P.D. Mechanisms of cellular uptake of free fatty acids. *Ann. Rev. Nutr.* 9: 253-270; 1989.

27. Richieri, G.V., Anel, A., and Kleinfeld, A.M. Interactions of long-chain fatty acids and albumin: determination of free fatty acid levels using the fluorescent probe ADIFAB. *Biochem.* 32: 7574-7580; 1993.

28. Schaffer, J.E., and Lodish, H.F. Expression cloning and characterization of a novel adipocyte long chain fatty acid transport protein. *Cell.* 79: 427-436; 1994.

29. Schnitzer, J.E., Sung, A., Horvat, R., and Bravo, J. Preferential interaction of albumin-binding proteins, gp30 and gp18, with conformationally modified albumins. Presence in many cells and tissues with a possible role in catabolism. *J. Biol. Chem.* 267: 24,544-24,553; 1992.

30. Schwieterman, W., Sorrentino, D., Potter, B.J., Rand, J., Kiang, C.-L., Stump, D., and Berk, P.D. Uptake of oleate by isolated rat adipocytes is mediated by a 40-kDa plasma membrane fatty acid binding protein closely related to that in liver and gut. *Proc. Natl. Acad. Sci. USA.* 85: 359-363; 1988.

31. Sorrentino, D., Stump, D., Potter, B.J., Robinson, R.B., White, R., Kiang, C.-L., and Berk, P.D. Oleate uptake by cardiac myocytes is carrier mediated and involves a 40-kD plasma membrane fatty acid binding protein similar to that in liver, adipose tissue and gut. *J. Clin. Invest.* 82: 928-935; 1988.

32. Spector, A.A., Fletcher, J.E., and Ashbrook, J.D. Analysis of long-chain free fatty acid binding to bovine serum albumin by determination of stepwise equilibrium constants. *Biochem.* 10: 3229-3232; 1971.

33. Stremmel, W., Lotz, G., Strohmeyer, G., and Berk, P.D. Identification, isolation, and partial characterization of a fatty acid binding protein from rat jejunal microvillous membranes. *J. Clin. Invest.* 75: 1068-1076; 1985a.

34. Stremmel, W., Strohmeyer, G., Borchard, F., Kochwa, S., and Berk, P.D. Isolation and partial characterization of a fatty acid binding protein in rat liver plasma membranes. *Proc. Natl. Acad. Sci. USA.* 82: 4-8; 1985b.

35. Stremmel, W., and Berk, P.D. Hepatocellular influx of [^{14}C] oleate reflects membrane transport rather than intracellular metabolism or binding. *Proc. Natl. Acad. Sci. USA.* 83: 3086-3090; 1986.

36. Stremmel, W. Fatty acid uptake by isolated rat heart myocytes represents a carrier-mediated transport process. *J. Clin. Invest.* 81: 844-852; 1988.

37. Stremmel, W. Transmembrane transport of fatty acids in the heart. *Mol. Cell. Biochem.* 88: 23-29; 1989.

38. Stremmel, W., Diede, H.-E., Rodilla-Sala, E., Vyska, K., Schrader, M., Fitscher, B., and Passarella, S. The membrane fatty acid-binding protein is not identical to mitochondrial glutamic oxaloacetic transaminase (mGOT). *Mol. Cell. Biochem.* 98: 191-199; 1990.

39. Stump, D.D., Zhou, S.-L., and Berk, P.D. Comparison of plasma membrane FABP and mitochondrial isoform of aspartate aminotransferase from rat liver. *Am. J. Physiol.* 265: G894-G902; 1993.

40. Trigatti, B.L., Mangroo, D., and Gerber, G.E. Photoaffinity labeling and fatty acid permeation in 3T3-L1 adipocytes. *J. Biol. Chem.* 266: 22,621-22,625; 1991.

41. Turcotte, L.P., Kiens, B., and Richter, E.A. Saturation kinetics of palmitate uptake in perfused skeletal muscle. *FEBS Lett.* 279: 327-329; 1991.

42. Turcotte, L.P., Richter, E.A., and Kiens, B. Increased plasma FFA uptake and oxidation during prolonged exercise in trained vs. untrained humans. *Am. J. Physiol.* 262: E791-E799; 1992.

43. Turcotte, L.P., Petry, C., Kiens, B., and Richter, E.A. Electrical stimulation increases the V_{max} for the uptake and oxidation of palmitate in perfused skeletal muscle. *Med. Sci. Sports & Exerc.* 24: S178; 1992.

44. Turcotte, L.P., Hespel, J.L.P., Graham, T.E., and Richter, E.A. Impaired plasma FFA oxidation imposed by extreme CHO deficiency in contracting rat skeletal muscle. *J. Appl. Physiol.* 77: 517-525; 1994.

45. Turcotte, L.P., Kiens, B., and Richter, E.A. Lipid metabolism during exercise. In *Exercise Metabolism.* Ed. M. Hargreaves, 99-130. Human Kinetics: Champaign, IL; 1995.

46. Turcotte, L.P., and Zavitz, M.D. Training-induced increase in FABP$_{PM}$ is associated with elevations in palmitate uptake and oxidation by contracting muscle. *Med. Sci. Sports & Exerc.* 28: S77; 1996.

47. Turcotte, L.P., and Zavitz, M.D. Plasma membrane FABP content is increased during conditions of increased fat oxidation. *The Physiologist.* 39: A13; 1996.

48. Turcotte, L.P., Srivastava, A.K., and Chiasson, J.-L. Fasting increases plasma membrane fatty acid binding protein (FABP$_{PM}$) in red skeletal muscle. *Mol. Cell. Biochem.* 166: 153-158; 1997.

49. Turcotte, L.P., Swenberger, J.R., Tucker, M.Z., and Bonen, A. Palmitate transport and bindings in rat muscle are saturable and inhibited by antibodies to $FABA_{PM}$. *FASEB Journal* 12(4): A231; 1998.
50. van der Vusse, G.J., and Roemen, T.H.M. Gradient of fatty acids from blood plasma to skeletal muscle in dogs. *J. Appl. Physiol.* 78(5): 1839-1843; 1995.
51. van der Vusse, G.J., and Reneman, R.S. Lipid metabolism in muscle. In *Handbook of Physiology. Section 12: Exercise: Regulation and Integration of Multiple Systems.* Ed. L.B. Rowell and J.T. Shepherd, 952-994. New York: Oxford Press; 1996.
52. Van Nieuwenhoven, F.A., Verstijnen, C.P.H.J., Abumrad, N.A., Willemsen, P.H.M., Van Eys, G.J.J.M., Van Der Vusse., G.J., and Glatz, J.F.C. Putative membrane fatty acid translocase and cytoplasmic fatty acid binding protein are co-expressed in rat heart and skeletal muscles. *Biochem. Biophys. Res. Comm.* 207: 747-752; 1995.
53. Zhou, S.-L., Stump, D., Sorrentino, D., Potter, B.J., and Berk, P.D. Adipocyte differentiation of 3T3-L1 cells involves augmented expression of a 43-kDa plasma membrane fatty acid-binding protein. *J. Biol. Chem.* 267: 14,456-14,461; 1992.

CHAPTER 17

Skeletal Muscle Amino Acid Transport and Metabolism

Anton J.M. Wagenmakers

Department of Human Biology & Stable Isotope Research Centre, Maastricht University, Maastricht, The Netherlands

The body of a 70 kg man contains about 12 kg of protein (amino acid polymers) and over 200 gram of free amino acids. About 7 kg of protein and some 120 gram of the free amino acids are present intracellularly in skeletal muscle. Exercise physiologists traditionally consider the amino acid pool in skeletal muscle as an inert reservoir from which the building blocks are obtained for the synthesis of contractile proteins and enzymes. It is expected though that changes in the concentration of most amino acids within the free amino acid pool would have important modulatory effects on their metabolism and possibly on the regulation of protein synthesis and degradation rates. For example, the catabolism of the branched-chain amino acids (BCAA) via transamination is accelerated as their concentration rises (e.g., after oral ingestion) due to the high K_m of the BCAA aminotransferase relative to their normal concentration in intracellular fluids (25). An increase in BCAA concentration also activates the oxidative decarboxylation step by a shift in the phosphorylation state of the branched-chain α-keto acid dehydrogenase, a key enzyme in the oxidative degradation route of the BCAA (42). Furthermore, increases in the intracellular concentration or availability (arterial blood concentration) of amino acids may stimulate protein synthesis and reduce protein degradation at whole-body level and in skeletal muscle (e.g., 4, 34, 53). Some of these effects may be specific for individual amino acids. For example, leucine and glutamine increase protein synthesis and reduce protein degradation in incubated and perfused skeletal muscles in vitro (29, 37); in vivo effects are less clear-cut though (36, 45).

For these reasons, control of amino acid uptake from the blood and of the size of the amino acid pools is important for the regulation of amino acid oxidation and protein metabolism in muscle and for the control of the planned release rate of end products of amino acid catabolism (alanine, glutamine, ammonia, etc.). Part of this control may be exerted at the muscle membrane by the amino acid transporters (12, 36). However, the role of amino acid transporters in the regulation of nitrogen metabolism in skeletal muscle probably is not as important as the glucose transporters are for carbohydrate metabolism (or potentially also fatty acid translocases for fat metabolism). There are seven known transporters for glucose, and one of them,

GLUT4, is translocated to the plasma membrane from microsomal membranes both by increased insulin concentrations (nutritional control) and by contractile activity (exercise control). As far as we know today, such complex control mechanisms leading to an extensive adaptability of glucose transport rates do not exist for amino acid transporters.

The Five Identified Amino Acid Transporters of the Muscle Membrane

Amino acid transporters are present not only in the intestine for absorption of nutrition-derived amino acids into the blood (portal vein) but also in most tissues. In the tissues the amino acid transporters are essential for uphill absorption into the tissue against a concentration gradient (12). Rennie (36) has published a state-of-the-art review of the five amino acid transporter systems that presently have been identified in the plasma membrane of skeletal muscle. Only a brief summary is presented here. System-A was the first to be discovered and identified (2). It transports small, neutral amino acids, particularly alanine (the A stands for alanine) and glycine and is a high-affinity, low-capacity transporter that is sodium dependent. A substantial part of the exchange of alanine between blood and muscle also appears to occur via Systems-ASC (sodium-dependent) and -L (sodium-independent), which are unresponsive to insulin in skeletal muscle. The combined action of these transporters (possibly in combination with high intramuscular production rates, see below) leads to a substantial concentration gradient between muscle and blood with the muscle concentration being sixfold to tenfold higher (5). System-ASC is a medium-affinity, medium-capacity, sodium-dependent transporter. Alanine (A), serine (S), and cysteine (C) have been identified as main substrates. System-L (L stands for leucine) in muscle is a low-affinity, high-capacity transporter (12, 24) that handles the branched-chain amino acids (BCAA) and aromatic amino acids. It is not sensitive to insulin and sodium-independent and effectively acts as a means of equalizing concentrations of its substrates across the muscle membrane. The distribution ratio between muscle and the extracellular space for the BCAA and aromatic amino acids is about 1.2:1 (5) due to a coupling of this system with the alanine gradient (inward transport of BCAA is coupled to outward transport of alanine). System-N^m is the muscle variant of the System-N transporter, which was first discovered in hepatocytes. It is unusual in muscle in being a high-capacity, low-affinity, sodium-dependent transporter taking only glutamine, asparagine, and histidine as major substrates (23). It is responsible for the large (thirtyfold to fortyfold) concentration difference for glutamine between muscle and blood (5) and appears to play a marked role in the controlled release of glutamine from muscle into the circulation (see below). Its activity has been shown to be elevated under conditions of acidosis, corticosteroid treatment, trauma, burns, and sepsis by mechanisms that presently are not understood but may involve the elevation of intracellular sodium (37). Glutamate, which is taken up by muscle for glutamine synthesis (see below), and aspartate are transported by System-X^-_{ac}.

In rat (and probably human skeletal muscle), the glutamate transporter has a high affinity and low capacity and is sodium-independent but H^+-dependent. In terms of the characteristics of this transporter it is not understood why a >50-fold concentration gradient is maintained between muscle and blood (5) and why muscle despite this gradient is able to extract substantial amounts of glutamate from the blood 24 hours a day (see below).

Muscle Amino Acid Metabolism at Rest

Arteriovenous Difference Studies in Postabsorptive Subjects

Muscle amino acid metabolism has been investigated in man in vivo by measuring the exchange of amino acids across a forearm or a leg (17, 19, 31, 43). After overnight fasting there is net breakdown of muscle proteins as protein synthesis is slightly lower than protein degradation (11). This implies that those amino acids that are not metabolized in muscle will be released in proportion to their relative occurrence in muscle protein, while a discrepancy will be found when amino acids are transaminated, oxidized, or synthesized. Human limbs release much more glutamine (48% of total amino acid release) and alanine (32%) than would be anticipated from the relative occurrence in muscle protein (glutamine 7% and alanine 9%). This implies that glutamine with two N-atoms per molecule is dominant for the amino acid N-release from human muscle. The BCAA (19% relative occurrence in muscle protein), glutamate (7%), aspartate and asparagine (together 9%), on the other hand, are not released or in lower amounts than their relative ocurrence. Glutamate, in fact, is constantly taken up from the circulation by skeletal muscle. This suggests that the BCAA, glutamate, aspartate, and asparagine originating from net breakdown of muscle proteins and glutamate taken up from the circulation are metabolized in muscle and used for de novo synthesis of glutamine and alanine after overnight starvation. All other amino acids are released in proportion to their relative occurrence in muscle protein, implying that little or no metabolism occurs in muscle. The same conclusion had been reached before in studies of rat and human skeletal muscle perfused and/or incubated in vitro (9, 10, 50).

The Source of Alanine and Glutamine Carbon and Nitrogen

Studies with [^{15}N]-leucine have shown that the amino group of the BCAA is indeed incorporated in humans in vivo in the α-amino nitrogen of alanine (21) and of glutamine (13). As glutamate is central in all aminotransferase reactions in muscle, this implies that the amino group of all six mentioned amino acids is interchangeable and can be incorporated in the α-amino nitrogen of alanine and of glutamine. The source of ammonia in glutamine synthesis (incorporated in the amide nitrogen) forms one of the puzzles in muscle amino acid metabolism remaining today. Between 5% and 10% is derived from the uptake of ammonia from the circulation (17, 43). Two intracellular enzymatic reactions are main candidates for the production of the remainder of the required ammonia: deamination of aspartate via the reactions of the purine nucleotide cycle (30) and glutamate dehydrogenase:

$$\text{glutamate} + NAD^+ \leftrightarrow \alpha\text{-ketoglutarate} + NH_4^+ + NADH.$$

The BCAA indirectly can also be deaminated by these reactions after transfer via transamination of the amino group to glutamate and aspartate. However, both the purine nucleotide cycle and glutamate dehydrogenase have been suggested to have very low activities in muscle both in vivo and in vitro (30). Estimates of limb production rates in the fed and fasted state nevertheless indicate that between 10 and 25 g of glutamine is synthesized in the combined human skeletal muscles per 24 h, that is much more than any other amino acid. This implies that there must be an equimolar rate of ammonia production in muscle.

In vitro muscle incubations and perfusions with [U-^{14}C]-amino acids have led to the consensus that the carbon skeletons of the six indicated amino acids are used for de novo synthesis of glutamine (10, 28, 50). No or very little radioactivity was found in lactate, pyruvate, and alanine during incubation of rat diaphragms (50) and perfusion of rat hindquarters (28) with [U-^{14}C]valine. This implies that there is no active pathway in muscle for conversion of TCA-cycle intermediates into pyruvate. It also implies that the carbon skeleton of the five amino acids that is converted to TCA-cycle intermediates cannot be used for complete oxidation or for pyruvate and alanine synthesis. Therefore, the only fate of these carbon skeletons is synthesis of TCA-cycle intermediates and glutamine. The question then is what is the source of the carbon atoms of alanine? The remaining sources are muscle glycogen and blood glucose converted by glycolysis into pyruvate. In agreement with this conclusion Chang and Goldberg (9) reported that over 97% of the carbons of the alanine, pyruvate, and lactate released by incubated diaphragms were derived from exogenous glucose.

The Glucose-Alanine Cycle: a Changing Concept

This conclusion slightly changes the concept of the glucose-alanine cycle (18), which by now has become generally accepted textbook knowledge. According to the original formulation of the glucose-alanine cycle the pyruvate used for alanine production in muscle was derived either from glycolysis of blood glucose or from pyruvate derived from metabolism of other muscle protein-derived amino acids. The alanine then is released to the blood and converted to glucose via gluconeogenesis in the liver. Muscle protein-derived carbon in this way was suggested to help and maintain blood glucose concentrations after overnight fasting and during prolonged starvation. The implication, however, of the above conclusions is that all pyruvate is either derived from glycolysis of blood glucose or from breakdown of muscle glycogen followed by glycolysis. In a recent tracer study in man (35) 42% of the alanine released by muscle was reported to originate from blood glucose. This implies that more than half of the alanine released by muscle is formed from pyruvate derived from muscle glycogen. This route provides a mechanism to slowly mobilize the sitting muscle glycogen stores during starvation, such that these stores can be used to help and maintain the blood glucose concentration. The amino acids liberated during starvation by increased net rates of protein degradation instead are converted to glutamine, which is a precursor for gluconeogenesis in both liver and kidney (55), though renal gluconeogenesis only starts to be significant (>10% of total glucose output) in man after 60 h (6) and prolonged (4–6 weeks) starvation (33). Protein-derived amino acids metabolized in muscle thus still can help and maintain blood glucose concentration during starvation but by a different route than suggested in the original formulation of the glucose-alanine cycle. Recent tracer studies in man also suggest that glutamine is more important than alanine as a gluconeogenic precursor after overnight starvation (32), and that glutamine is more important than alanine as a vehicle for transport of muscle protein-derived carbon and nitrogen from muscle through plasma to the sites of further metabolism (35).

Effect of Ingestion of Protein-Containing Meals

Following ingestion of a mixed protein-containing meal, small amounts of most amino acids are taken up by most tissues, as there is net protein deposition in the fed state (protein synthesis > protein degradation), which compensates for the net losses in the overnight fasting period. An excessively large uptake of BCAA and glutamate is seen in the 4 h period after ingestion of a mixed meal (16) and after ingestion of a large steak (15). BCAA and glutamate then together cover more than 90% of the muscle amino acid uptake. The BCAA originate from dietary protein. After digestion of dietary protein most of the resulting BCAA escape from uptake and metabolism in gut and liver due to the low BCAA aminotransferase activity in these tissues (22, 46). The source of the glutamate is not clear today. Marliss et al. (31) showed that the splanchnic area (gut and liver) in man constantly produces glutamate both after overnight and after prolonged starvation. After ingestion of a large steak the muscle release of glutamine more than doubled, while the alanine release was reduced to 10% of the overnight fasted value. In the 4 h period after ingestion of a mixed meal (16) the dominance of glutamine in carrying N out of skeletal muscle was even more clear than after overnight fasting. Glutamine then accounted for 71% of the amino acid release and 82% of the N-release from muscle. In summary, these data suggest that after consumption of protein-containing meals, BCAA and glutamate are taken up by muscle and their carbon skeletons are used for de novo synthesis of glutamine.

The Function of Muscle Glutamine Synthesis and Release

In the previous paragraphs it has become clear that glutamine is the main end product of muscle amino acid metabolism both in the overnight fasted state and during feeding. Alanine only serves to export part of the amino groups. Glutamine is the most abundant amino acid in human plasma (600–700μM) and in the muscle free amino acid pool (20 mM; 60% of the intramuscular pool excluding the nonprotein amino acid taurine). The synthesis rate of glutamine in muscle is higher than that of any other amino acid. Extrapolations of limb production rates in the fed and fasted state show that between 10 and 25 g of glutamine are synthesized in the combined human skeletal muscles per day. Tracer dilution studies even indicate that 80 gram of glutamine are produced per day (14), but this may be a methodological over-estimation due to slow mixing of the glutamine tracer with the large endogenous glutamine pool in muscle (41).

The reason for this high rate of glutamine production in muscle probably is that glutamine plays an important role in human metabolism in other organs. Glutamine has been shown to be an important fuel for cells of the immune system (3) and for mucosal cells of the intestine (40, 54), and is a precursor for DNA and RNA synthesis and plays a role in the maintenance of acid-base balance in prolonged starvation (26, 27). Low muscle and plasma glutamine concentrations are observed in patients with sepsis and trauma (37, 44), conditions that also are attended by mucosal atrophy, loss of the gut barrier function (bacterial translocation), and a weakened immune response. Though the link between the reduced glutamine concentrations and these functional losses has not been underpinned fully by experimental evidence, the possibility should seriously be considered that it is a causal relationship. Due to its numerous metabolic key functions and a potential shortage in patients with sepsis and trauma glutamine recently has been proposed to be a conditionally essential amino acid (27), which should be added especially to the nutrition of long-term hospitalized, critically-ill, and depleted patients. These patients have a reduced muscle mass due to continuous muscle wasting and, therefore, probably also a reduced capacity for glutamine production.

The Glutamine-Glutamate Cycle

The existence of the glutamine-glutamate cycle first was demonstrated by Marliss et al. (31). In muscle there is a continuous glutamate uptake and glutamine release, with the glutamate uptake accounting for about half of the glutamine release. Most of the glutamine produced by muscle is extracted by the splanchnic bed, most probably part by the gut (40) and part by the liver (38). This glutamine is converted to glutamate and ammonia by glutaminase. When generated in the gut the ammonia is transported via the portal vein to the liver and disposed of as urea; the same holds for ammonia generated in the liver. About half of the glutamate is retained in the splanchnic area and used as a fuel in the gut (40) or for gluconeogenesis in the liver (38), and the other half is released and transported back to the muscle. This glutamine-glutamate cycle provides a means to transport ammonia produced in muscle in the form of a nontoxic carrier (glutamine) through the blood to the splanchnic area where it can be removed as urea.

Muscle Amino Acid and Protein Metabolism During Exercise

The Anaplerotic Role of the Alanine Aminotransferase Reaction

During cycling exercise at intensities between 50% and 70% of W_{max} only two amino acids change substantially in concentration in the muscle free amino acid pool, i.e., glutamate and alanine (5, 39, 43). Glutamate decreases by 50–70% within 10 min of exercise, while alanine at that point in time is increased by 50–60%. The low concentration of glutamate is maintained when exercise is continued for periods up to 90 min or until exhaustion, while alanine slowly returns to resting levels. Substantial amounts of alanine, furthermore, are released into the circulation during the first 30 min of exercise (43). Alanine release is reduced again when exercise is continued and the muscle glycogen stores are gradually emptied (43). The functionality of the rapid fall in muscle glutamate concentration most likely is conversion of its carbon skeleton into α-ketoglutarate and TCA-cycle intermediates. The sum concentration of the most abundant TCA-cycle intermediates in skeletal muscle has been shown to increase about tenfold rapidly after the start of exercise (39). This increase most probably is needed to increase the flux in the TCA cycle during exercise (by substrate activation) and meet the increased energy demand for contraction.

The high rate of alanine production during the first 30 min of exercise (43) and the temporary increase in muscle alanine concentration after 10 min of exercise indicate that the alanine aminotransferase reaction is used for the rapid conversion of glutamate carbon into TCA-cycle intermediates (glutamate + pyruvate ↔ α-ketoglutarate + alanine). The alanine aminotransferase reaction is a near equilibrium reaction. At the start of exercise the rate of glycolysis and thus of pyruvate formation is high, as indicated by a temporary increase of the muscle pyruvate concentration (39). The increase in muscle pyruvate automatically forces the alanine aminotransferase reaction toward a new equilibrium with production of α-ketoglutarate and alanine. Felig and Wahren (19) have shown that the rate of release of alanine from muscle depended on the exercise intensity and suggested a direct relation between the rate of formation of pyruvate from glucose and alanine release. This led to the suggestion that the glucose-alanine cycle also operated during exercise: glucose taken up by muscle from the blood via glycolysis is converted to pyruvate and then via transamination to alanine to subsequently serve as substrate for gluconeogenesis in the liver and help to maintain blood glucose concentration during exercise. Here (also see above) we propose that the alanine aminotransferase reaction primarily functions for de novo synthesis of α-ketoglutarate and TCA-cycle intermediates at the start of exercise.

The Carbon Drain of the BCAA Aminotransferase Reaction in Glycogen Depleted Muscles: Its Potential Role in Fatigue Mechanisms

After the early increase of the concentration of TCA-cycle intermediates during exercise Sahlin et al. (39) observed a subsequent gradual decrease in human subjects exercising until exhaustion at 75% $\dot{V}O_2$max. It has been hypothesized that the increased oxidation of the BCAA plays an important role in that subsequent decrease (42, 43, 47, 48). The branched-chain α-keto acid dehydrogenase (BCKADH, the enzyme catalyzing the rate-determining step in the oxidation of BCAA in muscle) is increasingly activated during prolonged exercise, leading to glycogen depletion (42, 47). After prolonged exercise the muscle also begins to extract BCAA from the circulation in gradually increasing amounts (1, 42, 43). Ahlborg et al. (1) showed that these BCAA were released from the splanchnic bed. An increase in oxidation of the BCAA by definition will increase the flux through the BCAA aminotransferase step. In case of leucine this reaction will put a net carbon drain on the TCA cycle as the carbon skeleton of leucine is oxidized to 3 acetyl-CoA molecules and the aminotransferase step uses α-ketoglutarate as amino group acceptor. Increased oxidation of valine and isoleucine will not lead to net removal of TCA-cycle intermediates as the carbon skeleton of valine is oxidized to succinyl-CoA and that of isoleucine to both succinyl-CoA and acetyl-CoA. Net removal of α-ketoglutarate via leucine transamination can be compensated by regeneration of α-ketoglutarate in the alanine aminotransferase reaction as long as muscle glycogen is available and the muscle pyruvate concentration is kept high. However, as activation of the BCKADH complex is highest in glycogen depleted muscle (42) this mechanism eventually is expected to lead to a decrease in the concentration of TCA-cycle intermediates. This again may lead to a reduction of the TCA-cycle activity, inadequate ATP turnover rates, and via subsequent increases in the known cellular mediators to muscle fatigue (20).

BCAA after oral ingestion escape from hepatic uptake and are rapidly extracted by the leg muscles (42), and this is accompanied by activation of the BCKADH complex at rest and increased activation during exercise (42). This could imply that the indicated carbon drain on the TCA cycle is larger after BCAA ingestion and that BCAA ingestion by this mechanism leads to premature fatigue during prolonged exercise leading to glycogen depletion. Evidence in support of this hypothesis has been obtained (48) in patients with McArdle's disease, who have no access to muscle glycogen due to glycogen phosphorylase deficiency and, therefore, can be regarded as an "experiment of nature" from which we can learn what happens during exercise with glycogen depleted muscles. BCAA ingestion increased heart rate and led to premature fatigue during incremental exercise in these patients. This may contain the message that BCAA supplementation has a negative effect on performance by the proposed mechanism in healthy subjects in conditions where the glycogen stores have been emptied completely by highly demanding endurance exercise.

Importance of the Alanine Aminotransferase Reaction for the Maximal Rate of Substrate Oxidation During Exercise

Muscle glycogen is the primary fuel during prolonged high-intensity exercise such as practiced by elite marathon runners. High running speeds (≥ 20 km per h) are maintained by these athletes for periods of 2 h. However, the pace has to be reduced when the muscle glycogen concentration is falling and glycolytic rates cannot be maintained. This indicates that either there is a limit in the maximal rate at which fatty acids can be mobilized and oxidized or that there is a limitation in the maximal rate of the TCA cycle when glycolytic rates are falling as a consequence of glycogen depletion. It is proposed here that the decrease in muscle pyruvate concentration that occurs when the glycogen stores are reduced leads to a decrease of the anaplerotic capacity of the alanine aminotransferase reaction (reduced conversion of glutamate carbon to TCA-cycle intermediates) and thus leads to a decrease in the concentration of TCA-cycle intermediates. This again will lead to a reduction of TCA-cycle activity and the need to reduce the pace (fatigue). The following observation seems to support this hypothesis. Patients with McArdle's disease cannot substantially increase the glycolytic rate during exercise due to the glycogen breakdown defect in muscle, and they, therefore, do not increase muscle pyruvate (48). The arterial alanine concentration does not increase in these patients during exercise, and the muscle only produces alanine by means of protein degradation and not via the alanine aminotransferase reaction (48). This implies that the anaplerotic capacity of these patients is substantially reduced in comparison with healthy subjects. The maximal work rate and oxygen consumption of these patients during cycling exercise is between 40% and 50% of the maximum predicted for their age and build. In ultra-endurance exercise without carbohydrate ingestion healthy subjects have to reduce the work rate to about the same level when the glycogen stores have been emptied, suggesting that muscle glycogen indeed is needed to maintain high work rates potentially by means of its ability to establish and maintain high concentrations of TCA-cycle intermediates.

Alternative Anaplerotic Reactions Involving Amino Acids in Glycogen Depleted Muscles

From the previous sections it has become clear that the alanine aminotransferase reaction and the glutamate pool play an important role in the establishment and maintenance of adequate concentrations of TCA-cycle intermediates during exercise. In the glycogen depleted state, glucose released from the liver by glycogenolysis and gluconeogenesis and glucose absorbed from the gut following oral ingestion of carbohydrates may provide another source of pyruvate to serve as a driving force for synthesis of TCA-cycle intermediates via the alanine aminotransferase reaction. This, in fact, may explain why higher exercise intensities can be maintained for prolonged periods when athletes ingest carbohydrates during exercise. Other mechanisms that may generate TCA-cycle intermediates are increased deamination rates of amino acids in muscle. Increased deamination of amino acids indeed has been observed during prolonged one-leg exercise by Van Hall et al. (43). Deamination of valine, isoleucine, aspartate, asparagine, and glutamate in contrast to transamination does not use α-ketoglutarate as amino group acceptor. Deamination, therefore, leads to net production of ammonia and net synthesis of TCA-cycle intermediates. During prolonged one-leg exercise at 60-65% W_{max} we also observed an excessive net breakdown rate of muscle protein (51). During one-leg exercise the workload per kg muscle in the small muscle group used (maximally 3 kg) is exceedingly high, and this may be the reason why one-leg exercise leads to net protein degradation (protein synthesis < protein degradation) in muscle. The amino acid exchange observed under these conditions indicated that BCAA, glutamate, aspartate, and asparagine released by the net breakdown of muscle protein and BCAA and glutamate taken up from the circulation were used for net synthesis of TCA-cycle intermediates and glutamine (52). Removal of amino groups from muscle in the form of glutamine provides another mechanism for net synthesis of TCA-cycle intermediates as illustrated by the following net reactions:

$$2 \text{ glutamate} \rightarrow \text{glutamine} + \alpha\text{-ketoglutarate}$$

$$\text{valine} + \text{isoleucine} \rightarrow \text{succinyl-CoA} + \text{glutamine}$$

$$\text{aspartate} + \text{isoleucine} \rightarrow \text{oxaloacetate} + \text{glutamine}$$

An excessive release of ammonia and glutamine and excessive net breakdown of muscle protein (several-fold more than in one-leg exercise in healthy subjects) also was observed during two-legged cycling in patients with McArdle's disease (48), indicating that deamination of amino acids and synthesis of glutamine and TCA-cycle intermediates from glutamate and BCAA also provided alternative mechanisms of TCA-cycle anaplerosis in this muscle disease with zero glycogen availability and low pyruvate concentrations. The fact that high exercise intensities cannot be maintained by these patients and in glycogen depleted muscles seems to indicate that these alternative anaplerotic reactions involving amino acids derived from muscle protein and taken up from the circulation are not as effective as the alanine aminotransferase reaction and only allow muscular work at 40–50% of W_{max}.

It is far from clear whether dynamic whole-body exercise as practiced by athletes during competition (cycling or running) leads to net protein breakdown in muscle and helps to provide carbon skeletons for synthesis of TCA-cycle intermediates. Different stable isotope tracers used to measure protein synthesis and degradation in laboratory conditions gave different answers; different answers also were obtained for changes observed at whole body level and at muscle level (for a detailed review, see 36). Whole body measurements with L-[1-^{13}C]leucine suggest that there is increased net protein breakdown during 1–6 h of cycling exercise at intensities of 30–50% $\dot{V}O_2$max, but with other tracers (urea and [^{2}H$_5$-ring]phenylalanine) no such increases have been observed (7, 49). Carraro et al. (8) did not find an effect of cycling exercise at 40% $\dot{V}O_2$max on muscle protein synthesis. Furthermore, carbohydrate ingestion as practiced by endurance athletes during competition reduces net protein breakdown and amino acid oxidation and upgrades the relative importance of the alanine aminotransferase reaction for TCA-cycle anaplerosis.

Main Conclusions

There are five amino acid transporters in the muscle membrane. The characteristics of these transporters in part can explain the maintenance of the amino acid gradients between muscle and blood and seem to direct the interorgan exchange and muscle metabolism of amino acids. Six amino acids are metabolized in resting muscle: leucine, isoleucine, valine, asparagine, aspartate, and glutamate. They provide the amino groups and ammonia required for synthesis of glutamine and alanine, which are released in excessive amounts. Only leucine and part of the isoleucine molecule can be oxidized in muscle. The other carbon skeletons are used solely for de novo synthesis of TCA-cycle intermediates and glutamine. The carbon atoms of the released alanine originate primarily from glycolysis of blood glucose and of muscle glycogen (about half each in resting conditions). After consumption of a protein-containing meal BCAA and glutamate are taken up by muscle and their carbon skeletons are used for de novo synthesis of glutamine. About half of the glutamine release from muscle originates from glutamate taken up from the blood both in fasted conditions and after ingestion of a mixed meal. Glutamine produced by muscle is an important fuel and regulator of DNA and RNA synthesis in mucosal cells and immune system cells and fullfills several other important functions in human metabolism.

The alanine aminotransferase reaction functions to establish and maintain high concentrations of TCA-cycle intermediates in muscle during the first 10 min of exercise. The increase in concentration of TCA-cycle intermediates probably is needed to increase the rate of the TCA cycle and meet the increased energy demand of exercise. A gradual increase in leucine oxidation subsequently leads to a carbon drain on the TCA cycle in glycogen depleted muscles and thus may reduce the maximal flux in the TCA cycle and lead to fatigue. Deamination of amino acids and glutamine synthesis present alternative anaplerotic mechanisms in glycogen depleted muscles but only allow exercise at 40-50% of W_{max}. One-leg exercise leads to net breakdown of muscle protein. The liberated amino acids are used for synthesis of TCA-cycle intermediates and glutamine. Today it is not clear whether and how important this process is in endurance exercise in the field (running or cycling) in athletes who ingest carbohydrates. It is proposed that the maximal flux in the TCA cycle is reduced in glycogen depleted muscles due to insufficient TCA-cycle anaplerosis and that this presents a limitation for the maximal rate of fatty acid oxidation. Interactions between the amino acid pool and the TCA cycle, in other words, are suggested to play a central role in the energy metabolism of contracting muscle.

References

1. Ahlborg, G., Felig, P., Hagenfeldt, L., Hendler, R., and Wahren, J. Substrate turnover during prolonged exercise in man—Splanchnic and leg metabolism of glucose, free fatty acids, and amino acids. *Journal of Clinical Investigation* 53: 1080-1090; 1974.
2. Akedo, H., and Christensen, H.N. Nature of insulin action on amino acid uptake by isolated diaphragm. *Journal of Biological Chemistry* 237: 118-127; 1962.
3. Ardawi, M.S.M., and Newsholme, E.A. Glutamine metabolism in lymphocytes of the rat. *Biochemical Journal* 212: 835-842; 1983.

4. Bennet, W.M., Connacher, A.A., Scrimgeour, C.M., Smith, K., and Rennie, M.J. Increase in anterior tibialis muscle protein synthesis in healthy man during mixed amino acid infusion: studies of incorporation of [1-^{13}C]leucine. *Clinical Science* 76: 447-454; 1989.

5. Bergström, J., Fürst, P., and Hultman, E. Free amino acids in muscle tissue and plasma during exercise in man. *Clinical Physiology* 5: 155-160; 1985.

6. Björkman, O., Felig, P., and Wahren, J. The contrasting responses of splanchnic and renal glucose output to gluconeogenic substrates and to hypoglucagonemia in 60h fasted humans. *Diabetes* 29: 610-616; 1980.

7. Carraro, F., Kimbrough, T.D., and Wolfe, R.R. Urea kinetics in humans at two levels of exercise intensity. *Journal of Applied Physiology* 75: 1180-1185; 1993.

8. Carraro, F., Stuart, C.A., Hartl, W.H., Rosenblatt, J., and Wolfe, R.R. Effect of exercise and recovery on muscle protein synthesis in human subjects. *American Journal of Physiology* 259: E470-E476; 1990.

9. Chang, T.W., and Goldberg, A.L. The origin of alanine produced in skeletal muscle. *Journal of Biological Chemistry* 253: 3677-3684; 1978a.

10. Chang, T.W., and Goldberg, A.L. The metabolic fates of amino acids and the formation of glutamine in skeletal muscle. *Journal of Biological Chemistry* 253: 3685-3695; 1978b.

11. Cheng, K.N., Pacy, P.J., Dworzak, F., Ford, G.C., and Halliday, D. Influence of fasting on leucine and muscle protein metabolism across the human forearm determined using L-[1-^{13}C,^{15}N]leucine as the tracer. *Clinical Science* 73: 241-246; 1987.

12. Christensen, H.N. Role of amino acid transport and countertransport in nutrition and metabolism. *Physiological Reviews* 70: 43-77; 1990.

13. Darmaun, D., and Déchelotte, P. Role of leucine as a precursor of glutamine α-amino nitrogen in vivo in humans. *American Journal of Physiology* 260: E326-E329; 1991.

14. Darmaun, D., Matthews, D., and Bier, D. Glutamine and glutamate kinetics in humans. *American Journal of Physiology* 251: E117-E126; 1986.

15. Elia, M., and Livesey, G. Effects of ingested steak and infused leucine on forearm metabolism in man and the fate of amino acids in healthy subjects. *Clinical Science* 64: 517-526; 1983.

16. Elia, M., Schlatmann, A., Goren, A., and Austin, S. Amino acid metabolism in muscle and in the whole body of man before and after ingestion of a single mixed meal. *American Journal of Clinical Nutrition* 49: 1203-1210; 1989.

17. Eriksson, L.S., Broberg, S., Björkman, O., and Wahren, J. Ammonia metabolism during exercise in man. *Clinical Physiology* 5: 325-336; 1985.

18. Felig, P., Pozefsky, T., Marliss, E., and Cahill, G.F. Alanine: a key role in gluconeogenesis. *Science* 167: 1003-1004; 1970.

19. Felig, P., and Wahren, J. Amino acid metabolism in exercising man. *Journal of Clinical Investigation* 50: 2703-2714; 1971.

20. Fitts, R.H. Cellular mechanisms of muscle fatigue. *Physiological Reviews* 74: 49-94; 1994.

21. Haymond, M.W., and Miles, J.M. Branched-chain amino acids as a major source of alanine nitrogen in man. *Diabetes* 31: 86-89; 1982.

22. Hoerr, R.A., Matthews, D.E., Bier, D.M., and Young, V.R. Leucine kinetics from [^{2}H$_3$]- and [^{13}C]leucine infused simultaneously by gut and vein. *American Journal of Physiology* 260: E111-E117; 1991.

23. Hundal, H.S., Rennie, M.J., and Watt, P.W. Characteristics of L-glutamine transport in perfused rat hindlimb. *Journal of Physiology (London)* 393: 283-305; 1987.

24. Hundal, H.S., Rennie, M.J., and Watt, P.W. Characteristics of acidic, basic and neutral amino acid transport in perfused rat hindlimb. *Journal of Physiology (London)* 408: 93-114; 1989.

25. Krebs, H.A. Regulation of fuel supply in animals. *Advances in Enzyme Regulation* 10: 406-413; 1972.

26. Krebs, H.A. The role of chemical equilibria in organ function. *Advances in Enzyme Regulation* 15: 449-472; 1975.

27. Lacey, J.M., and Wilmore, D.W. Is glutamine a conditionally essential amino acid? *Nutrition Reviews* 48: 297-309; 1990.

28. Lee, S.-H.C., and Davis, E.J. Amino acid catabolism by perfused rat hindquarter. The metabolic fates of valine. *Biochemical Journal* 233: 621-630; 1986.

29. Li, J.B., and Odessey, R. Regulation of protein turnover in heart and skeletal muscle by branched-chain amino acids and the keto acids. In *Problems and Potential of Branched-Chain Amino Acids in Physiology and Medicine*. Ed.R. Odessey, 83-106. Amsterdam: Elsevier Science; 1986.

30. Lowenstein, J.M., and Goodman, M.N. The purine nucleotide cycle in skeletal muscle. *Federation Proceedings* 37: 2308-2312; 1978.

31. Marliss, E.B., Aoki, T.T., Pozefsky, T., Most, A.S., and Cahill, G.F. Muscle and splanchnic glutamine and glutamate metabolism in postabsorptive and starved man. *Journal of Clinical Investigation* 50: 814-817; 1971.

32. Nurjhan, N., Bucci, A., Perriello, G., Stumvoll, N., Dailey, G., Bier, D.M., Toft, I., Jenssen, T.G., and Gerich, J.E. Glutamine: a major gluconeogenic precursor and vehicle for interorgan carbon transport in man. *Journal of Clinical Investigation* 95: 272-277; 1995.

33. Owen, O.E., Felig, P., Morgan, A.P., Wahren, J., and Cahill,G.F. Liver and kidney metabolism during prolonged starvation. *Journal of Clinical Investigation* 48: 574-583; 1969.

34. Pacy, P.J., Price, G.M., Halliday, D., Quevedo, M.R., and Millward, D.J. Nitrogen homeostasis in man: the diurnal responses of protein synthesis and degradation and amino acid oxidation to diets with increasing protein intakes. *Clinical Science* 86: 103-118; 1994.

35. Perriello, G., Jorde, R., Nurjhan, N., Stumvoll, N., Dailey, G., Jenssen, T.G., Bier, D.M., and Gerich, J.E. Estimation of gluocose-alanine-lactate-glutamine cycles in postabsorptive humans: role of skeletal muscle. *American Journal of Physiology* 269: E443-E450; 1995.

36. Rennie, M.J. Influence of exercise on protein and amino acid metabolism. In *Handbook of Physiology. Section 12: Exercise: Regulation and Integration of Multiple Systems*. Ed. L.B. Rowell and J.T. Shepherd, 995-1035. Oxford, UK: Oxford University Press; 1996.

37. Rennie, M.J., Babij, P., Taylor, P.M., Hundal, H.S., MacLennan, P., Watt, P.W., Jepson, M.M., and Millward, D.J. Characteristics of a glutamine carrier

in skeletal muscle have important consequences for nitrogen loss in injury, infection and chronic disease. *Lancet* 2: 1008-1012; 1986.

38. Ross, B.D., Hems, R., and Krebs, H.A. The rates of gluconeogenesis from various precursors in the perfused rat liver. *Biochemical Journal* 102: 942-951; 1967.

39. Sahlin, K., Katz, A., and Broberg, S. Tricarboxylic acid cycle intermediates in human muscle during prolonged exercise. *American Journal of Physiology* 259: C834-C841; 1990.

40. Souba, W.W. Glutamine: a key substrate for the splanchnic bed. *Annual Reviews of Nutrition* 11: 285-308; 1991.

41. Van Acker, B.A.C., Hulsewé, K.W.E., Wagenmakers, A.J.M., Deutz, N.E.P., Van Kreel, B.K., Soeters, P.B., and Von Meijenfeldt, M.F. Measurement of glutamine metabolism in gastrointestinal cancer patients. *Clinical Nutrition* 15 (Supplement) 1; 1996.

42. Van Hall, G., MacLean, D.A., Saltin, B., and Wagenmakers, A.J.M. Mechanisms of activation of muscle branched-chain α-keto acid dehydrogenase during exercise in man. *Journal of Physiology* 494: 899-905; 1996.

43. Van Hall, G., Saltin, B., Van der Vusse, G.J., Söderlund, K., and Wagenmakers, A.J.M. Deamination of amino acids as a source for ammonia production in human skeletal muscle during prolonged exercise. *Journal of Physiology* 489: 251-261; 1995.

44. Vinnars, E., Bergström, J., and Fürst, P. Influence of the postoperative state on the intracellular free amino acids in human muscle tissue. *Annals of Surgery* 182: 665-671; 1975.

45. Wagenmakers, A.J.M., and Van Hall, G. Branched-chain amino acids: Nutrition and metabolism in exercise. In *Biochemistry of Exercise IX*. Ed. R.J. Maughan and S.M. Shirreffs, 431-443. Champaign, IL: Human Kinetics; 1996.

46. Wagenmakers, A.J.M., and Soeters, P.B. Metabolism of branched-chain amino acids. In *Amino Acid Metabolism and Therapy in Health and Nutritional Disease*. Ed. L.A. Cynober, 67-83. New York: CRC Press; 1995.

47. Wagenmakers, A.J.M., Beckers, E.J., Brouns, F., Kuipers, H., Soeters, P.B., Van der Vusse, G.J., and Saris, W.H.M. Carbohydrate supplementation, glycogen depletion, and amino acid metabolism during exercise. *American Journal of Physiology* 260: E883-E890; 1991.

48. Wagenmakers, A.J.M., Coakley, J.H., and Edwards, R.H.T. Metabolism of branched-chain amino acids and ammonia during exercise: clues from McArdle's disease. *International Journal of Sports Medicine* 11: S101-S113; 1990.

49. Wagenmakers, A.J.M., Pannemans, D.L.E., Jeukendrup, A.E., Gijsen, A.P., Senden, J.M.G., Halliday, D., and Saris, W.H.M. The effect of prolonged exercise on protein metabolism in top triathletes during carbohydrate ingestion. *Proceedings of the Nutrition Society*. In press; 1997.

50. Wagenmakers, A.J.M., Salden, H.J.M., and Veerkamp, J.H. The metabolic fate of branched-chain amino acids and 2-oxo acids in rat muscle homogenates and diaphragms. *International Journal of Biochemistry* 17: 957-965; 1985.

51. Wagenmakers, A.J.M., Van Hall, G., and Saltin, B. Excessive muscle pro-
    teolysis during one leg exercise is exclusively attended by increased de novo
    synthesis of glutamine, not of alanine. *Clinical Nutrition* 15 (Supplement) 1;
    1996a.
52. Wagenmakers, A.J.M., Van Hall, G., and Saltin, B. High conversion rates of
    glutamate and branched-chain amino acids to glutamine during prolonged
    one leg exercise. An alternative mechanism for synthesis of tricarboxylic
    acid cycle intermediates. *Physiologist* 39:A73; 1996b.
53. Watt, P.W., Corbett, M.E., and Rennie, M.J. Stimulation of protein synthesis
    in pig skeletal muscle by infusion of amino acids during constant insulin
    availability. *American Journal of Physiology* 263: 453-460; 1992.
54. Windmueller, H.G., and Spaeth, A.E. Uptake and metabolism of plasma
    glutamine by the small intestine. *Journal of Biological Chemistry* 249: 5070-
    5079; 1974.
55. Wirthensohn, G., and Guder, W. Renal substrate metabolism. *Physiological
    Reviews* 66: 469-497; 1986.

CHAPTER 18

Molecular Biology of the Glucose Transporter, GLUT4

David E. James

Centre for Molecular and Cellular Biology, University of Queensland,
Brisbane, Australia

Insulin increases glucose transport in muscle and adipocytes by stimulating the translocation of the glucose transporter GLUT4 from intracellular vesicles to the cell surface. A major challenge now is to dissect the molecular mechanism by which the intracellular GLUT4 vesicles undergo docking and fusion with the cell surface in the hope of determining the point of intersection with the insulin signal transduction pathway. The SNARE hypothesis suggests that membrane proteins (v-SNAREs) found in transport vesicles bind in a highly specific way to membrane proteins (t-SNAREs) found in the cognate target membrane. Thus, an important goal is to identify v- and t-SNAREs that regulate the trafficking of GLUT4. Using immunoelectron microscopy and a compartment ablation technique we recently showed that GLUT4 resides in an intracellular compartment that is segregated from endosomes and the trans-Golgi network. This compartment is marked by the presence of GLUT4, an amino peptidase (IRAP) and the v-SNARE, VAMP2. A specific role for VAMP2 in the insulin-dependent movement of GLUT4 is implicated, because VAMP2-specific peptides inhibit insulin-stimulated GLUT4 translocation when added to permeabilized 3T3-L1 adipocytes. These GLUT4containing vesicles appear to bind directly to the cell surface via the t-SNARE, syntaxin4. We previously have identified three Sec1/Munc-18 isoforms in adipocytes, a family of proteins that bind to syntaxins. Using an in vitro binding assay we show that individual Munc-18 isoforms display a unique specificity for different syntaxins. In particular, Syntaxin4 interacts with Munc-18c and, in so doing, its ability to bind to VAMP2 is blocked. Other proteins that modulate the interaction between these different SNAREs in adipocytes currently are under investigation. The molecules described here are good targets for insulin action in facilitating increased glucose entry into fat and muscle cells.

Our ultimate goal is to understand the molecular basis for insulin-resistant disease states such as Type II diabetes. Studies in first-degree relatives of Type II diabetics implicate a defect in peripheral insulin action as the primary pathophysiological disorder in the onteny of this disease (11). One of the most important biological actions of insulin is to increase glucose transport into muscle and adipose tissue, and this represents the main focus of our work. Several pivotal observations have been made over the past decade. In 1980 it was shown that insulin stimulates the movement of a glucose transporter from an intracellular site to the cell surface in adipocytes (6). Almost a decade later, from work in several laboratories (2, 5, 8, 9), a fat/muscle cell-specific glucose transporter, now known as GLUT4, was identified. This enabled production of isoform-specific antibodies, which were used to localize the protein in insulin-sensitive tissues by immunoelectron microscopy (19, 20). These studies provided the most conclusive evidence that insulin stimulates glucose transport in these cells by triggering the translocation of GLUT4 from an intracellular compartment to the cell surface. In nonstimulated cells, GLUT4 is excluded from the cell surface and is located in tubulo-vesicular elements clustered either in the trans-Golgi reticulum or in the cytoplasm, in many cases proximal to endosomes. GLUT4 co-localizes with other specialized proteins in muscle and fat, including the amino peptidase vp165 (13), a secretory vesicle membrane protein (SCAMPs) (10) and a synaptic vesicle membrane protein (VAMP2) (4, 12, 25). We have recently shown, in collaboration with Dr. Gould's laboratory at the University of Glasgow, that GLUT4, vp165, and VAMP2 co-localize in a unique population of intracellular vesicles in adipocytes that are segregated from endosomal/TGR membranes (12). The identification of this specialized compartment has enabled a detailed search for molecules that dictate whether these intracellular vesicles should or should not dock and fuse with the plasma membrane. Over the past two years four new proteins that are involved in the regulation of GLUT4 trafficking in adipocytes have been described. A future goal is to establish how these molecules communicate with each other and with the GLUT4 compartment and to determine the point of intersection between these vesicular transport proteins and the insulin signal transduction pathway.

The SNARE Hypothesis

Deciphering the molecular regulation of vesicle transport is a challenging problem. There must be some type of address that specifies that a particular compartment should fuse with another. There must also be a set of factors that catalyze the fusion reaction, an otherwise energy-unfavorable process. Finally, particularly in the case of regulated vesicular transport events such as the insulin-stimulated movement of GLUT4, there must be a biochemical means for either slowing or expediting the entire process. Recent work in the secretory pathway of both mammalian cells and yeast cells has elucidated a generalized set of proteins that fulfill these requirements for many vesicle fusion events. The SNARE hypothesis suggests that membrane proteins in a transport vesicle and in the target membrane, with which it is destined to fuse, interact to form a docking/fusion complex (21). The components of this complex (SNAREs) include synaptobrevin or VAMP2 in the transport vesicle and syntaxin and SNAP25 in the target membrane. It is thought that these proteins initially interact to form a 7s docking complex, and that once assembled this complex acts as a high-affinity binding site for a set of soluble proteins, that include the NEM-sensitive fusion factor NSF and the NSF attachment proteins α/B, γ-SNAPs resulting in the formation of a 20s particle. It has been proposed that ATP hydrolysis by NSF disassembles this complex, catalyzing membrane fusion (1, 18).

In support of this model:

1. tetanus and botulinum toxins irreversibly block synaptic vesicle exocytosis by selectively proteolysing synaptobrevin, syntaxin, and SNAP25 (15);
2. homologues of synaptobrevin, syntaxin, SNAP25, NSF, and α-SNAP have been identified in yeast and mutations in these genes result in vesicular transport defects (18);
3. these proteins belong to large gene families, the individual members of which are located at distinct intracellular loci (1).

Hence, different combinations of SNAREs may provide the necessary specificity required to regulate the multitude of different transport reactions that occur within eucaryotic cells. While syntaxin/synaptobrevin/SNAP25 represent the core of the docking/fusion complex, several other proteins that interact either genetically or biochemically play a modulatory role in the assembly of a functional complex. In particular, Sec1/Munc-18 binds to syntaxin and inhibits its binding affinity toward SNAP25 and synaptobrevin (16). Synaptobrevin also binds to synaptophysin, another synaptic vesicle protein, and this interaction is thought to decrease the likelihood of synaptobrevin binding to syntaxin (7).

To completely understand the biology of eucaryotic vesicle transport it will be necessary to define the stage-specific changes in SNARE interactions. This probably will involve multiple conformational changes that may be triggered by protein/protein interactions, phosphorylation, nucleotide binding, and so on. Most vesicle transport events occur constitutively, however, and so it has been difficult to capture a glimpse of each individual step. Regulated exocytic events provide a unique advantage, because secretory products are segregated away from the constitutive pathway in secretory vesicles that are held in stasis until the appropriate signal triggers their exocytosis often in an explosive manner (3). Hence, by studying this transition it should be possible to capture SNAREs at various stages of assembly or disassembly.

SNAREs and Insulin Action-Recent Progress

Four SNAREs have been implicated in the insulin-dependent movement of GLUT4:

1. *Synaptobrevin.* Two synaptobrevins, cellubrevin and VAMP2, that appear to co-localize with GLUT4 in adipocytes have been identified (4, 12, 25). Cellubrevin is confined to recycling endosomes together with the transferrin receptor (12, 14), whereas VAMP2 co-localizes with GLUT4 in a specialized post-endocytic compartment (12). We recently have obtained evidence to suggest that this latter compartment fuses directly with the cell surface in response to insulin, suggesting that this represents a specialized

exocytotic compartment (unpublished data). Consistent with this we also have shown that synthetic peptides comprising the Pro-rich amino terminus of VAMP2, a region that is unique compared with cellubrevin, inhibit insulin-stimulated GLUT4 translocation to the cell surface in permeabilized adipocytes (unpublished data). These data highlight the important role of VAMP2, in particular its unique N-terminus, in GLUT4 trafficking in adipocytes.

2. *Syntaxin4*. Syntaxin4 appears to be the t-SNARE that facilitates docking of GLUT4 vesicles with the cell surface in adipocytes because

- it is expressed at high levels in adipocytes (23, 24);
- it is the major adipocyte t-SNARE that interacts with VAMP2 in adipocytes as determined by far-western blotting and affinity chromatography of adipocyte fractions using a recombinant syntaxin4-GST fusion protein as a probe (S. Rea and D. James, unpublished data);
- it is targeted to the plasma membrane in basal and insulin-treated adipocytes (23); and
- Syntaxin4 antibodies inhibit insulin-stimulated GLUT4 translocation in 3T3-L1 adipocytes (23, 24).

3. *SNAP23*. Molecules that modulate the interaction between VAMP2 and Syntaxin4 are likely targets of insulin action. Two candidates are members of the SNAP25 and Sec1 families, because previous studies have shown that SNAP25 increases the affinity of VAMP2 for the neuronal syntaxin (16) whereas the neuronal Munc-18/Sec1 isoform inhibits this interaction (16). We recently have identified members of each of these families that may regulate GLUT4 trafficking in adipocytes. SNAP23 is an homologue of SNAP25 (17) that is expressed at high levels in adipocytes (S. Rea and D. James, unpublished data). SNAP23 interacts with Syntaxin4 in vitro and has an identical subcellular distribution to Syntaxin4 in 3T3-L1 adipocytes. A functional link between SNAP23 and GLUT4 trafficking in adipocytes has not been shown yet.

4. *Munc-18c*. Three Munc-18/Sec1 isoforms (Munc-18a-c) have been iden- tified in 3T3-L1 adipocytes (22). Each displays a unique syntaxin binding specificity. Notably, Munc-18c is the only isoform that binds to Syntaxin4 and in so doing inhibits its interaction with VAMP2 (23). Munc-18c has an identical subcellular distribution to Syntaxin4 and SNAP23 in adipocytes being highly enriched in the plasma membrane (23). Hence, these data implicate an important role for Munc-18c in GLUT4 trafficking in adipocytes.

Hypothesis

Based on our current data and by analogy with synaptic vesicle exocytosis we propose the following model. In the absence of insulin, GLUT4 is sequestered within the cell by two distinct mechanisms both of which prevent vesicle docking with the cell surface: (1) VAMP2, which co-localizes with GLUT4 in exocytic transport vesicles, is bound to another protein, possibly homologous to synaptophysin, and this interaction prevents VAMP2 from binding to Syntaxin4 at the cell surface; (2) Syntaxin4 is bound to Munc-18c, and this interaction prevents binding of VAMP2 and the GLUT4 vesicles to the cell surface. Both of these mechanisms afford a low exocytic rate of GLUT4. Insulin disrupts both of these interactions, i.e., VAMP2 with its intracellular binding protein and Syntaxin4 with Munc-18c at the cell surface. This results in increased binding of SNAP23 to Syntaxin4, so that the affinity of this cell surface t-SNARE complex for VAMP2 is enhanced, resulting in increased GLUT4 exocytosis (figure 18.1). This model lays the groundwork for future investigations in this area.

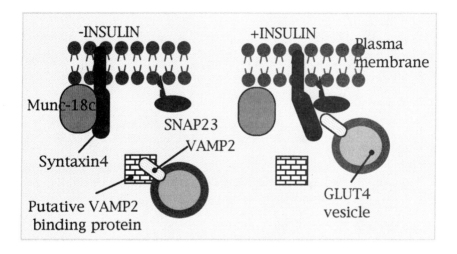

Figure 18.1   Proposed model for the effect of insulin on SNARE proteins in insulin sensitive cells. In the absence of insulin (-Insulin) Syntaxin4 may constitutively associate with Munc-18c resulting in decreased affinity of Syntaxin4 for the intracellular v-SNARE, VAMP2. VAMP2 may interact with intracellular proteins that provide a docikinglfusion clamp. Insulin may then result in two modulations: VAMP2 may disassociate from its putative binding protein in the intracellular GLUT4 containing vesicles and Munc-18c may disassociate from Syntaxin4 at the cell surface. Both of these modulations may result in increased affinity of Syntaxin4 for both SNAP23 and VAMP23, the end result being increased docking and fusion of intracellular GLUT4 containing vesicles with the plasma membrane of the adipocyte.

References

1. Bennett, M.K., and Scheller, R.H. The molecular machinery for secretion is conserved from yeast to neurons. *Proc. Natl. Acad. Sci. USA* 90: 2559-2563; 1993.

2. Birnbaum, M.J. Identification of a novel gene encoding an insulin-responsive glucose transporter protein. *Cell* 57: 305; 1989.

3. Burgess, T.L., and Kelly, R.B. Constitutive and regulated secretion of proteins. *Ann. Rev. Cell Biol.* 3: 243-293; 1987.

4. Cain, C.C., Trimble, W.S., and Lienhard, G.E. Members of the VAMP family of synaptic vesicle proteins are components of glucose transporter-containing vesicles from rat adipocytes. *J. Biol. Chem.* 267: 11,681-11,684; 1992.

5. Charron, M., Brosius, F.C., III, Alper, S.L., and Lodish, H.F. A glucose transport protein expressed predominantly in insulin-responsive tissues. *PNAS.* 86: 2535; 1989.

6. Cushman, S.W., and Wardzala, L.J. Potential mechanism of insulin action on glucose transport in the isolated rat adipose cell: apparent translocation of intracellular transport systems to the plasma membrane. *J. Biol. Chem.* 255: 4758-4762; 1980.

7. Edelmann, L., Hanson, P.I., Chapman, E.R., and Jahn, R. Synaptobrevin binding to synaptophysin: a potential mechanism for controlling the exocytotic fusion machine. *EMBO J.* 14: 224-231; 1995.

8. James, D.E., Strube, M., and Mueckler, M. Molecular cloning and characterization of an insulin regulatable glucose transporter. *Nature* 338: 83-87; 1989.

9. James, D.E., Brown, R., Navarro, J., and Pilch, P.F. Insulin-regulatable tissues express a unique insulin sensitive glucose transport protein. *Nature* 333: 183-185; 1988.

10. Laurie, S.M., Cain, C.C., Lienhard, G.E., Castle, J.D. The glucose transporter GLUT4 and secretory carrier membrane proteins (SCAMPs) colocalize in rat adipocytes and partially segregate during insulin stimulation. *J. Biol. Chem.* 268:19,110-19,117; 1993.

11. Martin, B.C., Warram, J.H., Krolewski, A.S., Bergman, R.N., Soeldner, J.S., and Kahn, C.R. Role of glucose and insulin resistance in development of Type 2 diabetes mellitus. *Lancet* 340: 925; 1992.

12. Martin, S., Tellam, J., Livingstone, C., Slot, J.W., Gould, G.W., and James, D.E. GLUT-4 and VAMP-2 are segregated from recycling endosomes in insulin-sensitive cells. *J. Cell Biol.* 134: 625-635; 1996.

13. Mastick, C.C., Aebersold, R., Lienhard, G.E. Characterisation of a major protein in GLUT4 vesicles. *J. Biol. Chem.* 269: 6089-6092; 1994.

14. McMahon, H.T., Ushkaryov, Y.A., Edelman, L., Link, E., Binz, T., Niemann, H., Jahn, R., and Sudhof, T.C. Cellubrevin is a ubiquitous tetanus-toxin substrate homologous to a putative synaptic vesicle fusion protein. *Nature* 364: 346-349; 1993.

15. Niemann, H., Blasi, J., and Jahn, R. Clostridial neurotoxins: new tools for dissecting exocytosis. *Trends Cell Biol.* 4: 179-185; 1994.

16. Pevsner, J., Hsu, S.C., Braun, J.E.A., Calakos, N., Ting, A.E., Bennett, M.K., and Scheller, R.H. Specificity and regulation of a synaptic vesicle docking complex. *Neuron* 13: 353-361; 1994.

17. Ravichandran, V., Chawla, A., and Roche, P.A. Identification of a novel syntaxin- and synaptobrevin/VAMP-binding protein, SNAP-23, expressed in non-neuronal tissues. *J. Biol. Chem.* 271: 13,300-13,303; 1996.

18. Rothman, J.E., and Warren, G. Implications of the SNARE hypothesis for intracellular membrane topology and dynamics. *Current Biology* 4: 220-233; 1994.
19. Slot, J.W., Geuze, H.J., Gigengack, S., Lienhard, G.E., and James, D.E. Immuno-localization of the insulin regulatable glucose transporter (GLUT4) in brown adipose tissue of the rat. *J. Cell Biol.* 113: 123-136; 1991.
20. Slot, J.W., Geuze, H.J., Gigengack, S., James, D.E., and Lienhard, G.E. Translocation of the glucose transporter GLUT4 in cardiac myocytes of the rat. *Proc. Natl. Acad. Sci. USA* 88: 7815-7819; 1991.
21. Sollner, T.,Whiteheart, S.W., Brunner, M., Erdjument-Bromage, H., Geromanos, S., Tempst, P., and Rothman, J.E. SNAP receptors implicated in vesicle targeting and fusion. *Nature* 362: 318-324; 1993.
22. Tellam, J.T., McIntosh, S., and James, D.E. Molecular identification of a family of mammalian Munc/Sec1 homologs. *J. Biol. Chem.* 270: 5857-5863; 1995.
23. Tellam, J.T., Macaulay, S.L., McIntosh, S., Hewish, D.R., Ward, C.W., and James, D.E. Characterization of Munc-18c and Syntaxin-4 in 3T3-L1 adipocytes. Putative role in insulin-dependent movement of GLUT-4. *J. Biol. Chem.* 272: 6179-6186; 1997.
24. Volchuk, A., Wang, Q., Ewart, S., Liu, Z., He, L., Bennett, M.K., and Klip, A. Syntaxin 4 in 3T3-L1 adipocytes: regulation by insulin and participation in insulin-dependent glucose transport. *Mol. Biol. Cell* 7: 1075-1082; 1996.
25. Volchuk, A., Sargeant, R., Sumitani, S., Liu, Z., He, L., and Klip, A. Cellubrevin is a resident protein of insulin-sensitive GLUT4 glucose transporter vesicles in 3T3-L1 adipocytes. *J. Biol. Chem.* 270: 8233-8240; 1995.

CHAPTER 19

Biochemical Regulation of Carbohydrate-Lipid Interaction in Skeletal Muscle During Low and Moderate Intensity Exercise

Lawrence L. Spriet and L. Maureen Odland

Department of Human Biology & Nutritional Sciences,
University of Guelph, Guelph, Ontario, Canada

It is well established that a mixture of carbohydrates (CHO) and lipids are used as fuels in skeletal muscle at rest and during exercise. Classic studies using the respiratory exchange ratio (RER) demonstrated that both CHO and lipid were used during exercise and that their relative contribution changed as a function of exercise intensity and duration and the pre-exercise diet (for review, see Asmussen 1). Low power outputs were associated with greater oxidation of lipid, while higher exercise intensities resulted in greater reliance on CHO.

Current controversy exists regarding the mechanisms that regulate substrate choice for oxidation in human skeletal muscle during aerobic exercise (for review, see Spriet and Dyck 53). There are several sites in the pathways of CHO and lipid metabolism where intramuscular fuel selection during exercise may be regulated. These include regulation at sites controlling CHO metabolism and oxidation as described in the classic glucose-fatty acid (G-FA) cycle, regulation of muscle glycogen degradation via glycogen phosphorylase (PHOS), and regulation of long-chain fatty acid transport into the mitochondria via the carnitine palmitoyl transferase (CPT) system (figure 19.1).

The concept of the G-FA cycle originally was introduced by Randle et al. (40, 41) to explain the interaction between CHO and lipid metabolism in disease states. It was proposed that increased delivery of free fatty acids (FFA) to muscle tissue enhanced the rate of fat oxidation, which led to increased acetyl-CoA and citrate production and resulted in down-regulation of key CHO metabolizing enzymes, pyruvate dehydrogenase (PDH), and phosphofructokinase (PFK), respectively (figure 19.1). Decreased glycolytic activity also resulted in glucose-6-phosphate (G6P) accumulation, which in turn decreased hexokinase activity and reduced glucose uptake (41).

Much of the original support for G-FA cycle was obtained from contracting heart muscle perfused with high FFA or resting diaphragm muscle bathed in a medium with high FFA (18, 19, 40, 41). Heart and resting diaphragm muscles are

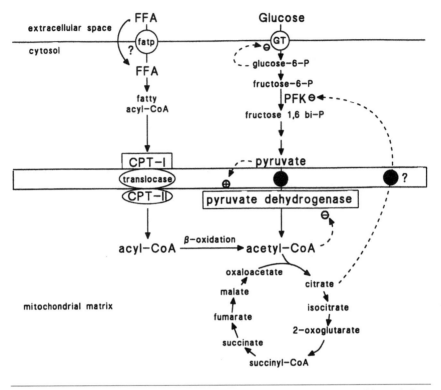

Figure 19.1. Schematic of the glucose-fatty acid cycle and the transport of long-chain fatty acids into the mitochondrion.

constantly active in a regular cycle, which ultimately requires that the majority of oxidizable substrate must come from exogenous sources. Conversely, in skeletal muscles other than diaphragm, the higher energy demands of moderate and high intensity exercise dictate that fuel must also come from the endogenous glycogen store. Several studies in humans have shown that CHO use is reduced when the provision of FFA is increased (12, 16, 17, 56). However, it was demonstrated recently that the classic explanation for the regulation between CHO and lipid interaction did not exist in skeletal muscle during 15 min of intense aerobic cycling (80–85% maximal oxygen uptake [$\dot{V}O_2$max]) (16, 17). High FFA provision did not alter muscle acetyl-CoA and citrate levels and had no effect on the transformation of PDH to the active a form. Instead, down-regulation of CHO metabolism occurred at the level of glycogen PHOS, not by altering the transformation to the more active a form but via post-transformational regulation. Enhanced FFA oxidation appeared to provide a better match between energy demand and energy provision at the onset of exercise, which led to reduced accumulations of free AMP (an allosteric activator of PHOS a) and P_i (a substrate for PHOS) and a lower flux through PHOS (17).

However, since smaller alterations in the energy status of the cell and therefore the requirement for glycogenolysis would be expected to occur during lower exercise intensities, it is possible that the regulation of CHO/lipid interaction at lower power outputs (i.e., <85% $\dot{V}O_2$max) may occur as classically proposed.

The rate-limiting step in the mitochondrial beta-oxidation of long-chain fatty acids is believed to be transport through the mitochondrial membranes (29, 31, 32, 55) (figure 19.1). The transport is L-carnitine dependent and catalyzed by the CPT enzyme system, which consists of CPT I, a carnitine-acylcarnitine translocase and CPT II (7, 31, 32). CPT I is located in the outer mitochondrial membrane and catalyzes the conversion of acyl-CoA to acylcarnitine, which then is transported through the inner mitochondrial membrane via the carnitine-acylcarnitine translocase. The translocase acts as an antiport, allowing the simultaneous export of L-carnitine from the mitochondrial matrix. The acylcarnitine is reconverted to acyl-CoA via CPT II on the matrix side of the inner mitochondrial membrane.

In vitro assays clearly have established, in a variety of tissues and species, that the activity of the rate-limiting enzyme, CPT I, can be reversibly inhibited by malonyl-CoA (M-CoA), while CPT II is unaffected (29, 30, 47). M-CoA is produced in the cytoplasm by the action of acetyl-CoA carboxylase (ACC) and is the first committed intermediate produced during fatty acid synthesis. It is a well-established regulator of fatty acid oxidation in the lipogenic tissues, adipose tissue, and liver (30). When CHO supply is abundant and lipid is synthesized in these tissues, high levels of M-CoA inhibit CPT I activity and the transport of lipid into the mitochondria.

The inhibition of CPT I by M-CoA is not restricted to lipogenic tissues (47). Measurable amounts of M-CoA have been detected in rat skeletal and heart muscles (46, 48, 57, 58). A muscle isoform of ACC also has been found in heart and skeletal muscle (4), and it appears to be regulated differently than hepatic ACC (46). Several recent investigations have promoted M-CoA as a regulator of fatty acid oxidation in isolated cardiac myocytes (2), perfused working heart muscle (2, 46), and contracting rat skeletal muscle (14, 58).

It is believed that resting levels of M-CoA in rodent muscle are sufficiently high to inhibit excessive entry of FFA into the mitochondria. During exercise, when the demand for energy from fat oxidation increases, rat skeletal muscle M-CoA content decreased during treadmill running and in response to electrical stimulation (14, 57, 48). The decrease in M-CoA was postulated to be instrumental for releasing the inhibition on CPT I and increasing FFA oxidation during prolonged exercise. Although increased M-CoA content has been related directly to decreased fatty acid oxidation rates in perfused working heart muscle (46), only correlational data are currently available regarding the role of M-CoA in skeletal muscle. Until recently, there had been no reported attempts to measure M-CoA content in human skeletal muscle. Consequently, the importance of M-CoA in the regulation of CPT I activity in contracting human skeletal muscle is not known.

The purpose of this chapter is to review recent studies examining the understanding of the regulation between CHO and lipid metabolism during low and moderate intensity exercise in humans. The reader is encouraged to consult other recent papers for more detailed reviews of the G-FA cycle (53, 55) and CPT I-malonyl-CoA interactions (32).

Effects of High FFA Provision on Muscle Metabolism During Low and Moderate Intensity Exercise

Odland et al. (35) increased the plasma [FFA] during low and moderate intensity cycle exercise in men to determine 1) if elevated FFA provision to the working muscles would decrease muscle glycogenolysis and glucose uptake, and 2) if decreases in CHO metabolism would be regulated as classically proposed in the G-FA cycle. Intralipid (20% triacylglycerol solution) and heparin were infused to artificially elevate the plasma [FFA] before and during 10 min of cycling at 40% $\dot{V}O_2$max and 60 min at 65% $\dot{V}O_2$max, as compared with a low FFA control condition. The radial artery and femoral vein were cannulated and leg blood flow was estimated from the leg arterial-venous (a-v) O_2 content difference and the pulmonary $\dot{V}O_2$ (27) to estimate substrate and metabolite exchange across the working leg. Muscle biopsies were sampled from the vastus lateralis muscle at rest and following 10, 20, and 70 min of cycling.

Intralipid infusion increased plasma arterial FFA from 0.25 ± 0.03 to 0.69 ± 0.04 mM at rest. During exercise, FFA remained high at all time points and peaked at 0.89 ± 0.05 mM at 50 min. In the control trial, FFA remained between 0.2-0.3 mM throughout the entire exercise period.

Whole-body $\dot{V}O_2$ was unaffected by Intralipid, but the RER was lower with elevated FFA; 0.92 ± 0.02 vs. 0.89 ± 0.01 at 40% $\dot{V}O_2$max and 0.94 ± 0.01 and 0.91 ± 0.01 at 65% $\dot{V}O_2$max. The blood a-v O_2 content differences and estimated leg blood flows were not different between trials. Calculated RQ across the working muscle bed was consistently lower with high FFA (40%, 0.91 ± 0.05 vs. 0.87 ± 0.05; 65%, 1.00 ± 0.02 vs. 0.94 ± 0.02).

Arterial blood glucose levels decreased at all exercise time points, and blood lactate increased above rest levels at both power outputs to the same extent in both conditions. Elevated arterial plasma FFA increased leg FFA uptake ~four-fold, decreased lactate efflux at 18 and 34 min of exercise (figure 19.2a), and had no effect on glucose uptake compared to control (figure 19.2b).

Resting glycogen levels were not different between conditions (figure 19.3). During exercise, muscle glycogen degradation was 23% lower in the Intralipid trial (297 ± 36 vs. 230 ± 29 mmol/min/kg dry muscle [dm]). Muscle ATP remained constant throughout exercise in both conditions (table 19.1). Phosphocreatine (PCr) decreased significantly from rest during exercise at 65% $\dot{V}O_2$max and muscle lactate increased above rest levels at 65% $\dot{V}O_2$max but both variables were unaffected by high FFA (table 19.1). Muscle acetyl-CoA and acetyl-carnitine increased from rest to 40% $\dot{V}O_2$max, and from 40% to 65% $\dot{V}O_2$max, but were unaffected by Intralipid infusion (table 19.1). High FFA provision increased muscle citrate content at rest and throughout exercise (figure 19.4). Calculated values for free ADP, AMP, and P_i were similar between conditions at rest, 10, and 20 min exercise, but were significantly reduced at 70 min during exercise in the presence of elevated FFA. The transformation of PDH to the active a form (PDH<u>a</u>) was significantly reduced at all exercise time points during the Intralipid trial (figure 19.5).

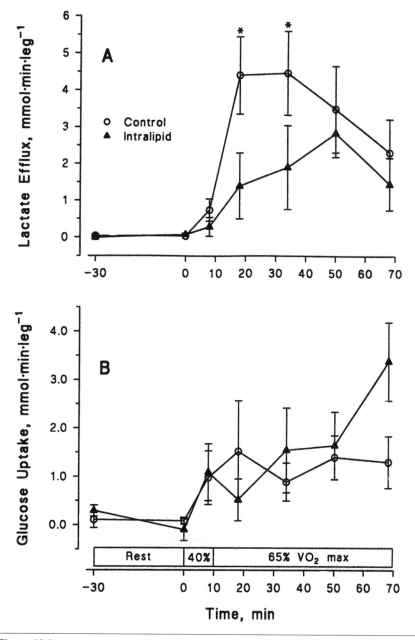

Figure 19.2. Net leg lactate efflux (A) and glucose uptake (B) during 70 min of cycling at 40% and 65% VO$_2$max with Intralipid infusion or Control. Values are ± SE. *, significantly different from Intralipid.

Reprinted, by permission, from Odland et al. 1998.

Table 19.1　Muscle Metabolites During Cycle Exercise (10 min at 40% and 60 min at 65% $\dot{V}O_2$max)

	Trial	Pre-exercise	10 min 40% $\dot{V}O_2$max	20 min	70 min
				65% $\dot{V}O_2$max	
ATP	CON	26.6 ± 1.2	26.1 ± 3.0	24.3 ± 2.9	24.6 ± 1.3
	INT	24.1 ± 2.9	25.4 ± 3.1	25.9 ± 0.6	24.6 ± 0.9
PCr	CON	83.4 ± 1.4	76.6 ± 8.8	54.4 ± 6.5#,¥	43.5 ± 4.6#,¥
	INT	81.7 ± 9.4	74.3 ± 9.2	50.1 ± 3.4#,¥	55.5 ± 4.6#,¥
Lactate	CON	6.2 ± 1.7	7.2 ± 2.4	29.0 ± 5.7#,¥	25.6 ± 6.1#,¥
	INT	8.2 ± 1.6	10.1 ± 2.4	30.2 ± 6.2#,¥	25.2 ± 4.3#,¥
Acetyl-CoA	CON	8.8 ± 0.9	14.8 ± 1.8#	23.2 ± 3.3#,¥	26.7 ± 1.9#,¥
	INT	8.6 ± 1.2	16.8 ± 2.8#	27.0 ± 2.1#,¥	26.6 ± 2.3#,¥
Acetyl-carn	CON	1.2 ± 0.2	5.4 ± 0.9#	11.4 ± 1.7#,¥	13.3 ± 1.0#,¥
	INT	1.1 ± 0.2	5.4 ± 0.9#	12.2 ± 0.8#,¥	13.6 ± 1.0#,¥

Mean ± SE, mmol/kg dry muscle (Acetyl-CoA, μmol/kg dry muscle). Acetyl-carn, acetyl-carnitine; PCr, phosphocreatine. #, significantly different from pre-exercise. ¥, significantly different from 10 min.

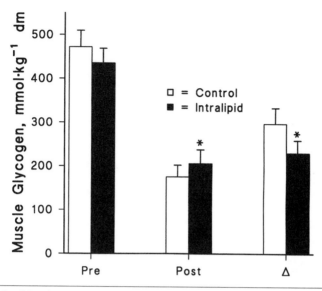

Figure 19.3.   Muscle glycogen content prior to and following 10 min at 40 and 60 min at 65% V̇O₂max with Intralipid infusion and control. Values are mean ± SE. * , significantly different from Control.

Reprinted, by permission, from Odland et al. 1998.

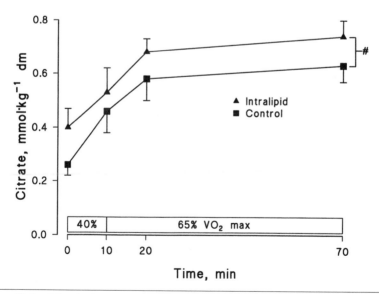

Figure 19.4.   Muscle citrate content at rest, and accumulation during 70 min of cycling at 40% and 65% V̇O₂max with Intralipid infusion or Control. Values are ± SE. #, significant main effect between trials.

Reprinted, by permission, from Odland et al. 1998.

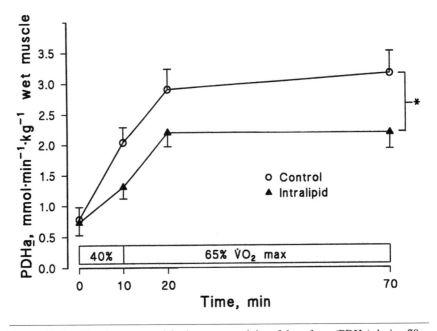

Figure 19.5.   Muscle pyruvate dehydrogenase activity of the <u>a</u> form (PDH<u>a</u>) during 70 min of cycling at 40% and 65% $\dot{V}O_2$max with Intralipid infusion or Control. Values are ± SE. *, significant main effect between trials.

Reprinted, by permission, from Odland et al. 1998.

The 23% reduction in glycogenolysis during 70 min of low and moderate intensity cycle exercise with high [FFA] is consistent with previous investigations in which subjects exercised at 70% $\dot{V}O_2$max for 30 and 60 min and used 40% and 28% less muscle glycogen in the high-fat compared with low-fat trials (12, 56). In the Odland et al. (35) study, the pulmonary RER and leg RQ data also indicated reduced CHO use and increased lipid oxidation in the Intralipid trial.

In addition to glycogen sparing, Intralipid infusion resulted in an elevated muscle citrate content and reduced transformation of PDH to the active <u>a</u> form. These results are consistent with the classic description of the G-FA cycle (40). However, there was no concomitant increase in muscle acetyl-CoA content and no reduction in glucose uptake associated with increased FFA availability. Lastly, calculated free ADP and AMP contents were significantly less in Intralipid compared with CON at 70 min of exercise. Therefore, although down-regulation of CHO metabolism in the presence of high [FFA] was partially regulated as classically proposed, other regulatory factors also appeared to contribute. The reduced flux through PHOS during Intralipid may have been due to reduced free AMP and P_i contents during exercise at 65% $\dot{V}O_2$max, and PFK may have been inhibited by increased cytoplasmic citrate. The mechanism for reduced transformation of PDH to the more active <u>a</u> form (which was unrelated to acetyl-CoA) is unknown but suggests reduced flux through PDH.

Carbohydrate Sparing

Few researchers have investigated glucose uptake during exercise in humans. Romijn et al. (45) reported that blood glucose disappearance (stable-isotope tracer technique) was unaffected by elevated plasma FFA during cycle exercise at 85% $\dot{V}O_2$max. A significant reduction in glucose uptake, however, was reported during 60 min of leg extension exercise at 80% knee-extensor work capacity in response to elevated FFA (22). The discrepancy in results is difficult to explain, but may be due to differences in experimental protocols. In the Hargreaves et al. (22) study, leg blood flow was measured directly via the thermodilution method, the control trials were performed immediately prior to Intralipid trials (opposite legs), and the subjects were fasted overnight, leading to high resting plasma [FFA] in the control trial (0.60 ± 0.07 vs. 0.25 ± 0.04 mM, Odland et al. [35]). Interestingly, no increase in FFA uptake was reported, and no reduction in muscle glycogen use occurred during knee-extensor exercise in response to elevated FFA. The knee-extensor model has increased muscle blood flow relative to power output compared to conventional dynamic exercise (i.e., cycling or running) and, combined with the relatively high control FFA levels, may account for the fact that Intralipid did not increase FFA uptake or decrease glycogen utilization.

Hargreaves et al. (22) did not measure muscle citrate, but no accumulation of G6P or glucose was associated with the reduced glucose uptake, and leg citrate release was similar between trials. The authors suggested that reduced glucose uptake was due to direct inhibition of the high FFA on glucose transport rather than by the classical G-FA cycle. If elevated plasma FFA exert a direct effect on glucose transport, this effect was not observed by Odland et al. (35) or during a similar protocol at a higher power output (85% $\dot{V}O_2$max) (45).

Control of Glycogen Phosphorylase

Independent of FFA-induced inhibition at the level of PFK or PDH (as suggested by the G-FA cycle), glycogen PHOS activity was lower, in order to account for the lower glycogenolysis in the Intralipid trial. PHOS exists in two interconvertible forms: a more active a and a less active b form. Transformation from PHOS b to a at the onset of exercise occurs primarily due to stimulation of PHOS kinase by increased cytoplasmic [Ca^{2+}] (8) and possibly to a minor extent by epinephrine (9). PHOS a then is regulated post-transformationally by free AMP, an allosteric modulator, and substrate availability (P_i and glycogen) (44).

The transformational state of PHOS was not measured in this study, but the levels of cytosolic Ca^{2+} and epinephrine were expected to be similar between trials as the power output was identical. In addition, previous measurements from this laboratory reported no change in PHOS transformation, despite a 45% sparing of muscle glycogen with elevated FFA during exercise at 85% $\dot{V}O_2$max (16, 17). The authors suggested that the reduction of PHOS activity was due largely to post-transformational regulation.

No difference existed in pre-exercise glycogen content between trials in the Odland et al. (35) study. The decreased accumulation of AMP during Intralipid and the tendency for decreased P_i may have reduced PHOS activity sometime between 20 and 70 min of exercise. The time course of the reduced glycogenolysis in the INT trial was not determined, since glycogen measurements were made only at rest and at the completion of exercise. As previously mentioned, RER and PDH transformation were reduced by 10 min of exercise, and lactate efflux was reduced during Intralipid at 18 and 34 min of exercise. These results suggest that reduced glycogenolysis and CHO use occurred, both early in exercise and well beyond 20 min during Intralipid. The mechanisms that may have reduced PHOS activity early in exercise are unknown. It may be that the magnitude of changes in the regulators of PHOS a were too small to detect.

Muscle Citrate and Phosphofructokinase Activity

The classic G-FA cycle suggests that increased fat oxidation should result in citrate-mediated inhibition of PFK, accumulation of G6P, and subsequent inhibition of glucose uptake (40, 41). In this study, glucose uptake was unaffected by Intralipid, but muscle citrate was slightly (but significantly) elevated at rest, and this increase remained during exercise. Since muscle G6P and glucose levels were not measured, it is difficult to assess whether PFK activity was down-regulated by the accumulation of muscle citrate or lower due to the reduced PHOS a activity higher up in the pathway. The lack of reduction in glucose uptake, however, suggests that either PFK and hexokinase activities were not inhibited, or, if enzyme activities were reduced by citrate, this reduction had no effect on glucose uptake.

Previous investigations of muscle citrate accumulation in response to increased FFA availability during exercise have produced mixed results. Studies involving high intensity exercise (75–85% $\dot{V}O_2max$) have consistently reported no effect of FFA elevation on muscle citrate levels, despite the occurrence of significant glycogen sparing (16, 17, 39, 52). Two earlier studies reported increased muscle citrate and decreased muscle glycogenolysis during cycle exercise at 65% $\dot{V}O_2max$ in response to five days of high-fat diet and aerobic training, respectively (24, 25). Research at lower power outputs (<65% $\dot{V}O_2max$) has failed to measure muscle citrate (28, 42), while several studies have reported increased muscle citrate with elevated FFA levels at rest (17, 24, 39, 52).

Citrate is produced in the mitochondria and in order to affect PFK must move to the cytoplasm. Unfortunately, muscle measurements of citrate are performed on whole tissue, and the proportion of citrate in the different cellular compartments is unknown. Therefore, an increase in total muscle citrate may not necessarily reflect increased citrate in the cytoplasmic compartment.

Early in vitro investigations apparently overestimated the potency of citrate to inhibit PFK activity during exercise, since increases in positive regulators may override its inhibitory effect (6, 37). Recent in vitro work reported that the citrate-mediated inhibition of PFK appears to be most powerful at rest, and that additional increases that occur in response to increased FFA availability are not expected to increase the inhibition of PFK at rest or during exercise (37). In the present study, we cannot conclusively state whether the small increases in citrate with high [FFA] decreased CHO metabolism. However, the recent in vitro data suggest that the elevated muscle citrate content would have little effect on PFK.

Control of Pyruvate Dehydrogenase Activity

In response to Intralipid infusion, the transformation of PDH to the active $\underline{a}$ form was reduced at all exercise time points, while the acetyl-CoA increase was similar in both conditions. PDH activity is regulated by a complex phosphorylation cycle. It is activated (to PDH$\underline{a}$) when dephosphorylated by PDH phosphatase, and inactivated by PDH kinase (phosphorylated). Several metabolites are purported to be regulators of the PDH complex (for review, see Reed [43]). Ca^{2+} is a potent activator of PDH phosphatase, while PDH kinase can be inhibited by pyruvate and $NAD^+/NADH$, and activated by ATP/ADP and acetyl-CoA/CoA.

The concept of the G-FA cycle is based partially on the premise that increased fat oxidation will result in accumulation of acetyl-CoA that in turn will activate PDH kinase and down-regulate PDH (40). Acetyl-CoA control of PDH activity appears to exist in humans at rest (5, 39), but several recent investigations have suggested that this potential regulatory effect is overridden by other factors during exercise (10, 16, 17, 39). Since acetyl-CoA increased at a time when PDH was transformed to the $\underline{a}$ form during exercise in both trials, our results support this concept.

There are other potential regulators of PDH that may account for the lower transformation of PDH to the active $\underline{a}$ form during INT. PDH phosphatase is known to be activated by Ca^{2+} (13), but Ca^{2+} was believed to be similar between conditions, since power outputs were the same in both trials. PDH kinase activity can be inhibited by ADP, which may accumulate in the mitochondria during exercise. The calculated accumulation of free ADP was reduced during INT only at 70 min of exercise and, therefore, unlikely to contribute early in exercise. Also, the intramitochondrial [ADP] is unknown, making it difficult to assess the potential effect of ADP on PDH transformation.

PDH kinase also can be inhibited by pyruvate and $NAD^+/NADH$ (38). Pyruvate levels were not measured, but the reduced glycogenolysis that occurred with elevated FFA would suggest that less pyruvate/unit time should have been available to PDH during INT. Potentially, a feed-forward mechanism could exist whereby reduced glycogenolytic flux may be directly linked to down-regulation of PDH$\underline{a}$. Evidence to support this proposal is found in a study that utilized dietary manipulations to alter plasma FFA levels and found reduced glycogenolysis, pyruvate content, and PDH$\underline{a}$ during cycle exercise at 75% $\dot{V}O_2max$ following a low CHO diet (39). Direct measurements of muscle pyruvate in the two conditions will be needed to examine this possibility.

Evidence from in vitro studies suggest that a decrease in the $NAD^+/NADH$ concentration ratio would result in a decrease in transformation of PDH to the active $\underline{a}$ form (38). Direct measurement of the mitochondrial reduction-oxidation (redox) state of the $NAD^+/NADH$ couple in skeletal muscle during contraction, however, has produced inconsistent results. Using the glutamate dehydrogenase reaction, Graham and Saltin (20) reported a substantial increase in estimated mitochondrial redox state during cycle exercise in humans (75% and 100% $\dot{V}O_2$max). In contrast, Duhaylongsod et al. (15) monitored redox state in canine gracilis muscle using near-infrared spectroscopy during electrical stimulation and found the redox state to decrease at all workloads. Furthermore, NADH measured directly in human muscle has been shown to decrease at 45% $\dot{V}O_2$max, and to increase at higher power outputs (49). No attempt was made to estimate the mitochondrial redox state in the present study, but it is possible that NADH accumulation may be increased by elevated availability of fat during Intralipid. This would result in reduced PDH<u>a</u> transformation if NADH exerts a significant regulatory effect on PDH kinase.

Thus, recent investigations confirm previous findings that elevated FFA levels result in significant sparing of muscle glycogen during moderate intensity exercise. Although regulation of the CHO/lipid interaction was partially as classically proposed, other regulatory factors must also exist. No difference in glucose uptake was present between trials, thus regulation of CHO use occurred at the level of glycogen PHOS, PFK, and/or PDH. Reduced flux through PHOS with high [FFA] may have been due to reduced free AMP and P_i accumulation that occurred later in exercise. PFK flux may have been reduced by increased cytoplasmic citrate, but more likely was the result of decreased substrate provision. PDH transformation to the more active <u>a</u> form during exercise was reduced following Intralipid infusion, but the decrease was not due to increased acetyl-CoA accumulation. The mechanisms for the reduced transformation of PDH are unknown but may be due to reduced pyruvate and/or increased NADH accumulation during exercise with Intralipid infusion.

Skeletal Muscle Malonyl-CoA Content During Low and Moderate Intensity Exercise

Odland et al. (34, 36) measured the M-CoA content in human skeletal muscle of men and women to determine the time course of changes during low and moderate intensity cycle exercise. Pulmonary gases were measured to determine RER and estimate lipid and CHO fuel use during exercise. In study I, muscle biopsies were taken at rest and following 10 min at 40% $\dot{V}O_2$max and 10 and 60 min at 65% $\dot{V}O_2$max. In study II, muscle biopsies were taken at rest and 1 and 10 min of exercise in one trial at 35% $\dot{V}O_2$max and at the same time points in a second trial at 65% $\dot{V}O_2$max.

The resting M-CoA content in study I was 1.53 ± 0.18 μmol/kg/dm (figure 19.6). During exercise, it was unchanged at 1.39 ± 0.21 μmol/kg dm after 10 min at 40% $\dot{V}O_2$max. Following 10 and 60 min at 65% $\dot{V}O_2$max, M-CoA was 1.46 ± 0.14 and 1.22 ± 0.15 μmol/kg/dm, respectively. These values were not different from the resting or 40% $\dot{V}O_2$max M-CoA contents. ATP levels remained constant throughout exercise, while PCr decreased during cycling at 65% $\dot{V}O_2$max. There was no change in muscle lactate during exercise at 40% $\dot{V}O_2$max, while it increased after 10 min at 65% $\dot{V}O_2$max and decreased to near resting level by the end of exercise.

RER increased significantly from 4–6 min (40% $\dot{V}O_2$max) to 14–16 min (65% $\dot{V}O_2$max) concomitant with the increase in power output at 10 min (table 19.2).

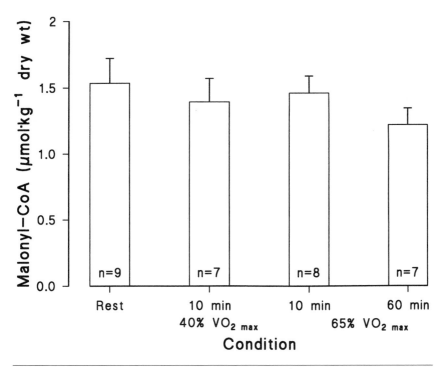

Figure 19.6.   Malonyl-CoA content in human skeletal muscle at rest, and after 10 min at 40% $\dot{V}O_2$max and 60 min at 65% $\dot{V}O_2$max.
Reprinted, by permission, from Odland et al. 1996.

Table 19.2   Power Output (Percentage of $\dot{V}O_2$max), RER and Fat Utilization at Various Time Points During Prolonged Submaximal Cycle Exercise

Time (min)	% $\dot{V}O_2$max	RER	Fat (g/min)
4–6	41.1 ± 0.7	0.88 ± 0.02*	0.33 ± 0.05
14–16	64.3 ± 2.2	0.95 ± 0.02	0.20 ± 0.10
30–32	67.8 ± 2.5	0.91 ± 0.02*‡	0.39 ± 0.09*
46–48	69.2 ± 2.1	0.91 ± 0.02*‡	0.42 ± 0.08*
64–66	69.1 ± 2.7	0.89 ± 0.02*	0.50 ± 0.13*‡

Values are means ± SE. Exercise consisted of 10 min at 40% $\dot{V}O_2$max followed by 60 min at 65% $\dot{V}O_2$max.
*, significant differences from the 14–16 min time point; ‡, significant differences from the 4–6 min time point.

RER decreased during 60 min at 65% $\dot{V}O_2$max such that the 30–32, 46–48, and 64–66 min values were lower than at 14–16 min. Similarly, the calculated rate of fat oxidation increased during the 65% $\dot{V}O_2$max power output (table 19.2). Fat oxidation rates at 30–32, 46–48, and 64–66 min were greater than the 14–16 min rate and the 64–66 min rate was also greater than at 4–6 min.

In study II, resting muscle M-CoA varied among subjects and ranged from 0.78 to 3.41 µmol/kg dm, but the mean data was similar at 35% and 65% $\dot{V}O_2$max, 1.85 ± 0.29 and 1.85 ± 0.46 µmol/kg/dm (figure 19.7). M-CoA decreased from rest at 1 min during exercise at 35% $\dot{V}O_2$max but returned to the rest level by 10 min. No decrease in M-CoA content occurred at 1 or 10 min of exercise at 65% $\dot{V}O_2$max. Muscle ATP remained constant during exercise at 35% and 65% $\dot{V}O_2$max. PCr was decreased after 1 and 10 min of cycling at both power outputs, and lactate was increased following 1 and 10 min at 65% $\dot{V}O_2$max.

RER was 0.84 ± 0.02 at 35% and 0.92 ± 0.02 at 65% $\dot{V}O_2$max. The calculated fat oxidation rates were 0.24 ± 0.04 g/min at 35% and 0.18 ± 0.03 g/min at 65% $\dot{V}O_2$max.

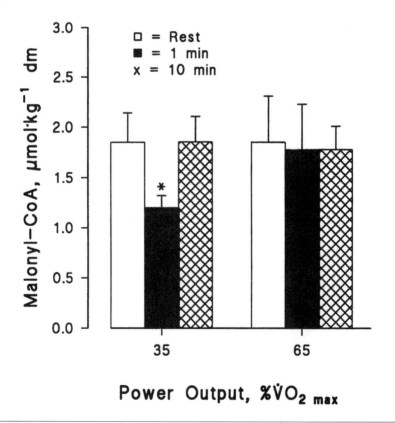

Figure 19.7.   Malonyl-CoA content in human skeletal muscle at rest and during exercise at 35% and 65% $\dot{V}O_2$max. Values are means ± SE. *, significantly different from rest and 10 min. Reprinted, by permission, from Odland et al. 1998.

The primary purpose of these investigations was to measure M-CoA in human skeletal muscle biopsies at rest and during low and moderate intensity exercise. The RER data demonstrated that exercise at 35–40% and 65% $\dot{V}O_2$max was associated with a considerable contribution of fat utilization to total energy production. M-CoA ranged from 0.68 to 3.41 μmol/kg dm at rest, but generally did not change in response to exercise at varying time points and power outputs. An exception was at 35% $\dot{V}O_2$max, where M-CoA fell significantly at 1 min but returned to resting level by 10 min. M-CoA also decreased to 84% of the initial content during 60 min of cycling at 65% $\dot{V}O_2$max, but this decrease was not statistically significant.

Malonyl-CoA and Intramuscular Fuel Selection

Observed decreases in M-CoA content during exercise have led to the suggestion that it is a regulator of fat-CHO interaction in rodent skeletal muscle at rest and during exercise (32, 48, 57). It has also been postulated to play the same role in contracting human skeletal muscle, although M-CoA was not measured in these investigations (26, 51). Two of the main regulatory enzymes responsible for the choice of oxidative substrates are PDH and CPT I, which control the entry of CHO and FFA into the mitochondria, respectively. M-CoA has been shown to be a potent inhibitor of CPT I activity in rat and human skeletal muscle in vitro (3, 30, 47). Because its concentration may be influenced by acetyl-CoA, which is an end product of the PDH reaction and a substrate for M-CoA formation, alterations in M-CoA could provide a mechanism to control the balance between fat and CHO oxidation. During low and moderate submaximal exercise, when fat provides a significant proportion of fuel for contraction, a reduction in M-CoA may relieve the inhibition of CPT I and allow more FFA into the mitochondria for oxidation.

The human muscle M-CoA data do not support the correlation between exercise-induced decreases in M-CoA content and increases in fatty acid transport and oxidation. Consequently, the data suggest that M-CoA is not a primary regulator of fatty acid oxidation during exercise and/or M-CoA levels are not predictive of the fatty acid oxidation rate in human skeletal muscle during low and moderate intensity exercise. It should be noted that there was a 16% decrease in M-CoA from 0 to 60 min of cycling at 65% $\dot{V}O_2$max. While this decrease was not statistically significant, it is not known whether this change is biologically significant. In addition, the time course of this 16% decrease is unknown, except that it occurred sometime between 0 and 60 min. A more detailed time-course analysis is required to answer this question.

The early decline in M-CoA at 35% $\dot{V}O_2$max may be explained by fiber-type recruitment. In rodent skeletal muscle, M-CoA content was reported to vary among different fiber types and to correlate with mitochondrial content of muscle fibers (58). In resting rats, [M-CoA] in the white region of the quadriceps (fast-glycolytic) was only 28% of the deep red quadriceps content (slow-oxidative). Furthermore, during treadmill running at 21 m/min up a 15% grade, the most rapid decrease in rodent muscle M-CoA occurred in the red quadriceps, while the slowest decrease occurred in white quadriceps (58). While regions of rat skeletal muscle are homogeneous with respect to fiber type, human skeletal muscle is comprised of ~50% type I, 40% IIa, and 10% type IIb fibers (50). The M-CoA content within specific human fiber types is not known, but the contents of many metabolites in human type I and II fibers are similar (11, 21, 54). However, if there was specific recruitment of mainly type I fibers during exercise at 35% $\dot{V}O_2$max, a decrease in the M-CoA content of ~67% in these fibers would be required to account for the early ~33% decrease in total muscle M-CoA at this power output.

Fiber-type recruitment, however, cannot explain the return of M-CoA to resting levels by 10 min of exercise at 35% $\dot{V}O_2$max. It is possible that the early decline at 35% "primes" the fatty acid oxidation machinery at the onset of exercise, and that the mechanism responsible for the early decline at 35% $\dot{V}O_2$max is overridden by other regulators over time and at higher power outputs. In addition, there is the possibility that the M-CoA-induced inhibition of CPT I is maximal at rest such that further increases are not necessary during high intensity exercise.

Malonyl-CoA Interaction With CPT I

M-CoA content did not correlate with fat utilization during exercise, but the possibility of M-CoA-induced inhibition of CPT-I and subsequent regulation of fatty acid oxidation cannot be dismissed. McGarry et al. (30) determined that the I_{50} (concentration of M-CoA required for 50% inhibition of enzyme activity) for human semitendinosis muscle CPT I was 0.025 μM. The calculated concentration of M-CoA measured in the human studies ranged between 0.2 and 0.9 μM which, if located entirely in the cytoplasm, should be sufficient to completely inhibit CPT I. However, it is possible that only a fraction of the total M-CoA content measured is available to interact with CPT I in the functioning muscle cell. While it is expected that the bulk of hepatic M-CoA is present extramitochondrially (30), the distribution of M-CoA in skeletal muscle is uncertain. It has been established that propionyl-CoA carboxylase (present in muscle mitochondria) can nonspecifically convert acetyl-CoA to M-CoA within the mitochondria (23), suggesting that some of the M-CoA present in human skeletal muscle may be mitochondrial. Thus, the possibility exists that overall measurements of tissue M-CoA are not predictive of the concentration of M-CoA that interacts with CPT I.

It must also be remembered that the I_{50} determination was in vitro (30), and it is highly probable that other regulators are present in vivo, which could reduce the M-CoA-induced inhibition of CPT I independent of a change in M-CoA content. For example, cytoplasmic acetyl-CoA, free CoA, acetyl-carnitine, and pH have been suggested as potential regulators that affect the sensitivity of CPT I to inhibition by M-CoA (33, 59).

In summary, human skeletal muscle M-CoA content remained remarkably constant during exercise at varying time points and power outputs. Regardless of the mechanism(s) regulating M-CoA formation and/or degradation, M-CoA content was not correlated with fatty acid oxidation rate. Therefore, if M-CoA is involved in CPT I regulation during exercise at 35-40% and 65% $\dot{V}O_2$max, it is not primarily due to decreases in its concentration. In conclusion, other regulators may be present during exercise that interact with the inhibition of CPT I activity by M-CoA in human skeletal muscle.

Summary

Intramuscular fuel selection during low and moderate exercise in humans appeared to be regulated at several sites in the pathways of CHO and lipid metabolism. In the presence of increased FFA provision, the following occurred:

1. Muscle glycogen utilization was reduced but glucose uptake was not affected.
2. Reduced accumulations of free AMP and P_i during prolonged moderate intensity exercise were correlated with reduced flux though PHOS.
3. A reduction in substrate provision could explain the decreased flux through PFK, although small increases in muscle citrate also may have contributed.
4. The reduction in PDH transformation to the more active a form was reduced, implying a lower PDH flux.

The mechanism(s) for the reduced transformation is unknown but was not due to increased acetyl-CoA accumulation.

In response to exercise of low and moderate intensity, human skeletal muscle M-CoA content generally was unaffected and not related to the fatty acid oxidation rate, suggesting that the regulation of skeletal muscle CPT I activity must be more complex than simply to control via malonyl-CoA.

References

1. Asmussen, E. Muscle metabolism during exercise in man. A historical survey. In *Muscle Metabolism During Exercise*. Ed. B. Pernow and B. Saltin, 1-2. New York: Plenum Press; 1971.
2. Awan, M.M., and Saggerson, E.D. Malonyl-CoA metabolism in cardiac myocytes and its relevance to the control of fatty acid oxidation. *Biochemical Journal* 295: 61-66; 1993.
3. Berthon, P.M., Howlett, R.A., Heigenhauser, G.J.F., and Spriet, L.L. Human skeletal muscle carnitine palmitoyltransferase I activity determined in intact isolated mitochondria. *Journal of Applied Physiology* 85: In press; 1998.
4. Bianchi, A., Evans, J.L., Iverson, A.J., Norlund, A., Watts, T.D., and Witters, L.A. Identification of an isozymic form of acetyl-CoA carboxylase. *Journal of Biological Chemistry* 265: 1502-1509; 1990.
5. Boden, G., Jadali, F., White, J., Liang, Y., Mozzoli, M., Chen, X., Coleman, E., and Smith, C. Effects of fat on insulin-stimulated carbohydrate metabolism in normal men. *Journal of Clinical Investigation* 88: 960-966; 1991.
6. Bosca, L., Aragon, J.J., and Sols, A. Modulation of muscle phosphofructokinase at physiological concentration of enzyme. *Journal of Biological Chemistry* 260: 2100-2107; 1985.

7. Bremer, J. Carnitine metabolism and functions. *Physiological Reviews* 63: 1420-1480; 1983.

8. Chasiotis, D., Sahlin, K., and Hultman, E. 1982. Regulation of glycogenolysis in human skeletal muscle at rest and during exercise. *Journal of Applied Physiology* 53: 708-715; 1982.

9. Chesley, A., Hultman, E., and Spriet, L.L. Effects of epinephrine infusion on muscle glycogenolysis during intense aerobic exercise. *American Journal of Physiology* 268: E127-E134; 1995.

10. Constantin-Teodosiu, D., Cederblad, G., and Hultman, E. Acetyl group accumulation and pyruvate dehydrogenase activity in human muscle during incremental exercise. *Acta Physiologica Scandinavica* 143: 367-372; 1991.

11. Constantin-Teodosiu, D., Howell, S., and Greenhaff, P.L. Carnitine metabolism in human muscle fiber types during submaximal dynamic exercise. *Journal of Applied Physiology* 80: 1061-1064; 1996.

12. Costill, D.L., Coyle, E., Dalsky, G., Evans, W., Fink, W., and Hoops, D. Effects of elevated plasma FFA and insulin on muscle glycogen usage during exercise. *Journal of Applied Physiology* 43: 695-699; 1977.

13. Denton, R.M., Randle, P.J., and Martin, B.R. Stimulation by calcium ions of pyruvate dehydrogenase phosphate phosphatase. *Biochemical Journal* 128: 161-163; 1972.

14. Duan, C., and Winder, W.W. Nerve stimulation decreases malonyl-CoA in skeltal muscle. *Journal of Applied Physiology* 72: 901-904; 1992.

15. Duhaylongsod, F.G., Griebel, J.A., Bacon, D.S., Wolfe, W.G., and Piantadosi, A. Effects of muscle contraction on cytochrome a,a3 redox state. *Journal of Applied Physiology* 75: 790-797; 1993.

16. Dyck, D.J., Putman, C.T., Heigenhauser, G.J.F., Hultman, E., and Spriet, L.L. Regulation of fat-carbohydrate interaction in skeletal muscle during intense aerobic cycling. *American Journal of Physiology* 265: E852-E859; 1993.

17. Dyck, D.J., Peters, S.A., Wendling, P.S., Chesley, A., Hultman, E., and Spriet, L.L. Regulation of muscle glycogen phosphorylase activity during intense aerobic cycling with elevated FFA. *American Journal of Physiology* 265: E116-E125; 1996.

18. Garland, P.B., Randle, P.J., and Newsholme, E.A. Citrate as an intermediary in the inhibition of phosphofructokinase in rat heart by fatty acids, ketone bodies, pyruvate, diabetes, and starvation. *Nature* 200: 169-170; 1963.

19. Garland, P.B., and Randle, P.J. Regulation of glucose uptake by muscle: 10. Effects of alloxan-diabetes, starvation, hypophysectomy and adrenolectomy, and of fatty acids, ketone bodies and pyruvate on the glycerol output and concentrations of free fatty acids, long-chain fatty acyl-coenzyme A, glycerol phosphate and citrate-cycle intermediates in rat heart and diaphragm. *Biochemical Journal*: 678-687; 1964.

20. Graham, T.E., and Saltin, B. Estimation of mitochondrial redox state in human skeletal muscle during exercise. *Journal of Applied Physiology* 66: 561-566; 1989.

21. Greenhaff, P.L., Ren, J.-M., Soderlund, K., and Hultman, E. Energy metabolism in single human muscle fibers during contraction without and with epinephrine infusion. *American Journal of Physiology* 260: E713-E718; 1991.

22. Hargreaves, M., Kiens, B., and Richter, E.A. Effect of plasma free fatty acid concentration on muscle metabolism in exercising men. *Journal of Applied Physiology* 70: 194-210; 1991.

23. Hulsmann, W.C. On the synthesis of malonyl-Coenzyme A in rabbit-ear sarcomeres. *Biochimica Biophysica Acta* 125: 398-400; 1966.

24. Jannson, E., and Kaijser, L. Leg citrate metabolism at rest and during exercise in relation to diet and substrate utilization in man. *Acta Physiologica Scandinavica* 122: 145-153; 1984.

25. Jannson, E., and Kaijser, L. Substrate utilization and enzymes in skeletal muscle of extremely endurance-trained men. *Journal of Applied Physiology* 62: 999-1005; 1987.

26. Jeukendrup, A.E., Saris, W.H.M., Brouns, F., Halliday, D., and Wagenmakers, A.J.M. Effects of carbohydrate (CHO) and fat supplementation on CHO metabolism during prolonged exercise. *Metabolism* 45: 915-921; 1996.

27. Jorfeldt, L., and Wahren, J. Leg blood flow during exercise in man. *Clinical Science* 41:459-473; 1971.

28. Knapik, J.J., Meridith, C.N., Jones, B.H., Suek, L., Young, V.R., and Evans, W.J. Influence of fasting on carbohydrate and fat metabolism during rest and exercise in men. *Journal of Applied Physiology* 64: 1923-1929; 1988.

29. McGarry, J.D., Leatherman, G.F., and Foster, D.W. 1978. Carnitine palmitoyltransferase I. *Journal of Biological Chemistry* 253: 4128-4136; 1978.

30. McGarry, J.D., Mills, S.E., Long, C.S., and Foster, D.W. Observations on the affinity for carnitine, and malonyl-CoA sensitivity, of carnitine palmitoyltransferase-I in animal and human tissues. *Biochemical Journal* 214: 21-28; 1983.

31. McGarry, J.D., Sen, A., Esser, V., Woeltje, K.F., Weis, B.C., and Foster, D.W. New insights into the mitochondrial carnitine palmitoyltransferase enzyme system. *Biochimie* 73: 77-84; 1991.

32. McGarry, J.D., and Brown, N.F. The mitochondrial carnitine palmitoyltransferase system. From concept to molecular analysis. *European Journal of Biochemistry* 244: 1-14; 1997.

33. Mills, S.E., Foster, D.W., and McGarry, J.D. Effects of pH on the interaction of substrates and malonyl-CoA with mitochondrial carnitine palmitoyltransferase I. *Biochemical Journal* 219: 601-608; 1984.

34. Odland, L.M., Heigenhauser, G.J.F., Lopaschuk, G.D., and Spriet, L.L. Human skeletal muscle malonyl-CoA at rest and during prolonged submaximal exercise. *American Journal of Physiology* 270: E541-E544; 1996.

35. Odland, L.M., Heigenhauser, G.J.F., Wong, D., Hollidge-Horvat, M.G., and Spriet. L.L. Effects of increased fat availability on fat-carbohydrate interaction during prolonged aerobic exercise in humans. *American Journal of Physiology* 275: R894-R902; 1998.

36. Odland, L.M., Howlett, R.A., Heigenhauser, G.J.F., Hultman, E., and Spriet, L.L. Skeletal muscle malonyl-CoA at the onset of exercise at varying power outputs in humans. *American Journal of Physiology* 275: E1080-E1085; 1998.

37. Peters, S.J., and Spriet. L.L. Skeletal muscle phosphofructokinase activity examined under physiological conditions in vitro. *Journal of Applied Physiology* 78: 1853-1858; 1995.

38. Pettit, F.H., Pelley, J.W., and Reed, L.J. Regulation of pyruvate dehydrogenase kinase and phosphatase by acetyl-CoA/CoA and NADH/NAD+ ratios. *Biochemical and Biophysical Research Communications* 65: 575-582; 1975.

39. Putman, C.T., Spriet, L.L., Hultman, E., Lindinger, M.I., Lands, L.C., McKelvie, R.S., Cederblad, G., Jones, N.L., and Heigenhauser, G.J.F. Pyruvate dehydrogenase activity and acetyl-group accumulation during exercise after different diets. *American Journal of Physiology* 265: E752-E760; 1993.

40. Randle, P.J., Hales, C.N., Garland, P.B., and Newsholme, E.A. The glucose fatty-acid cycle. Its role in insulin sensitivity and the metabolic disturbances of diabetes mellitus. *Lancet* 1: 785-789; 1963.

41. Randle, P.J., Newsholme, E.A., and Garland, P.B. Regulation of glucose uptake by muscle: 8. Effects of fatty acids, ketone bodies and pyruvate, and of alloxan-diabetes and starvation, on the uptake and metabolic fate of glucose in rat heart and diaphragm muscles. *Biochemical Journal* 93: 652-665; 1964.

42. Ravussin, E., Bogardus, C., Scheidegger, K., LaGrange, B., Horton, E.D., and Horton, E.S. Effect of elevated FFA on carbohydrate and lipid oxidation during prolonged exercise in humans. *Journal of Applied Physiology* 60: 893-900; 1986.

43. Reed, L.J. Regulation of pyruvate dehydrogenase complex by a phosphorylation-dephosphorylation cycle. *Current Topics of Cellular Regulation* 8: 95-106; 1981.

44. Ren, J.M., and Hultman, E. Regulation of phosphorylase-a activity in human skeletal muscle. *Journal of Applied Physiology* 69: 919-923; 1990.

45. Romijn, J.A., Coyle, E., Sidossis, L.S., Zhang, X.J., and Wolfe, R.R. Relationship between fatty acid delivery and fatty acid oxidation during strenuous exercise. *Journal of Applied Physiology* 79: 1939-1945; 1995.

46. Saddik, M., Gamble, J., Witters, L.A., and Lopaschuk, G.D. Acetyl-CoA carboxylase regulation of fatty acid oxidation in the heart. *Journal of Biological Chemistry* 268: 25,836-25,845; 1993.

47. Saggerson, E.D., and Carpenter, C.A. Carnitine palmitoyltransferase and carnitine octanoyltransferase activities in liver, kidney cortex, adipocyte, lactating mammary gland, skeletal muscle and heart. *FEBS Letters* 129: 229-232; 1981.

48. Saha, A.K., Kurowski, T.G., and Ruderman, N.B. A malonyl-CoA fuel-sensing mechanism in muscle: effects of insulin, glucose, and denervation. *American Journal of Physiology* 32: E283-E289; 1995.

49. Sahlin, K., Katz, A., and Henriksson, J. Redox state and lactate accumulation in human skeletal muscle during dynamic exercise. *Biochemical Journal* 245: 551-556; 1987.

50. Schantz, P., Randall-Fox, E., Hutchison, W., Tyden, A., and Astrand, P.-O. Muscle fibre type distribution, muscle cross-sectional area and maximal voluntary strength in humans. *Acta Physiologica Scandinavica* 117: 219-226; 1983.

51. Sidossis, L.S., Gastaldelli, A., Klein, S., and Wolfe, R.R. Regulation of plasma fatty acid oxidation during low- and high-intensity exercise. *American Journal of Physiology* 272: E1065-E1070; 1997.

52. Spriet, L.L., MacLean, D.A., Dyck, D.J., Hultman, E., Cederblad, G., and Graham, T.E. Caffeine ingestion and muscle metabolism during prolonged exercise in humans. *American Journal of Physiology* 262: E891-E898; 1992.
53. Spriet, L.L., and Dyck, D.J. The glucose-fatty acid cycle in skeletal muscle at rest and during exercise. In *Biochemistry of Exercise IX*. Ed. R. Maughan, A. Huxley, and C. Williams, 127-155. Champaign, IL: Human Kinetics; 1996.
54. Tesch, P., and Karlsson, J. Effects of exhaustive, isometric training on lactate accumualtion in different muscle fiber types. *International Journal of Sports Medicine* 5: 89-91; 1984.
55. van der Vusse, G.J., and Reneman, R.S. Lipid metabolism in muscle. In *Handbook of Physiology: Exercise: Regulation and integration of multiple systems*. (sec.12, pp. 952-994) New York: Oxford Press; 1996.
56. Vukovich, M.D., Costill, D.L., Hickey, M.S., Trappe, S.W., Cole, K.J., and Fink, W.J. Effect of fat emulsion and fat feeding on muscle glycogen utilization during cycle exercise. *Journal of Applied Physiology* 75: 1513-1518; 1993.
57. Winder, W.W., Arogyasami, J., Barton, R.J., Elayan, I.M., and Vehrs, P.B. 1989. Muscle malonyl-CoA decreases during exercise. *Journal of Applied Physiology* 67: 2230-2233; 1989.
58. Winder, W.W., Arogyasami, J., Elayan, I.M., and Cartmill, D. Time course of the exercise-induced decline in malonyl-CoA in different muscle types. *American Journal of Physiology* 259: E266-E271; 1990.
59. Zierz, S., and Engel, A.G. Different sites of inhibition of carnitine palmitoyltransferase by malonyl-CoA, and by acetyl-CoA and CoA, in human skeletal muscle. *Biochemical Journal* 245: 205-209; 1987.

CHAPTER 20

Fat Oxidation During Exercise: Role of Lipolysis, FFA Availability, and Glycolytic Flux

Edward F. Coyle

The Human Performance Laboratory, Department of Kinesiology and Health Education, The University of Texas at Austin, Austin, TX, USA

Although it long has been recognized that fat and carbohydrate are the two primary substrates oxidized during exercise, much remains to be learned about the factors that regulate fat oxidation. Fat oxidation by skeletal muscle during exercise is derived predominantly from either the vast store of triglyceride in adipose tissue or from triglyceride contained directly within muscle fibers (i.e., intramuscular triglyceride; IMTG) (11). Another potential source of fat, plasma triglyceride seem to contribute little to fat oxidation during exercise because chylomicron triglyceride hydrolysis by lipoprotein lipase and uptake by muscle during exercise appears relatively low (24, 26). However, plasma triglyceride probably serve primarily to replenish IMTG after exercise.

This review will discuss the conditions during which the delivery of fatty acids (FA) into the mitochondria of skeletal muscle may limit fat oxidation during exercise. The amount of FA in plasma or in the sarcoplasm is small, and thus their turnover must be high in order to deliver significant FA to the mitochondria for oxidation. Therefore, triglyceride hydrolysis (i.e., lipolysis) theoretically establishes the upper limit for fat oxidation during exercise. Not all of the FA liberated in lipolysis is oxidized due to re-esterification or limitations at various steps in transporting FA. This includes the transport of FA out of adipose tissue, through the circulation of plasma or from capillary to mitochondria (38). Finally, the transport of FA through the mitochondrial membrane can limit fat oxidation (8). Evidence is accumulating that this latter process is sensitive to carbohydrate metabolism in general and glycolytic flux in particular (8).

This review supports the idea that fat oxidation in skeletal muscle during exercise is largely regulated by carbohydrate metabolism. It is interesting that situations that shift substrate oxidation from fat toward carbohydrate, such as pre-exercise carbohydrate meals and increasing exercise intensity, appear to exert well-coordinated effects both on adipose tissue and skeletal muscle (8, 21, 18, 29, 34). It seems that the various systems of the body coordinate to minimize the appearance of FA that cannot be oxidized by skeletal muscle during exercise.

Substrate Mobilization and Oxidation
During Exercise of Increasing Intensity

Figure 20.1 describes the pattern with which endurance-trained subjects oxidize carbohydrate and fat during exercise of increasing intensity, performed after an overnight fast (29). Constant rate infusion of stable isotopes and determination of isotopic ennrichment, allowed calculation of the rates of plasma FFA appearance (RaFFA) and disappearance (RdFFA). The FA oxidation in excess of plasma FFA disappearance is calculated to be that derived from IMTG with the assumption that oxidation of other triglyceride stores during exercise (i.e., plasma triglyceride) is relatively small and that RdFFA is fully oxidized. As shown (figure 20.1), the turnover (Rd) of FFA in plasma declines as the intensity of exercise increases from low (i.e., 25% $\dot{V}O_2$max, comparable to walking), to moderate (i.e., 65% $\dot{V}O_2$max, comparable to a running pace for 2–4 hours), to high intensity (i.e., 85% $\dot{V}O_2$max, the highest pace maintained for 30–60 min) (figure 20.1). Furthermore, this decline in plasma FFA turnover (i.e., both Ra and Rd) was directly reflected in plasma FFA concentrations after 30 min of exercise (figure 20.2). Measurements were not made at 45% $\dot{V}O_2$max for comparison to 25% $\dot{V}O_2$max to identify the exercise intensity with the highest rate of RdFFA. Interestingly, RaFFA and plasma FFA concentrations increased abruptly with the cessation of exercise at 85% $\dot{V}O_2$max and to a lesser extent after 65% $\dot{V}O_2$max with little perturbation after 25% $\dot{V}O_2$max (figure 20.2). In that this influx of FFA into plasma upon exercise cessation was not associated with increased lipolysis, it would seem to reflect entry into plasma of FA trapped in adipose tissue during exercise, possibly due to inadequate adipose tissue blood flow (2, 33). If indeed the magnitude of the post-exercise increase in plasma FFA concentration reflects the relative FA trapping, it appears to be directly related to exercise intensity. The assumed greater trapping of FA in adipose tissue during high-intensity compared with low-intensity exercise (i.e., 85% vs. 25% $\dot{V}O_2$max), despite higher rates of lipolysis during high-intensity exercise (29), seems to prevent the mobilization into plasma of FA under conditions during which their oxidation by skeletal muscle is reduced, as discussed below.

Figure 20.1 indicates that the increase in total fat oxidation when intensity is increased from 25% to 65% $\dot{V}O_2$max appears to be derived from increased IMTG oxidation in these endurance-trained subjects. At 25% $\dot{V}O_2$max, total FA oxidation is matched to RdFFA, whereas at 65% $\dot{V}O_2$max, RdFFA can account for approximately one-half of total FA oxidation, with IMTG presumably providing the remainder. Total fat oxidation is reduced at 85% compared to 65% $\dot{V}O_2$max in association with large increases in glycolytic flux, which, as discussed below, may be a mechanism by which carbohydrate metabolism actively regulates FA oxidation.

Lipolysis During Exercise

At rest the "triglyceride-FA substrate cycle" is high, with only about 30% of the FA resulting from lipolysis appearing in plasma, whereas the remainder is re-esterified

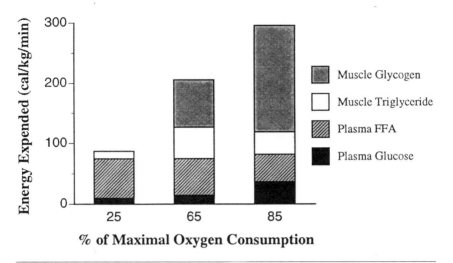

Figure 20.1.   Contribution of the four major substrates to energy expenditure after 30 min of exercise at 25%, 65%, and 85% of maximal oxygen uptake when fasted.
Reprinted, by permission, from Romijn et al. 1993.

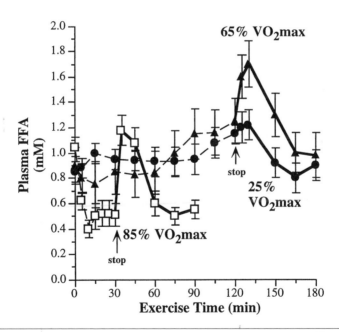

Figure 20.2.   Plasma FFA concentration during and following exercise at 25%, 65%, and 85% of maximal oxygen uptake when fasted. Note the magnitude of increase in plasma FFA when exercise is stopped, especially after exercise at 85% and 65% of maximal oxygen uptake.
Reprinted, by permission, from Romijn et al. 1993.

(40). Therefore, resting lipolysis greatly exceeds RaFFA and total FA oxidation (3, 20, 40). However, re-esterification is reduced markedly during exercise, and thus the difference between lipolysis and fat oxidation is reduced (40). During exercise in the fasted state, lipolysis exceeds fat oxidation by 10–30% depending upon the intensity of exercise and oxidative ability of the subjects (18, 22).

Lipolysis in adipose tissue as well as IMTG is controlled by hormone sensitive lipase, which is activated by beta-adrenergic receptor stimulation during exercise, largely via the action catecholamines in general, particularly epinephrine (1, 6). Lipolysis of IMTG also may be influenced by metabolic processes within the contracting muscle fibers (26, 36). These processes may be responsible for the apparent increase in IMTG oxidation during exercise at 65% compared to 25% $\dot{V}O_2$max (figure 20.1). Insulin is by far the most potent antipolytic hormone (4, 6). The lowering of insulin below resting basal levels during exercise may serve to further release inhibition of lipolysis. At rest, alpha-adrenergic receptor stimulation inhibits adipose tissue lipolysis, but this system exerts little influence on lipolysis during exercise (1). Therefore, in a simplistic sense, lipolysis during exercise appears to be activated largely through catecholamines and intramuscular metabolism and inhibited by the action of insulin.

Lipolysis Can Limit Fat Oxidation During Exercise

During exercise in the fasted state, which is accompanied by low insulin levels, whole-body lipolysis exceeds and thus does not appear to limit fat oxidation. Studies that have given subjects nicotinic acid, an antilipolytic agent in adipose tissue, have observed reduced fat oxidation during exercise (15). This observation indicates that reductions in lipolysis are associated with reduced fat oxidation, but it did not establish quantitatively that fat oxidation was reduced to the level of lipolysis (15). Until recently, studies have not directly measured lipolysis during exercise following administration of antilipolytic agents to quantitatively compare lipolysis and fat oxidation. Recognizing that lipolysis is very sensitive to insulin, Horowitz et al. (18) recently directly determined that lipolysis is reduced markedly during exercise after a carbohydrate meal to a level where it appears to equal and thus limits fat oxidation. As shown in figure 20.3, during exercise (44% $\dot{V}O_2$peak) in the fasted state, lipolysis exceeded fat oxidation by 13%. However, when exercise was performed 60 min after ingesting glucose to increase insulin, fat oxidation was suppressed to the point where it equaled lipolysis (figure 20.3). During a third trial, also performed after glucose ingestion (i.e., GLUCOSE + HIGH LIPOLYSIS), lipolysis was exogenously increased by intravenous infusion of triglycerides and heparin; the latter releases LPL to hydrolyze the infused triglyceride in plasma. Interestingly, this stimulation of lipolysis increased fat oxidation by 30% during exercise after carbohydrate ingestion (figure 20.3), suggesting that lipolysis did indeed limit fat oxidation. However, it should be noted that the increase in lipolysis did not restore fat oxidation to the levels observed during exercise in the fasted state. This implies that carbohydrate ingestion has additional effects on skeletal muscle that directly reduce its ability to oxidize fat. As discussed below, one factor appears to be a reduction in the transport of FA into the mitochondrial.

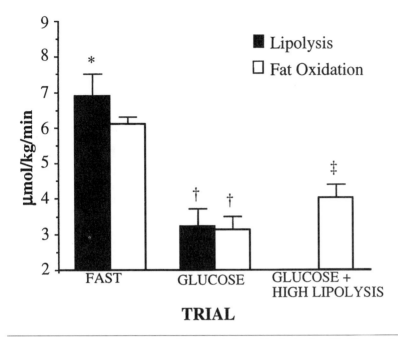

Figure 20.3.   Comparison of fat oxidation relative to lipolysis after 20–30 min of exercise at 44% of peak oxygen consumption. During exercise in the fasted state, lipolysis exceeded fat oxidation (* indicates $P < 0.05$). During exercise after ingesting glucose 60 min. before exercise, both lipolysis and fat oxidation are significantly reduced († indicates $P < 0.05$). The increase in lipolysis via intravenous infusion of Intralipid and heparin, after glucose ingestion, resulted in a significant increase in fat oxidation compared to glucose (‡ indicates $P < 0.05$), but not a restoration to the fasted levels. Reprinted, by permission, from Horowitz et al. 1997.

Plasma FFA Availability

Under conditions when lipolysis exceeds fat oxidation during exercise, fat oxidation still might be limited by impaired entry of the liberated FA to the mitochondria. As discussed above, high-intensity exercise seems to be a good example of this situation with high lipolysis but very low RaFFA and plasma FFA concentrations (9, 21, 30). We have found well-trained cyclists to oxidize FA at 43 μmol/kg/min after 30 min at 65% $\dot{V}O_2$max compared with only 27–30 μmol/kg/min when exercising at 85% $\dot{V}O_2$max (29, 30). In that the restoration of plasma FFA concentration to 1–2 mM during exercise at 85% $\dot{V}O_2$max increased fat oxidation to 34 μmol/kg/min during exercise at 85% $\dot{V}O_2$max, it seems that approximately one-half of the normal decline in fat oxidation when intensity is increased from 65% to 85% $\dot{V}O_2$max is due to limited availability of plasma FFA for oxidation by muscle. However, the failure of fat oxidation to reach the higher levels capable by muscle (i.e., 43 μmol/kg/min at 65% $\dot{V}O_2$max), indicates that fat oxidation at 85%

$\dot{V}O_2$max is not merely limited by plasma FFA supply. As discussed below, intracellular factors such as increased glycolytic flux may directly impair FA oxidation in muscle.

During exercise at both 85% $\dot{V}O_2$max and at 44% $\dot{V}O_2$max (after glucose ingestion), we have seen that the restoration of plasma FFA concentration to 1–2 mM reduced carbohydrate oxidation due solely to a sparing of muscle glycogen utilization, with no apparent reduction in blood glucose disappearance (18, 30). This glycogen sparing effect agrees with the original findings of Costill et al. (7), who had subjects run at 70% $\dot{V}O_2$max, 5–6 hours after a carbohydrate meal, which resulted in a very low plasma FFA concentration during the control trial (i.e., 0.2 mM). Ingestion of heavy cream and heparin injection raised fat oxidation 32% and reduced muscle glycogen use by 40%. Therefore, it seems that fat oxidation is somewhat impaired and muscle glycogen use increased during moderate- to high-intensity exercise when plasma FFA concentration is only 0.2–0.3 mM and that total fat oxidation can be increased somewhat by raising plasma FFA concentration (7, 9, 30). However, in studies with plasma FFA during the control trial in the range of 0.5–1.0 mM, it seems that further elevation does not raise fat oxidation or spare muscle glycogen use (16, 28, 30). This is probably because plasma FFA availability did not limit total fat oxidation in the control conditions.

Mitochondrial Factors and Endurance Training

Fat oxidation occurs in mitochondria, which, of course, have great potential to influence fat oxidation during exercise. The increased mitochondrial density characteristic of endurance training results in increased fat oxidation and reduced glycolytic flux from both muscle glycogen and blood glucose (5, 17, 19, 35). The reduction in glycogenolysis during exercise at a given absolute intensity is associated with a lesser disturbance of cellular homeostasis (i.e., lesser increase in AMP and ADP) and increased oxidation of IMTG (17, 19, 25). The extent to which endurance training increases fat oxidation directly by increasing mitochondrial density, as opposed to the indirect effect of increased mitochondrial density on reducing glycolytic flux, is not clear.

During low-intensity exercise eliciting an oxygen consumption of 20 ml/kg/min, endurance-trained and untrained subjects have comparable levels of whole-body lipolysis as well as plasma FFA Ra and Rd (22). Despite this comparable stimulation of FA mobilization, the untrained subjects display lower fat oxidation. Thus, during exercise at this low intensity, the endurance-trained subjects show a close matching between RdFFA and total fat oxidation, whereas the untrained subjects display reduced fat oxidation despite high plasma FFA Rd as well as a higher plasma FFA concentration. This suggests that during low-intensity exercise in untrained subjects, FA availability does not limit fat oxidation. It is likely that mitochondrial factors limit fat oxidation. In contrast to low-intensity exercise at a given absolute intensity, exercise at 70% $\dot{V}O_2$max in endurance-trained compared to untrained subjects elicits higher rates of lipolysis and total fat oxidation (23). The higher lipolytic rate probably is due, for the most part, to greater IMTG lipolysis (19, 25,

29), and it is not surprising that total fat oxidation is increased because total oxidative metabolism is increased.

Glycolytic Flux and Fat Oxidation

We tested the hypothesis that fat oxidation is regulated by carbohydrate metabolism in general and muscular glycolytic flux in particular (8). Pre-exercise carbohydrate feedings were used as a tool to produce hyperglycemia and hyperinsulinemia and thus to increase glycolytic flux during exercise (8). As shown in figure 20.4, FA oxidation rates were measured during constant rate intravenous infusion of trace amounts of a long-chain fatty acid (1-^{13}C-palmitate) vs. a medium-chain fatty acid (1-^{13}C-octanoate). Octanoate oxidation is not as limited by transport into mitochondria as is palmitate oxidation (14, 32). We observed that the increased glycolytic flux from glucose ingestion significantly reduced palmitate oxidation, whereas it had no effect on octanoate oxidation. This suggests that glycolytic flux regulates long-chain fatty acid oxidation in skeletal muscle during exercise, possibly by inhibiting its transport into the mitochondria. A similar experimental approach, using increased exercise intensity to raise glycolytic flux, found similar results (34).

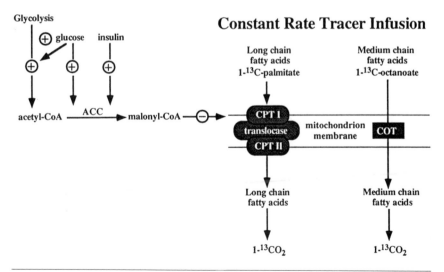

Figure 20.4. Scheme of the experimental approach used by Coyle et al. (8) to determine if long-chain fatty acid oxidation (i.e., palmitate) is reduced more than medium-chain fatty acid oxidation (i.e., octanoate) by elevations in plasma glucose and insulin. It is hypothesized that the formation of malonyl-CoA from acetyl-CoA through the activity of acetyl-CoA carboxylase (ACC) is a potent inhibitor of carnitine palmitoyltransferase (CPT) with much less of an effect on carnitine octanoyltransferase.

Glycolytic flux may regulate fat oxidation within exercising skeletal muscle via the formation of malonyl-CoA (10, 39). Malonyl-CoA is formed from acetyl-CoA through the activity of acetyl-CoA carboxylase (ACC), and it is a potent inhibitor of carnitine palmitoyltransferase (CPT) with much less of an effect on carnitine octanoyltransferase (COT) (31, 32) (figure 20.4). It is hypothesized that malonyl-CoA concentration can reflect the availability of carbohydrate as a substrate with increased glycolytic flux producing more pyruvate and thus increasing acetyl-CoA/CoA and malonyl-CoA, which then reduces fatty acid oxidation by reducing transport of long-chain fatty acids into mitochondria via inhibition of CPT.

It seems that increases in glycolytic flux increase carbohydrate oxidation while directly and actively reducing FA oxidation. If the reduction in FA oxidation following glucose ingestion were simply a passive phenomenon from increased glycolytic flux and acetyl-CoA production from glucose, the oxidation of palmitate versus octanoate should have been equal in response to increased glycolytic flux. This did not occur. Therefore, it seems that the preference for carbohydrate oxidation when both carbohydrate and fat are made available to muscle is mediated, in part, by the active inhibition of fat oxidation via a process that appears to involve transport of FA into the mitochondria.

Summary

Situations that shift substrate oxidation from fat toward carbohydrate, such as pre-exercise carbohydrate meals, and increasing exercise intensity, appear to exert well-coordinated effects both on adipose tissue and skeletal muscle. High-intensity exercise (85% vs. 65% $\dot{V}O_2$max) reduces plasma FFA mobilization in concert with a direct reduction in FA oxidation within skeletal muscle (figures 20.1 and 20.2) (29). Furthermore, lipolysis appears to limit fat oxidation when exercise is performed following a pre-exercise meal, most likely from the inhibition of lipolysis by insulin (18). However, increased lipolysis (via intralipid and heparin infusion) after pre-exercise carbohydrate ingestion only partially restores fat oxidation toward fasting conditions (figure 20.3) (18). Increased glycolytic flux from pre-exercise carbohydrate ingestion also appears to actively and directly inhibit the oxidation of FA in skeletal muscle, possibly by inhibiting FA transport into mitochondria (figure 20.4) (8). It appears that various systems of the body somehow act in concert to minimize the appearance of FA that cannot be oxidized by skeletal muscle during exercise when glycolytic flux is increased.

References

1. Arner, P., Kriegholm, E., Engfeldt, P., and Bolinder, J. Adrenergic regulation of lipolysis in situ at rest and during exercise. *J. Clin. Invest.* 85: 893-898; 1990.

2. Bülow, J. Subcutaneous adipose tissue blood flow and triacylglycerol-mobilization during prolonged exercise in dogs. *Pflügers Arch* 392: 230-234; 1982.

3. Bonadonna, R.C., Groop, L.C., Zych, K., Shank, M., and DeFronzo, R.A. Dose-dependent effect of insulin on plasma free fatty acid turnover and oxidation in humans. *Am. J. Physiol.* 259: E736-E750; 1990.

4. Campbell, P.J., Carlson, M.G., Hill, J.O., and Nurjhan, N. Regulation of free fatty acid metabolism by insulin in humans: role of lipolysis and reesterification. *Am. J. Physiol.* 26: E1063-E1069; 1992.

5. Coggan, A.R., Kohrt, W.M., Spina, R.J., Bier, D.M., and Holloszy, J.O. Endurance training decreases plasma glucose turnover and oxidation during moderate-intensity exercise in men. *J. Appl. Physiol.* 68: 990-96; 1990.

6. Coppack, S.W., Jense, M.D., and Miles, J.M. In vivo regulation of lipolysis in humans. *J. Lipid Res.* 35: 77-193; 1994.

7. Costill, D.F., Coyle, E.F., Dalsky, G., Evans, W., Fink, W., and Hoopes, D. Effects of elevated plasma FFA and insulin on muscle glycogen usage during exercise. *J. Appl. Physiol.* 43: 695-699; 1977.

8. Coyle, E.F., Jeukendrup, A.E., Wagenmakers, A.J.M., and Saris, W.H.M. Fatty acid oxidation is directly regulated by carbohydrate metabolism during exercise. *Am. J. Physiol.* 273 (Endocrinol. Metab. 36): E268-E275; 1997.

9. Dyck, D.J., Peters, S.A., Wendling, P.S., Chesley, A., Hultman, E., and Spriet, L.L. Regulation of muscle glycogen phosphorylase activity during intense aerobic cycling with elevated FFA. *Am. J. Physiol.* 265 (Endocrinol. Metab. 33): E116-E125; 1996.

10. Elayan, I.M., and Winder, W.W. Effect of glucose infusion on muscle malonyl-CoA during exercise. *J. Appl. Physiol.* 70(4): 1495-1499; 1991.

11. Essen, B., Hagenfeldt, L., and Kaijser, L. Utilization of blood-borne and intramuscular substrates during continuous and intermittent exercise in man. *J. Physiol.* 265: 489-506; 1977.

12. Garland, P.B., and Randle, P.J. Regulation of glucose uptake by muscle. Effects of alloxan-diabetes, starvation, hypophysectomy and adrenalectomy, and of fatty acids, ketone bodies and pyruvate, on the glycerol output and concentrations of free fatty acids, long-chain fatty acyl-coenzyme A, glycerol phosphate and citrate-cycle intermediates in rat heart and diaphragm muscles. *Biochem. J.* 93: 678-687; 1964.

13. Garland, P.B., and Randle, P.J. Control of pyruvate dehydrogenase in the perfused rat heart by the intracellular concentration of acetyl-coenzyme A. *Biochem. J.* 91: 6C-7C; 1964.

14. Gollnick, P.D., Pernow, B., Essen, B., Jansson, E., and Saltin, B. Availability of glycogen and plasma FFA for substrate utilization in leg muscle of man during exercise. *Clin. Physiol.* 1: 27-42; 1981.

15. Groot, P.H., and Hulsmann, W.C. The activation and oxidation of octanoate and palmitate by rat skeletal muscle mitochondria. *Biochim Biophys Acta* 316(2): 124-135; 1973.

16. Hargreaves, M., Kiens, B., and Richter, E.A. Effect of increased plasma free fatty acid concentrations on muscle metabolism in exercising men. *J. Appl. Physiol.* 70(1): 194-201; 1991.

17. Holloszy, J.O., and Coyle, E.F. Adaptations of skeletal muscle to endurance exercise and their metabolic consequences. *J. Appl. Physiol.* 56(4): 831-838; 1984.

18. Horowitz, J.F., Mora-Rodriguez, R., Byerley, L.O., and Coyle, E.F. Lipolytic suppression following carbohydrate ingestion limits fat oxidation during exercise. *Am. J. Physiol.* 273 (Endocrinol. Metab. 36): E768-E775; 1997.

19. Hurley, B.F., Nemeth, P.M., Martin, W.H., Hagberg, J.M., Dalsky, G.P., and Holloszy, J.O. Muscle triglyceride utilization during exercise: effect of training. *J. Appl. Physiol.* 60: 562-567; 1986.

20. Jensen, M.D., Caruso, M., Heiling, V., and Miles, J.M. Insulin regulation of lipolysis in nondiabetic and IDDM subjects. *Diabetes* 38: 1595-1601; 1989.

21. Jones, N.L., Heigenhauser, G.J.F., Kuksis, A., Matsos, C.G., Sutton, J.R., and Toews, C.J. Fat metabolism in heavy exercise. *Clin. Sci.* 59: 469-478; 1980.

22. Klein, S., Coyle, E.F., and Wolfe, R.R. Fat metabolism during low-intensity exercise in endurance-trained and untrained men. *Am. J. Physiol.* 267 (Endocrinol. Metab. 30): E934-E940; 1994.

23. Klein, S., Weber, J.-M., Coyle, E.F., and Wolfe, R.R. Effect of endurance training on glycerol kinetics during strenuous exercise in humans. *Metab.* 45: 357-361; 1996.

24. Mackie, B.G., Dudley, G.A., Kaciuba-Uscilko, H., and Terjung, R.L. Uptake of chylomicron triglycerides by contracting skeletal muscle in rats. *J. Appl. Physiol.* 49: 851-855; 1980.

25. Martin, W.H., Dalsky, G.P., Hurley, B.F., Matthews, D.E., Bier, D.M., Hagberg, J.M., and Holloszy, J.O. Effect of endurance training on plasma FFA turnover and oxidation during exercise. *Am. J. Physiol.* 265 (Endocrinol. Metab. 28): E708-E714; 1993.

26. Oasci, L.B., Essig, D.A., and Palmer, W.K. Lipase regulation of muscle triglyceride hydrolysis. *J. Appl. Physiol.* 69: 1571-1577; 1990.

27. Randle, P.J., Newsholme, E.A., and Garland, P.B. Regulation of glucose uptake by muscle. Effects of fatty acids, ketone bodies and pyruvate, and of alloxan-diabetes and starvation, on the uptake and metabolic fate of glucose in rat heart and diaphragm muscles. *Biochem. J.* 93: 652-664; 1964.

28. Ravussin, E., Bogardus, C., Scheidegger, K., LaGrange, B., Horton, E.D., and Horton, E.S. Effect of elevated FFA on carbohydrate and lipid oxidation during prolonged exercise in humans. *J. Appl. Physiol.* 60: 893-900; 1986.

29. Romijn, J.A., Coyle, E.F., Sidossis, L.S., Gastaldelli, A., Horowitz, J.F., Endert, E., and Wolfe, R.R. Regulation of endogenous fat and carbohydrate metabolism in relation to exercise intensity. *Am. J. Physiol.* 265 (Endocrinol. Metab. 28): E380-E391; 1993.

30. Romijn, J.A., Coyle, E.F., Sidossis, L.S., Zhang, X.J., and Wolfe, R.R. Relationship between fatty acid delivery and fatty acid oxidation during strenuous exercise. *J. Appl. Physiol.* 79: 1939-1945; 1995.

31. Saddik, M., Gamble, J., Witters, L.A., and Lopaschuk, G.D. Acetyl-CoA carboxylase regulation of fatty acid oxidation in the heart. *J. Biol. Chem.* 268: 25,836-25,845; 1993.

32. Saggerson, E.D., and Carpenter, C.A. Carnitine palmitoyltransferase and carnitine octanoyltransferase activities in liver, kidney cortex, adipocyte, lactating mammary gland, skeletal muscle and heart. *FEBS Letters* 129: 229-232; 1981.

33. Samra, J.S., Simpson, E.J., Clark, M.L., Forster, C.D., Humphreys, S.M., Macdonald, I.A., and Frayn, K.N. Effects of epinephrine infusion on adipose tissue: interactions between blood flow and lipid metabolism. *Am. J. Physiol.* 271: E834-E839; 1996.

34. Sidossis, L.S., Gastaldelli, A., Klein, S., and Wolfe, R.R. Regulation of plasma fatty acid oxidation during low- and high-intensity exercise. *Am. J. Physiol.* 272: E1065-1070; 1997.

35. Spina, R.J., Chi, M.M., Hopkins, M.G., Nemeth, P.M., Lowry, O.H., and Holloszy, J.O. Mitochondrial enzymes increase in muscle in response to 7-10 days of cycle exercise. *J Appl Physiol.* 80(6): 2250-2254; 1996.

36. Spriet, L.L., Heigenhauser, G.J., and Jones, N.L. Endogenous triacylglycerol utilization by rat skeletal muscle during tetanic stimulation. *J. Appl. Physiol.* 60(2): 410-415; 1986.

37. Terjung, R.L., Mackie, B.G., Dudley, G.A., and Kaciuba-Uscilko, H. Influence of exercise on chylomicron triacylglycerol metabolism: plasma turnover and muscle uptake. *Med. Sci. Sports Exerc.* 15: 340-347; 1983.

38. Van der Vusse, G.J., and Reneman, R.S. Lipid metabolism in muscle. In *Handbook of Physiology. Section 12: Exercise: Regulation and Integration of Multiple Systems,* chapter 21, 952-994. Bethesda, MD: Am. Physiol. Soc.; 1996.

39. Winder, W.W., Arogyasami, J., Barton, R.J., Elayan, I.M., and Vehrs, P.R. Muscle malonyl-CoA decreases during exercise. *J. Appl. Physiol.* 67(6): 2230-2233; 1989.

40. Wolfe, R.R., Klein, S., Carraro, F., and Weber, J.M.. Role of triglyceride-fatty acid cycle in controlling fat metbolism in humans during and after exercise. *Am. J. Physiol.* 258 (Endocrinol. Metab. 21): E382-E389; 1990.

CHAPTER 21

Energy Metabolism of Skeletal Muscle Fiber Types and the Metabolic Basis of Fatigue in Humans

Paul L. Greenhaff, Anna Casey, Dumitru Constantin-Teodosiu, and Kostas Tzintzas*

Department of Physiology and Pharmacology, The Medical School, University of Nottingham, Nottingham, UK; *Department of Physical Education, Sports Science and Recreation Management, University of Loughborough, Loughborough, UK

The vast majority of studies to date aimed at investigating the metabolic basis of fatigue in human skeletal muscle have involved biochemical analysis of homogenates of muscle biopsy samples or the collection of 31^P magnetic resonance spectroscopy scans before, during, and after fatiguing exercise. Based on the results of such studies, conclusions have been drawn about the principal metabolic factors responsible for fatigue under most exercise conditions. However, it is also clear from the literature that the propulsive muscles of man are composed of at least three major fiber types, which differ widely in their physiological, metabolic, and contractile characteristics (5, 15, 17), a point that cannot be considered using the techniques described above. It is perhaps therefore premature to believe that conclusions drawn from the analysis of such heterogeneous tissue will offer a clear insight into the metabolic basis of fatigue in exercising man. For example, we have demonstrated that during 30 s of intense electrically evoked intermittent isometric contraction, the rate of glycogenolysis measured in mixed-fiber biopsy samples obtained from the vastus lateralis muscle was 1.47 + 0.66 mmol.s^{-1}.kg^{-1} dry material (dm) (18), which is in good agreement with other literature describing the metabolic responses of mixed-fibered skeletal muscle during maximal exercise under a variety of conditions (9, 25). However, closer inspection, by undertaking quantitative measurements on fragments of muscle fibers dissected (and characterized) from the same biopsy samples, revealed marked differences between fiber types, with the rate of glycogenolysis being negligible (0.18 + 0.14 mmol.s^{-1}.kg^{-1} dm) in type I fibers but extremely rapid (3.54 + 0.53 mmol.s^{-1}.kg^{-1} dm) in type II fibers. Indeed, the rate of glycogenolysis was close to the V_{max} of glycogen phosphorylase measured in type II fibers by Harris et al. (22). This demonstrates clearly that the metabolic responses observed in a muscle biopsy sample as a whole are not representative of the metabolic responses occurring at the single-fiber level but reflect the average metabolic response occurring across fiber types. Even when circulation was occluded in an attempt to accelerate glycogenolysis in type I fibers, under the same experimental conditions the rate of glycogenolysis was still markedly lower in the type I fibers (2.05 + 0.70 mmol.s^{-1}.kg^{-1} dm) compared with type II fibers (4.32 mmol.s^{-1}.kg^{-1} dm), despite glycogenolysis proceeding maximally in both fiber types (19). This demonstrates that metabolic differences between fiber types will be dependent not only on such factors as recruitment but also on inherent metabolic and functional differences between fiber types.

Another example of mixed-fibered tissue analysis clouding data interpretation is illustrated in figure 21.1. The upper figure shows the change in mixed-fibered muscle acetylcarnitine during 30s of maximal isokinetic cycling exercise and during 4 min of subsequent recovery. The lower figure shows the same measurements made in different fiber types. As can be seen, during exercise there was an accumulation of

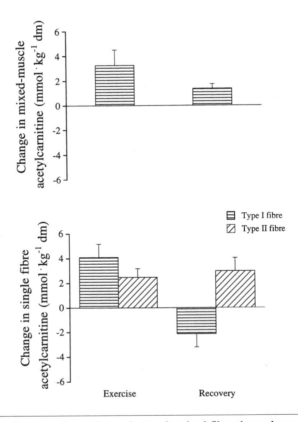

Figure 21.1. The upper figure shows changes in mixed-fibered muscle acetylcarnitine concentration during 30 s of maximal isokinetic cycling exercise and during 4 min of subsequent recovery. The lower figure shows the same measurements made in type I and type II muscle fibers dissected from the same biopsy samples used for mixed-fibered measurements. Values represent mean + SEM.

acetylcarnitine in both fiber types and, reassuringly, the mean of the increases equaled that observed in the mixed-fibered tissue. However, during recovery, contrasting responses were observed when comparing fiber types, which were not discernible from the mixed-fibered measurements. Clearly, therefore, the mixed-fibered measurements were not representative of the true metabolic responses.

In contrast to the above examples, scenarios also exist where one might predict that differences would exist between fiber type metabolic responses, but upon examination no differences are found. Such a scenario is illustrated in figure 21.2, which shows muscle-free carnitine, acetylcarnitine and total carnitine in type I and type II muscle fibers at rest and immediately following prolonged cycling exercise to exhaustion at ~75% of maximal oxygen consumption (10). One would have predicted that the total carnitine content of type I fibers would be somewhat greater than that of type II fibers, due to its widely reported role in mitochondrial fat transport (16). However, as figure 21.2 shows, no fiber type differences in total carnitine

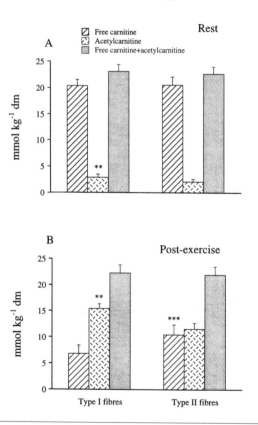

Figure 21.2. Free carnitine, acetylcarnitine and their sum in human type I and II muscle fibers at rest (upper figure) and after exhaustive exercise at ~75% of maximal oxygen consumption (lower figure). dm, dry muscle. Values represent mean + SEM. Significant differences between fiber types: ** P < 0.01; *** P < 0.001.

Reprinted, by permission, from Constantin-Teodosiu et al. 1996.

concentration were evident, which was probably attributable to carnitine also having an important role in buffering excess acetyl group accumulation in both fiber types during exercise, which has not been widely reported. As one might have predicted, acetylcarnitine accumulation was greater in the type I fibers during exercise, which was probably a reflection of the greater mitochondrial content of this fiber type.

Based on the above discussion, the intention of this review is to highlight recent studies from this laboratory that have been aimed at investigating the metabolic responses of different muscle fiber types in an attempt to offer further insight into the mechanisms responsible for the development of fatigue during exercise in human subjects.

Methods to Study Energy Metabolism in Fiber Types During Exercise

It is appropriate to question why so many previous studies have been devoted to investigating the metabolic responses of mixed-fibered muscle tissue during exercise, when such clear differences exist between fiberer-type responses to exercise. The answer to this question is simply that the generation of quantitative data relating to muscle fiber-type responses is labor intensive and time consuming. Attempts have been made to study fiber-type responses using histochemical techniques (26, 27, 35); the benefit being such methods are less labor intensive. However, it should be recognized that the array of substrate and metabolite measurements available using histochemical analysis is not extensive and that the methods that are available are at best semiquantitative. For example, Vollestad et al. (35) investigated the resynthesis of glycogen in different muscle fiber types following prolonged, submaximal exercise in man using photometric determination of PAS stain intensity. In brief, the authors demonstrated that glycogen resynthesis was significantly less during the initial three hours of recovery in type I fibers when compared with type II fibers. However, employing biochemical analysis of glycogen in fragments of muscle fibers, we have found the opposite response (6). This could not be attributed to notable differences between studies in the extent of glycogen depletion achieved during exercise or to either the timing or composition of the glucose solutions administered at the end of exercise, but did meet with known fiber-type differences in glucose uptake and disposal. We therefore concluded that a likely reason for the lack of agreement between studies was related to the analytical techniques employed. Namely, a relatively poor linear relationship ($r = 0.80$) has been shown to exist between PAS stain intensity and muscle glycogen concentration, with residual variance due to unmeasured or random factors accounting for 35% of the total variance (35). This contrasts with biochemical determination of fiber-type glycogen concentration, where excellent agreement is found between mixed-fibered muscle measurements and the mean concentrations determined in type I and type II fibers from the same biopsy sample ($r = 0.96$, 6).

Another widely used approach to investigate fiber-type responses during exercise is to sample muscle tissues that are composed of exclusively one type. In practice this is almost impossible to achieve in the case of human propulsive muscles, but it can be achieved using animal models. As table 21.1 shows, the human vastus lateralis muscle is composed of a mixture of fiber types. However, by sampling, for example, the rat extensor digitorum longus (EDL) and soleus muscles before and after electrically evoked contraction, metabolic data relating to fast and slow fibers can be obtained (13). A word of caution, however. While such experiments can offer an enormous amount of information, it would be presumptuous to relate this information directly to exercising humans. For example, table 21.2 shows there to be a number of differences between rat soleus and EDL muscles and human type I and type II fibers. In particular, the muscle glycogen store is at least threefold lower in the rat, and the phosphocreatine (PCr) concentration of rat soleus is about 50% of that measured in human type I fibers. Both of these observations could account for the relatively low anaerobic ATP production rates measured in rat slow (~1.8 mmol.s^{-1}.kg^{-1} dm), fast oxidative (~4 mmol.s^{-1}.kg^{-1} dm), and fast glycolytic (~7 mmol.s^{-1}.kg^{-1} dm) muscles when compared with human type I (~6 mmol.s^{-1}.kg^{-1} dm) and type II (~12 mmol.s^{-1}.kg^{-1} dm) fibers under similar

Table 21.1   Fiber Type Composition of Human and Rat Skeletal Muscle Samples

| Species | Muscle | Fiber types (%) | | | |
		I	IIa	IIb	IIc
Man	Vastus lateralis	41	39	20	0
Rat	EDL	2	24	74	0
Rat	Soleus	96	4	0	0

(From 13.)

Table 21.2   Resting Muscle Metabolite Concentrations in Human Vastus Lateralis and Rat Soleus and Extensor Digitorum Longus (EDL) Muscles

	Rat Soleus	Human I	Rat EDL	Human II
ATP	19.7 ± 2.3	23.7 ± 0.7	28.0 ± 1.0	25.2 ± 0.6
ADP	3.18 ± 0.25	4.25 ± 1.05	3.03 ± 0.33	3.70 ± 0.89
AMP	0.16 ± 0.07	0.32 ± 0.10	0.19 ± 0.09	0.33 ± 0.12
PCr	47.3 ± 6.6	79.4 ± 2.1	82.9 ± 6.5	89.6 ± 5.2
ATP/ADP	6.3 ± 0.8	5.52 ± 0.8	9.4 ± 1.4	6.1 ± 1.7
Glycogen	113 ± 42	364 ± 23	133 ± 4	480 ± 24

All data are mmol kg^{-1} dm.
(From 13, 19, 28, 31.)

experimental conditions of maximal electrically evoked contraction without circulation (12, 19, 32, 36). Indeed, we have been unable to find any data demonstrating that rat soleus muscle can accumulate anything but moderate amounts of lactate, even under the most extreme experimental conditions. Furthermore, it has also been demonstrated that the activities of many of the enzymes involved with the regulation of energy metabolism are markedly greater in rat slow and fast muscles when compared with the corresponding measurements in human muscle fiber types (table 21.3), which seems somewhat surprising given the lower anaerobic capacity of rat skeletal muscle outlined above.

There is an abundance of illustrations in the literature demonstrating differences between rat and human skeletal muscle metabolism. It is not, however, the intention of this review to highlight these differences but rather to illustrate that, although the rat model may be a useful experimental tool because it enables muscle energy metabolism to be studied from a mechanistic approach, it is somewhat flawed if attempts are then made to relate these findings to exercising humans.

Table 21.3   Phosphorylase (Phosph.), Phosphofructokinase (PFK), Citrate Synthase (CS), and Succinate Dehydrogenase (SDH) Activities Measured in Rat Slow Oxidative (SO), Fast Oxidative (FOG), and Fast Glycolytic (FG) Fibers and in Human Type I, Type IIa, and Type IIb Fibers

	Rat SO	Human I	Rat FOG	Human IIa	Rat FG	Human IIb
Phosph.	58.8	11.8	483.0	24.4	718.2	37.0
PFK	88.2	31.5	289.8	57.5	420.0	73.5
CS	96.6	45.4	172.2	36.1	37.8	27.3
SDH	29.4	29.8	37.8	20.2	21.0	10.5

All data are mmol min^{-1} kg^{-1} dm, from 30.

The Metabolic Basis of Fatigue During Maximal Exercise in Humans With Respect to Muscle Fiber Types

As previously stated, the quadriceps muscle group in man is not homogeneous, and it is therefore not unreasonable to suggest that the analysis of mixed-fibered muscle biopsy samples will not truly reflect the metabolic and physiological responses that occur in human skeletal muscle during exercise. In one particular study (31), muscle biopsy samples were obtained from the vastus lateralis at rest and after 10 and 20 s of intermittent electrical stimulation (1.6 s stimulation, 1.6 s rest, at a frequency of 50 Hz) with muscle blood flow intact. The concentrations of ATP and PCr then were measured in type I and II muscle fibers dissected from each biopsy sample and are depicted in figure 21.3. During the initial 10 s of contraction, the rate

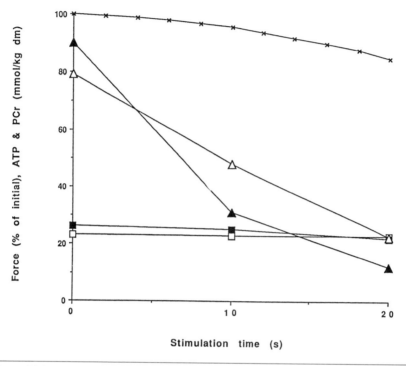

Figure 21.3.   Muscle isometric force production (x) and ATP (□) and PCr (△) concentrations in type I (open symbols) and type II (closed symbols) muscle fibers during 20 s of intense electrical stimulation (1.6 s stimulation, 1.6 s rest; 50 Hz) in man. Reprinted from Greenhaff et al. 1997.

of PCr degradation was 60% greater in type II fibers. However, during the remaining 10 s of contraction, PCr degradation fell in type II fibers by 60%, such that there was no difference in the rate of PCr degradation between fiber types in this period.

We also have investigated glycogenolysis in different muscle fiber types during contraction in man under a variety of experimental conditions, e.g., electrical stimulation (1.6 s stimulation, 1.6 s rest for 30 s) with open circulation (18), electrical stimulation with open circulation and adrenaline infusion (18), electrical stimulation with occluded circulation (19), and during maximal sprint exercise (21). The results are depicted in figure 21.4. In short, it would appear that during intense contraction with circulation intact, glycogenolysis was already occurring maximally in type II fibers, such that when attempts were made to stimulate glycogenolysis further using adrenaline infusion and circulatory occlusion no increase in glycogenolysis was observed. This suggestion is supported by the finding that the rates of glycogenolysis observed were close to the V_{max}. of phosphorylase measured in type II fibers (22). Glycogenolysis was negligible in type I fibers during electrically evoked contraction with blood flow intact. This probably was attributable to the comparatively low anaerobic ATP turnover rate of this fiber type during contraction and the oxidative resynthesis of ATP during each 1.6 s of recovery between each contraction. The latter suggestion

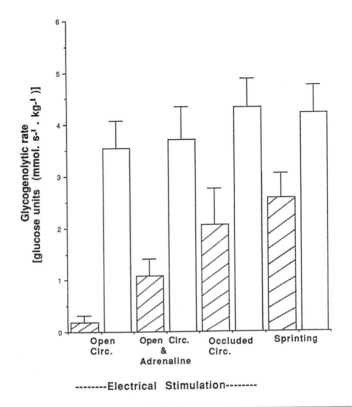

Figure 21.4.   Glycogenolytic rates in type I (hatched bars) and type II (open bars) fibers during 30 s of intermittent electrical stimulation at 50 Hz with open circulation (circ.), open circulation with adrenaline infusion, occluded circulation, and during 30 s of maximal sprint running in man.

Reprinted from Greenhaff et al. 1997.

has been confirmed recently by an experiment in which the interval between each contraction was reduced to 0.8 s, resulting in a rapid increase in type I fiber glycogenolysis, i.e., the oxidative rephosphorylation of ADP between each contraction was reduced resulting in AMP accumulation and the activation of glycogen phosphorylase (Greenhaff et al., unpublished observation). Similarly, when blood flow was occluded and sprint exercise was performed, glycogenolysis also was dramatically increased in type I fibers (figure 21.4), presumably by the same mechanism. This conclusion is supported by the dramatic decline in type I fiber ATP concentrations in the latter studies (19, 21) and in the unpublished study of Greenhaff et al. It is logical that the increase in glycogenolysis is in type I fibers following adrenaline infusion occurred as a direct consequence of the activation of phosphorylase by adrenaline (18).

The question remains whether the biochemical changes observed in the above series of studies offer any explanation for the development of fatigue during intense contraction in man. It is highly likely that fatigue development is a multi-factorial process. However, the findings presented above suggest that during near maximal intensity exercise ATP is supplied initially by maximal rates of PCr degradation and glycogenolysis, particularly in type II muscle fibers. As exercise progresses, it would appear that the rate of anaerobic ATP turnover declines due to lack of availability of PCr and a reduced rate of glycogenolysis. In short, therefore, an imbalance between the rate of ATP demand and ATP provision occurs that could be the mechanism responsible for the development of fatigue. Support for this suggestion comes in part from experiments where repeated bouts of maximal exercise have been performed with short recovery periods between exercise bouts (3, 7). These studies have shown a significant relationship exists between the extent of PCr resynthesis between exercise bouts and subsequent exercise performance. Indeed, we have recently demonstrated that when 2 bouts of 30 s maximal, isokinetic cycling exercise were performed, separated by 4 min of recovery, the extent of PCr resynthesis during recovery was positively correlated with work output during the second bout of exercise ($r = 0.8$, $n = 9$, $P < 0.05$, 7). Furthermore, in agreement with the suggestion that the depletion of PCr specifically in type II muscle fibers may be primarily responsible for fatigue, it was demonstrated that the rate of PCr hydrolysis during the first bout of exercise was 35% greater in type II fibers compared with type I fibers. However, during the second bout of exercise the rate of PCr hydrolysis declined by 33% in type II fibers (figure 21.5),

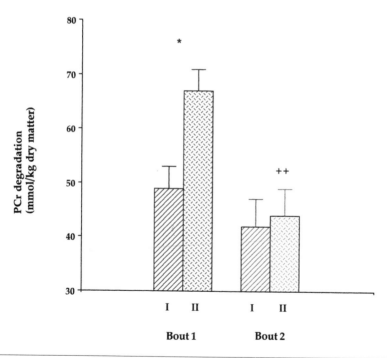

Figure 21.5.   Changes in phosphocreatine (PCr) in type I and type II muscle fibers during two bouts of 30 s maximal intensity, isokinetic cycling exercise in man. Each bout of exercise was performed at 80 rev/min and separated by 4 min of passive recovery. * indicates significant differences between fiber types ($p < 0.05$) and ++ indicates significantly different from exercise bout 1 ($p < 0.01$).
Reprinted, by permission, from Casey et al. 1996.

which was attributable to incomplete resynthesis of PCr in this fiber type during recovery. Conversely, PCr resynthesis was almost complete in type I fibers during recovery from exercise bout 1, and PCr utilization was unchanged in this fiber type during exercise bout 2 (figure 21.5). This led us to conclude that the reduction in work output during the second bout of exercise may have been related to a slower resynthesis and consequently a reduced availability of PCr in type II muscle fibers (7).

The importance of type II fiber PCr availability to exercise performance has been confirmed by recent experiments demonstrating that dietary creatine (Cr) supplementation can increase muscle-free Cr and PCr concentrations (23, 24) and thereby improve subsequent maximal exercise performance in man (8, 20). In particular, we have demonstrated that the increase in resting PCr in type II muscle fibers following Cr supplementation can be positively correlated with PCr degradation in this fiber type during subsequent exercise (r = 0.78, n = 8, P < 0.05) and with the increase in work production observed during exercise (r = 0.66, n = 8, P < 0.05; 8). No corresponding significant correlations were observed with respect to type I muscle fibers (r = 0.22, n = 8, P > 0.05; r = 0.32, n = 8, P > 0.05, respectively). These observations led us to suggest that the improvements in exercise performance resulted from the increase in pre-exercise PCr concentration in type II muscle fibers following Cr ingestion increasing ATP resynthesis during exercise.

The Metabolic Basis of Fatigue During Submaximal Exercise in Humans With Respect to Muscle Fiber Types

It is widely accepted that, providing hydration status is maintained, fatigue during prolonged submaximal exercise will occur as a result of carbohydrate depletion (1, 2). The exact mechanism by which carbohydrate depletion causes fatigue is currently open to debate but would appear to be related to an inability of muscle to rephosphorylate ADP to ATP at the required rate, thereby causing a rise in ADP and P_i concentrations and the inhibition of contraction coupling (4, 29). As might be expected, there is now considerable evidence to demonstrate that carbohydrate ingestion during exercise can delay the onset of fatigue. It would appear that this effect is mediated by carbohydrate supplementation delaying the rate of muscle glycogen utilization during exercise (14) or by maintaining the rate of carbohydrate oxidation toward the end of exercise when glycogen stores are limiting substrate delivery (11). However, because all of these studies have based their conclusions on the analysis of mixed-fibered biopsy samples, they are therefore open to the same criticism already raised concerning the biochemical analysis of whole-muscle homogenates.

With this latter point in mind, we have demonstrated recently that the ingestion of a 5.5% carbohydrate solution immediately prior to and during 60 min of treadmill running (at 70% maximal oxygen consumption) did have a glycogen sparing effect, but this effect was restricted to solely type I muscle fibers, where a 42% reduction in glycogen utilization was observed (33). Glycogen utilization was unaffected in type II fibers by carbohydrate ingestion. These findings were confirmed in a subsequent study (34), when it was demonstrated that carbohydrate ingestion under the same experimental conditions produced a 25% reduction of glycogen utilization in type I muscle fibers following 104 min of treadmill running but had no effect on type II fiber glycogen stores. In addition, carbohydrate ingestion enabled a further 30 min of exercise to be performed before exhaustion occurred. It was concluded that because at the point of exhaustion glycogen was depleted only in the type I fibers, irrespective of whether carbohydrate was ingested or not, carbohydrate ingestion improved endurance capacity by maintaining carbohydrate oxidation specifically in type I muscle fibers. Indeed, the glycogen stores of the type II fibers at exhaustion were still approximately threefold higher than those of the type I fibers from the same biopsy sample.

It is clear from these studies that the possibility exists that the development of fatigue that occurs as a consequence of glycogen depletion is restricted to type I muscle fibers. It is apparent therefore that future experiments aimed at elucidating the metabolic basis of fatigue during prolonged exercise will be improved by focusing at the muscle fiber level rather than the whole-muscle homogenate level.

In conclusion, although work based at the muscle fiber level is time consuming and labor intensive, it is the hope of these authors that the reader has been convinced that quantitative biochemical analysis of different muscle fiber types is the best method of gaining a clear insight into the metabolic basis of fatigue in exercising humans. Of course, in an ideal world our understanding of human exercise biochemistry could be improved via in vivo biochemical analysis at the subcellular level within different fiber types. At the present time, such techniques are not available, but with the rapid development of nuclear magnetic resonance spectroscopy, this may become a reality one day.

References

1. Bergstrom, J., Hermansen, L., Hultman, E., and Saltin, B. Diet, muscle glycogen and physical performance. *Acta Physiol. Scand.* 71: 140-150; 1967.
2. Bergstrom, J., and Hultman, E. A study of glycogen metabolism during exercise in man. *Scand. J. Clin. Lab. Invest.* 19: 218-228; 1967.
3. Bogdanis, G.C., Nevill, M.E., Boobis, L.H., and Lakomy, H.K.A. Contribution of phosphocreatine and aerobic metabolism to energy supply during repeated sprint exercise. *J. Appl. Physiol.* 80: 876-884; 1996.
4. Broberg, S., and Sahlin, K. Adenine nucleotide degradation in human skeletal muscle during prolonged exercise. *J. Appl. Physiol.* 67: 116-122; 1989.
5. Brooke, M.H., and Kaiser, K.K. Muscle fibre types: how many and what kind? *Archives in Neurology* 23: 369-379; 1970.
6. Casey, A., Short, A.H., Hultman, E., and Greenhaff, P.L. Glycogen resynthesis in human muscle fibre types following exercise-induced glycogen depletion. *J. Physiol.* 483: 265-271; 1995.
7. Casey, A., Constantin-Teodosiu, D., Howell, S., Hultman, E., and Greenhaff, P.L. The metabolic response of type I and II muscle fibres during repeated bouts of maximal exercise in humans. *Am. J. Physiol.* 271: E38-E43; 1996.

8. Casey, A., Constantin-Teodosiu, D., Howell, S., Hultman, E., and Greenhaff, P.L. Creatine supplementation favourably affects performance and muscle metabolism during maximal intensity exercise in humans. *Am. J. Physiol.* 271: E31-E37; 1996.

9. Cheetham, M.E., Boobis, L.H., Brooks, S.E., Williams, C. Human muscle metabolism during sprint running. *J. Appl. Physiol.* 61: 54-60; 1986.

10. Constantin-Teodosiu, D., Howell, S., and Greenhaff, P.L. Carnitine metabolism in human muscle fibre types during submaximal dynamic exercise. *J. Appl. Physiol.* 80: 1061-1064; 1996.

11. Coyle, E.F., Coggan, A.R., Hemmert, M.K., and Ivy, J.L. Muscle glycogen utilisation during prolonged strenuous exercise when fed carbohydrate. *J. Appl. Physiol.* 61: 165-172; 1986.

12. Dudley, G.A., and Terjung, R.L. Influence of aerobic metabolism on IMP accumulation in fast-twitch muscle. *Am. J. Physiol.* 248: C37-C42; 1985.

13. Edstrom, L., Hultman, E., Sahlin, K., and Sjoholm, H. The contents of high-energy phosphates in different fibre types in skeletal muscles from rat, guinea-pig and man. *J. Physiol.* 332: 47-58; 1982.

14. Erickson, M.A., Schwartzkopf, R.J., and McKenzie R.D. Effects of caffeine, fructose and glucose ingestion on muscle glycogen utilisation during exercise. *Med. Sci. Sports Exercise* 19: 579-583; 1987.

15. Faulkner, J.A., Claflin, D.R., and McCully, K.K. Power output of fast and slow fibres from human skeletal muscles. In *Human Power Output*. Ed. N.L. Jones, N. McCartney, and A.J. McComas, 81-89. Champaign, IL: Human Kinetics; 1986.

16. Fritz, I.B. Carnitine and its role in fatty acid metabolism. *Adv. Lipid Res.* 1: 285-334; 1963.

17. Green, H.J. Muscle power: fibre type recruitment, metabolism and fatigue. In *Human Power Output*. Ed. N.L. Jones, N. McCartney, and A.J. McComas, 65-79. Champaign, IL: Human Kinetics; 1986.

18. Greenhaff, P.L., Ren, J.-M., Soderlund, K., and Hultman, E. Energy metabolism in single human muscle fibres during contraction without and with epinephrine infusion. *Am. J. Physiol.* 260: E713-E718; 1991.

19. Greenhaff, P.L., Soderlund, K., Ren, J.-M., and Hultman, E. Energy metabolism in single human muscle fibres during intermittent contraction with occluded circulation. *J. Physiol.* 460: 443-453; 1993.

20. Greenhaff, P.L., Casey, A., Short, A.H., Harris, R.C., Soderlund, K., and Hultman, E. Influence of oral creatine supplementation on muscle torque during repeated bouts of maximal voluntary exercise in man. *Clin. Sci.* 84: 565-571; 1993.

21. Greenhaff, P.L., Nevill, M.E., Soderlund, K., Bodin, K., Boobis, L.H., Williams, C., and Hultman, E. The metabolic responses of human type I and II muscle fibres during maximal treadmill sprinting. *J. Physiol.* 478: 149-155; 1994.

22. Harris, R.C., Essen, B., and Hultman, E. Glycogen phosphorylase in biopsy samples and single muscle fibres of musculus quadriceps femoris of man at rest. *Scand. J. Clin. Lab. Invest.* 36: 521-526; 1976.

23. Harris, R.C., Soderlund, K., and Hultman, E. Elevation of creatine in resting and exercised muscle of normal subjects by creatine supplementation. *Clin. Sci.* 83: 367-374; 1992.

24. Hultman, E., Soderlund, K., Timmons, J., Cederblad, G., and Greenhaff, P.L. Muscle creatine loading in man. *J. Appl. Physiol.* 81: 232-237; 1996.

25. Jones, N.K., McCartney, N., Graham, T., Spriet, L.L., Kowalchuk, J.M., Heigenhauser, G.J.F., and Sutton, J.R. Muscle performance and metabolism in maximal isokinetic cycling at slow and fast speeds. *J. Appl. Physiol.* 59: 132-136; 1985.

26. Kirwan, J.P., Costill, D.L., Mitchell, J.B., Houmard, J.A., Flynn, M.G., Fink, W.J, and Beltz, J.D. Carbohydrate balance in competitive runners during successive days of intensive training. *J. Appl. Physiol.* 65: 2601-2606; 1988.

27. Norman, B., Sollevi, A., and Jansson, E. Increased IMP content in glycogen depleted muscle fibres during submaximal exercise in man. *Acta Physiol. Scand.* 11: 375-384; 1988.

28. Piehl, K. Time course for refilling of glycogen stores in human muscle fibres following exercise-induced glycogen depletion. *Acta Physiol. Scand.* 90: 297-302; 1974.

29. Sahlin, K., Katz, A., and Broberg, S. Tricarboxylic cycle intermediates in human muscle during submaximal exercise. *Am. J. Physiol.* 259: C834-C841; 1990.

30. Saltin, B., and Gollnick, P. *Handbook of Physiology. Section 10,* 555-632; 1983.

31. Soderlund, K., Greenhaff, P.L., and Hultman. E. Energy metabolism in type I and type II human muscle fibres during short term electrical stimulation at different frequencies. *Acta Physiol. Scand.* 144: 5-22; 1992.

32. Spriet, L.L. ATP utilisation and provision in fast-twitch skeletal muscle during tetanic contractions. *Am. J. Physiol.* 257: E595-E605; 1989.

33. Tsintzas, O-K., Williams, C., Boobis, L., and Greenhaff, P. Carbohydrate ingestion and glycogen utilisation in different muscle fibre types in man. *J. Physiol.* 489: 243-250; 1995.

34. Tsintzas, O.-K., Williams, C., Boobis, L., and Greenhaff, P. Carbohydrate ingestion and single muscle fibre glycogen metabolism during prolonged running in man. *J. Appl. Physiol.* 81: 801-809; 1996.

35. Vollestad, N.K., Blom, P.C.S., and Gronnerod, O. Resynthesis of glycogen in different muscle fibre types after prolonged exhaustive exercise in man. *Acta Physiol. Scand.* 137: 15-21; 1989.

36. Whitlock, D.M., and Terjung R.L. ATP depletion in slow-twitch red muscle of rat. *Am. J. Physiol.* 253: C426-C432; 1987.

PART V

Adaptation

CHAPTER 22

Contractile Activity and Skeletal Muscle Gene Expression

P. Darrell Neufer

John B. Pierce Laboratory, Yale University, New Haven, CT, USA

The ability of skeletal muscle to adapt to sustained increases in activity level represents one of the most fundamental and intriguing aspects of exercise biochemistry. The adaptive responses that occur within the myofiber are dedicated to enhancing the capacity for ATP resynthesis in an effort to efficiently match metabolic flux with functional demand. The objective of this brief review is to provide a perspective on the current understanding of how exercise training modulates gene expression in skeletal muscle and, in so doing, present some of the challenges and questions that have yet to be addressed, particularly at the molecular level.

What Genes Are Regulated by Contractile Activity

Studies conducted over the past 30 years have demonstrated that the concentrations of nearly all proteins involved in the transport, activation, and complete oxidation of metabolic substrates markedly increase in response to exercise training. These include transport proteins such as GLUT4, myoglobin, and fatty acid binding protein; proteins required for substrate activation such as carnitine palmityl transferase I and II; and proteins required for the terminal oxidation of substrates such as enzymes of the tricarboxylic acid cycle and electron transport chain (2, 3, 36). Of particular interest is the fact that the actual number of mitochondria increases with endurance training (35). Mitochondria contain their own circular DNA genome that encodes for a small number of the respiratory chain protein subunits, two ribosomal RNA subunits, and the complete set of transfer RNAs required for translation in the mitochondrial matrix (5). This implies, therefore, that the proliferation of mitochondria induced by contractile activity requires the coordinate regulation of mitochondrial DNA replication and transcription with the transcriptional regulation of nuclear–genes encoding proteins destined for the mitochondria.

In addition to proteins involved in energy metabolism, contractile activity also influences the activity and/or expression of various components of the Ca^{2+}-regulatory system and contractile apparatus. Continuous low-frequency stimulation of rabbit fast-twitch muscle via the motor nerve elicits a switch in sarcoplasmic reticulum (SR) Ca^{2+}-ATPase expression from the fast SERCA1a to the slow SERCA2a isoform, a response that accounts, at least in part, for the corresponding decrease in SR Ca^{2+}-ATPase activity (27, 28). Stimulation also activates transcription of the gene encoding for phospholamban, a Ca^{2+}-binding protein that is normally not expressed in fast-twitch muscle (11), while expression of parvalbumin, a Ca^{2+}-binding protein normally expressed at high levels in fast-twitch glycolytic muscle, is down-regulated with stimulation (12, 16). Transitions in myosin heavy chain and light chain isoforms as well as regulatory proteins of the thin filament also are induced by chronic contractile activity and generally proceed from fast glycolytic to slow oxidative isoforms (17, 18, 27, 28). The collective appearance of various protein isoforms that are either not normally expressed or present at very low levels suggests that prolonged contractile activity is triggering the transcriptional induction of these genes (11, 33).

It should be emphasized at this point that endurance training also induces the expression of slow isoforms of various Ca^{2+}-regulatory and contractile proteins, but in a less dramatic manner than evoked by continuous low-frequency motor nerve stimulation. Although qualitatively similar to endurance training, the gene switching events triggered by chronic stimulation are more complete and, thus, demonstrate the remarkable plasticity of skeletal muscle tissue. After several weeks of continuous low-frequency stimulation, nearly every aspect of the structural and biochemical makeup of a fast-twitch glycolytic muscle is transformed to that of a slow-twitch oxidative muscle. The reader is referred to several recent reviews for detailed information on the various models of exercise training and their impact on specific proteins and cellular processes (2, 27, 36).

Are Gene Regulatory Events Fiber-Type Specific?

In general, the magnitude of an adaptive response within the skeletal muscle is directly proportional to the intensity and duration of the exercise training program (6, 8, 31). These relationships, however, must be viewed in terms of the percent fiber-type distribution of the working muscle and the known order of fiber-type recruitment. The pattern of motor unit recruitment progressively involves the slow-oxidative type I fibers and the fast-oxidative type IIa fibers for exercise intensities below ~80% of maximal oxygen consumption. Fast-glycolytic type IIb/IId fibers do not begin to be recruited until exercise intensity exceeds ~80% of maximal oxygen consumption with additional motor units becoming activated through supramaximal exercise intensities. Importantly, the adaptive increases in mitochondrial enzymes at a given intensity of exercise training also display a fiber-type specific pattern that coincides with the pattern of fiber-type recruitment. For example, training induced increases in cytochrome c content peak in muscles with a high proportion of type I and IIa fibers when exercise intensity is ~80%, but are not evident in muscles with a high proportion of type IIb fibers until training intensity exceeds 80% of maximal oxygen consumption (32). Although there is certainly overlap in the progressive recruitment of different fiber-type populations, the general ordered pattern of recruitment and adaptation suggests that endurance training responses are induced within individual myofibers based on their relative participation during the activity.

As mentioned above, continuous low-frequency motor nerve stimulation elicits qualitatively similar, but quantitatively more rapid and dramatic adaptive responses to contractile activity. Since stimulation is delivered to the motor nerve innervating an entire muscle, it generally has been assumed that all myofibers are contracting and that the adaptive changes in gene expression measured in whole-muscle homogenates qualitatively represent all myofibers. However, data from two recent studies employing both immunohistochemistry and in situ hybridization suggest that gene regulatory events may indeed be fiber-type specific during the adaptive response to chronic stimulation (20, 23). Hsp70 and αB-crystallin, two members of the stress protein family, are dramatically induced within the first three days after the onset of stimulation but only in the type I and a subpopulation of type IIa myofibers (figure 22.1) (23).

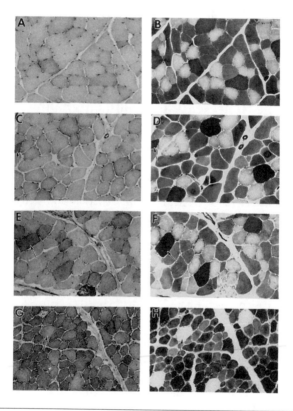

Figure 22.1. Representative photomicrographs (100X) of serially obtained transverse sections from control (A, B), 1 d (C, D), 3 d (E, F), and 21 d (G, H) stimulated rabbit TA muscle subjected to Hsp70 immunohistochemistry (A, C, E, G) and myosin ATPase staining (B, D, F, H) for fiber type determination. Note that for control, 1 d, and 3 d sections (B, D, F) type I slow-twitch/oxidative fibers ("|") stain dark, type IIa fast-twitch/oxidative fibers ("a") stain light, and type IId fast-twitch/glycolytic fibers ("d") stain intermediate (preincubation pH = 4.54), while for 21 d section (H) type I slow-twitch/oxidative fibers stain light, and type II fibers (mixed population) stain intermediate to dark (preincubation pH = 10.3). Reprinted, by permission, from Neufer et al. 1996.

Interestingly, both force output and electromyographic activity decline within minutes and remain depressed for several days after the onset of stimulation, indicating that a significant portion of the myofibers, most likely the fast-glycolytic type IId fibers, become refractory to motor nerve stimulation during the first few days (4, 9). Both twitch output and EMG activity gradually begin to recover within six days of stimulation, presumably reflecting renewed activity of the fast-glycolytic fibers (9). Likewise, expression of Hsp70 and αB-crystallin becomes evident in nearly all type II fibers after several weeks of stimulation but is completely absent in type I fibers (figure 22.1G and H) (20, 23). These findings, therefore, provide evidence that chronic low-frequency stimulation, in and of itself, is not sufficient to elicit a uniform adaptive response within all innervated myofibers. The mechanism by which fast-glycolytic type II fibers are able to resume contractile activity after several days of motor nerve stimulation has not been established but may involve adaptations in the excitation-contraction coupling process, such as increased in Na^+-K^+-ATPase content (9), or other adaptive events, such as the growth of new capillaries and subsequent delivery of O_2 and exchange of metabolites (13).

Energy Charge Represents a Potential Signaling Mechanism

How skeletal myofibers sense changes in metabolic demand and transduce this information to specific genes is a particularly challenging question in muscle biology, given the diversity of cell types within muscle tissue and the complex nature of the metabolic and physical stimuli experienced by myofibers during contractile activity (36). Three lines of evidence now indicate that, of the many potential primary signaling events arising within contracting myofibers, a disruption in energy charge may be a key intracellular perturbation that leads to specific gene regulatory events. Energy charge, defined as the ratio of ATP to the product of free ADP and inorganic phosphate ($ATP/ADP_{free} \times P_{i\,free}$), declines within 15 min after the initiation of low-frequency motor nerve stimulation and remains depressed as long as contractile activity is maintained (7, 10). Secondly, addition of a nonmetabolizable creatine analogue, β-guanidinopropionic acid (β-GPA), to the diet of rats lowers energy charge within skeletal muscle and, within several weeks, elicits an adaptive response that is qualitatively similar to the changes induced by endurance training (e.g., increased mitochondrial density, GLUT4 and slow isoforms of the myosin heavy and light chains) (15, 19, 29). Finally, mice carrying a null mutation of the muscle creatine kinase gene also display a marked reduction in muscle energy charge that is associated with a dramatic increase in the number and size of mitochondria, mitochondrial enzyme content, and oxidative capacity (34).

How may a shift in energy charge activate specific gene regulatory events? One possibility may be that, since ATP also represents the principal substrate for many intracellular kinase reactions, disruptions in energy charge may alter the kinetics of such reactions, thereby affecting the phosphorylation state and activity of key signaling or regulatory proteins. For example, phosphorylation plays a major role in regulating the nuclear translocation, DNA binding activity, and transactivation potential of many different transcription factors. A reduction in energy charge also may trigger a secondary signaling event, such as a rise in intracellular free Ca^{2+} via ADP-mediated inhibition of the sarcoplasmic reticulum Ca^{2+}-ATPase (14), resulting in the activation of Ca^{2+}-dependent signaling cascades.

Evidence that disruptions in energy charge may be linked directly to the activation of contractile responsive genes comes from recent experiments in which mitochondrial inhibitors were infused directly into resting or contracting tibialis anterior muscle of anesthetized rabbits (26). Infusion of the respiratory chain inhibitor, rotenone, in resting muscle elicited an ~5-fold induction of the c-*fos* immediately early gene after both 2 and 4 h of infusion (figure 22.2). In contrast, infusion of rotenone did not influence the timing or magnitude of c-*fos* mRNA accumulation during 4 h of contractile activity, suggesting that continuous motor nerve stimulation provides a maximal stimulus for expression of this immediate early gene. Infusion of the respiratory chain uncoupler, dinitrophenol, generated similar results, thus providing evidence that inhibition of ATP production in resting muscle is sufficient to induce an increase in c-*fos* expression similar to that seen with contractile activity.

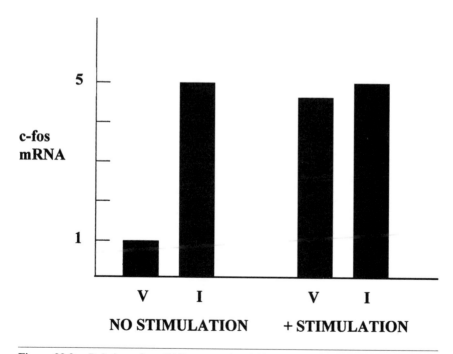

Figure 22.2.   Relative c-fos mRNA content in rabbit tibialis anterior muscle infused for 4 h with vehicle (V) or 200 uM rotenone (I), a metabolic inhibitor. Infusions were performed during no stimulation or during continuous contractile activity induced by low-frequency motor nerve stimulation.

Is Recovery From Exercise Associated With Gene Regulation?

In contrast to continuous low-frequency motor nerve stimulation, the stimulus experienced by myofibers during exercise training is intermittent. This raises the obvious fundamental question as to when the adaptive gene regulatory events to endurance training actually are induced (i.e., during exercise or after exercise). There is now a growing body of evidence to suggest that the expression of training-induced genes may be transiently activated during recovery from exercise. For example, the transcription rate and/or mRNA content of GLUT4, citrate synthase, hexokinase II, and lipoprotein lipase have been shown to increase by twofold to fourfold in skeletal muscle within the first 3–4 h of recovery from exercise, returning to baseline by 24 h after exercise (22, 24, 25, 30).

Recovery from motor nerve stimulation also is associated with striking changes in the expression of several genes. Transcript levels of the c-*fos*, αB-crystallin, and hsp70 genes shown to increase by greater than tenfold specifically during the first 2–4 h of recovery from 8 h of continuous contractile activity (239). The induction pattern is transient, however, as c-*fos* and hsp70 mRNA levels quickly return to baseline while αB-crystallin mRNA content remains elevated through 24 h of recovery. Similar expression profiles were found during recovery from 7 d and 14 d of intermittent (8 h/d) stimulation, providing evidence that the transient expression of immediate early and stress protein genes during recovery from contractile activity may be important for the adaptive responses evoked by endurance training.

When considering the adaptive process that takes place within a cell in response to a particular stimulus, it is helpful to keep in mind several basic principles of cell biology. Nearly all proteins within a cell have a characteristic turnover rate. This means that the absolute level of a particular protein is a function of the equilibrium that is established between that protein's rate of synthesis and rate of degradation. The synthetic process may be regulated at several steps including transcriptional activation of the gene that encodes for the particular protein, processing and stabilization of the nascent RNA transcript, translation of the mRNA to protein, and post-translational processing to yield the mature protein. Although not as well understood, the rate at which a mature protein degrades is also regulated and, under normal conditions, appears to be determined by factors such as structure, conformation, and type of N-terminal residue. Any disruption in the extracellular or intracellular environment of the cell that elicits an adaptive response does so by shifting the equilibrium of the synthetic and/or degradative processes to yield a new steady-state level of the target protein.

It is important to recognize, however, that the adaptive response to exercise training does not follow steady-state kinetics, since exercise represents an intermittent stimulus. Thus, this raises the possibility that the responses associated with endurance training may represent the cumulative effects of transient changes in gene expression. To understand the temporal aspects of an adaptive response to a stimulus that is intermittent, the response must be viewed in terms of the half-lives of the individual mRNAs or proteins. For example, mRNAs with relatively short half-lives (e.g., immediate early genes, stress protein genes, etc.) that are induced during recovery from exercise will decline rapidly as the stimulus diminishes such that no functional increase in mRNA remains by the time the cell experiences the next stimulus (next exercise/recovery period). Alternatively, mRNAs with longer half-lives (e.g., those mRNAs encoding mitochondrial proteins) will not decline as rapidly such that by the time the next stimulus is delivered there remains a small net increase in the mRNA (figure 22.3a). Thus, it follows that the long-term adaptive increases in mRNAs with relatively long half-lives are generated from the cumulative transient changes in expression that occur in response to each exercise/recovery stimulus (figure 22.3b) (for review, see 36).

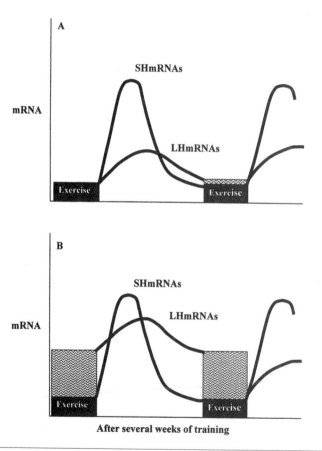

Figure 22.3. Kinetic features of mRNA content in skeletal muscle during recovery from exercise. A. Transient pattern of mRNA expression during recovery from exercise for mRNA species with relatively short half-lives (SHmRNAs), such as immediate early genes, and relatively long half-lives (LHmRNAs), such as mitochondrial enzymes. SHmRNAs return to baseline while LHmRNAs remain slightly elevated by the next exercise session. B. After several weeks of training, LHmRNAs display a cumulative increase (training effect). Based on data from (21, 24, 25, 30) and Neufer, Ordway, and Williams, unpublished observations.

Conclusion

Exercise represents one of the most important forms of preventative medicine; yet, very little is known about how the adaptive responses to exercise are initiated and maintained at the molecular level. This information is critical to the clinician's ability to properly assess the benefits of prescribed exercise in the context of other mitigating factors such as daily work activity, preexisting disease states, and genetic background. For example, although the benefits of exercise to cardiac health are well accepted, it is extremely difficult, based on epidemiological studies, to assess the impact of different types, intensities, frequencies, and durations of exercise on specific cardiovascular risk factors, even when studies are carried out over several months to years (1). Future efforts to delineate the primary molecular events triggering and maintaining the adaptive response of myofibers to exercise training likely will provide the necessary framework to more accurately and efficiently address these important questions.

References

1. Barinaga, M. How much pain for cardiac gain? *Science* 276: 1324-1327; 1997.
2. Booth, F. W., and Baldwin, K.M. Muscle plasticity: energy demand and supply processes. In *Handbook of Physiology. Section 12: Exercise: Regulation and Integration of Multiple Systems*. Ed. L.B. Rowell and J.T. Shepherd, 1075-1123. New York: Oxford Press; American Physiological Society; 1996.
3. Booth, F.W., and Thomason, D.B. Molecular and cellular adaptation of muscle in response to exercise: perspectives of various models. *Physiol. Rev.* 71: 541-585; 1991.
4. Cadefau, J.A., Parra, J., Cusso, R., Heine, G., and Pette, D. Responses of fatigable and fatigue-resistant fibres of rabbit muscle to low-frequency stimulation. *Pflugers Arch* 424: 529-537; 1993.
5. Clayton, D.A. Replication and transcription of vertebrate mitochondrial DNA. *Annu. Rev. Cell. Biol.* 7: 453-478; 1991.
6. Dudley, G.A., Abraham, W.M., and Terjung, R.L. Influence of exercise intensity and duration on biochemical adaptations in skeletal muscle. *J. Appl. Physiol.* 53: 844-850; 1982.
7. Green, H. J., Dusterhoft, S., Dux, L., and Pette, D. Metabolite patterns related to exhaustion, recovery and transformation of chronically stimulated rabbit fast-twitch muscle. *Pflugers Arch* 420: 359-366; 1992.
8. Harms, S.J., and Hickson, R.C. Skeletal muscle mitochondria and myoglobin, endurance, and intensity of training. *J. Appl. Physiol.* 54: 798-802; 1983.
9. Hicks, A., Ohlendieck, K., Gopel, S.O., and Pette, D. Early functional and biochemical adaptations to low-frequency stimulation of rabbit fast-twitch muscle. *Am. J. Physiol.* 273. In press.

10. Hood, D.A., and Parent, G. Metabolic and contractile responses of rat fast-twitch muscle to 10-Hz stimulation. *Am. J. Physiol.* 260: C832-C840; 1991.

11. Hu, P., Yin, C., Zhang, K.M., Wright, L.D., Nixon, T.E., Wechsler, A.S., Spratt, J.A., and Briggs, F.N. Transcriptional regulation of phospholamban gene and translational regulation of SERCA2 gene produces coordinate expression of these two sarcoplasmic reticulum proteins during skeletal muscle phenotype switching. *J. Biol. Chem.* 270: 11,619-11,622; 1995.

12. Huber, B., and Pette, D. Dynamics of parvalbumin expression in low-frequency-stimulated fast-twitch rat muscle. *Eur. J. Biochem.* 236: 814-819; 1996.

13. Hudlicka, O., Dodd, L., Renkin, E.M., and Gray, S.D. Early changes in fiber profile and capillary density in long-term stimulated muscles. *Am. J. Physiol.* 243: H528-H535; 1982.

14. Korge, P., Byrd, S.K., and Campbell, K.B. Functional coupling between sarcoplasmic-reticulum-bound creatine kinase and Ca(2+)-ATPase. *Eur. J. Biochem.* 213: 973-980; 1993.

15. Lai, M.M., and Booth, F.W. Cytochrome c mRNA and alpha-actin mRNA in muscles of rats fed beta-GPA. *J. Appl. Physiol.* 69: 843-848; 1990.

16. Leberer, E., Seedorf, U., and Pette, D. Neural control of gene expression in skeletal muscle. Calcium-sequestering proteins in developing and chronically stimulated rabbit skeletal muscles. *Biochem. J.* 239: 295-300; 1986.

17. Leeuw, T., and Pette, D. Coordinate changes in the expression of troponin subunit and myosin heavy-chain isoforms during fast-to-slow transition of low-frequency-stimulated rabbit muscle. *Eur. J. Biochem.* 213: 1039-1046; 1993.

18. Leeuw, T., and Pette, D. Coordinate changes of myosin light and heavy chain isoforms during forced fiber type transitions in rabbit muscle. *Dev. Genet.* 19: 163-168; 1996.

19. Moerland, T.S., Wolf, N.G., and Kushmerick, M.J. Administration of a creatine analogue induces isomyosin transitions in muscle. *Am. J. Physiol.* 257: C810-C816; 1989.

20. Neufer, P.D., and Benjamin, I.J. Differential expression of αB-crystallin and Hsp27 in skeletal muscle during continuous contractile activity. Relationship to myogenic regulatory factors. *J. Biol. Chem.* 271: 24,089-24,095; 1996.

21. Neufer, P.D., Carey, J.O., and Dohm, G.L. Transcriptional regulation of the gene for glucose transporter GLUT4 in skeletal muscle. Effects of diabetes and fasting. *J. Biol. Chem.* 268: 13,824-13,829; 1993.

22. Neufer, P.D., and Dohm, G.L. Exercise induces a transient increase in transcription of the GLUT-4 gene in skeletal muscle. *Am. J. Physiol.* 265: C1597-C1603; 1993.

23. Neufer, P.D., Ordway, G.A., Hand, G.A., Shelton, J.M., Richardson, J.A., Benjamin, I.J., and Williams, R.S. Continuous contractile activity induces fiber-type specific expression of Hsp70 in skeletal muscle. *Am. J. Physiol.* 271: C1828-C1837; 1996.

23a. Neufer, P.D., Ordway, G.A., and Williams, R.S. Transient regulation of c-fos, αB-crystallin, and hsp70 in muscle during recovery from contractile activity. *Am. J. Physiol.* 274: C342-C346; 1998.

24. O'Doherty, R.M., Bracy, D.P., Granner, D.K., and Wasserman, D.H. Transcription of the rat skeletal muscle hexokinase II gene is increased by acute exercise. *J. Appl. Physiol.* 81: 789-793; 1996.

25. O'Doherty, R.M., Bracy, D.P., Osawa, H., Wasserman, D.H., and Granner, D.K. Rat skeletal muscle hexokinase II mRNA and activity are increased by a single bout of acute exercise. *Am. J. Physiol.* 266: E171-E178; 1994.

26. Ordway, G.A., Williams, R.S., Crippens, D., and Neufer, P.D. Disruption of ATP production increases c-fos expression in skeletal muscle. *Med. Sci. Sports. Ex.* 27: S123; 1995.

27. Pette, D., and Staron, R.S. Mammalian skeletal muscle fiber type transitions. *Int. Rev. Cytol.* 170: 143-223; 1997.

28. Pette, D., and Vrbova, G. Adaptation of mammalian skeletal muscle fibers to chronic electrical stimulation. *Rev. Physiol. Biochem. Pharmacol.* 120: 115-202; 1992.

29. Ren, J.M., Semenkovich, C.F., and Holloszy, J.O. Adaptation of muscle to creatine depletion: effect on GLUT-4 glucose transporter expression. *Am. J. Physiol.* 264: C146-C150; 1993.

30. Seip, R.L., Mair, K., Cole, T.G., and Semenkovich, C.F. Induction of human skeletal muscle lipoprotein lipase gene expression by short-term exercise is transient. *Am. J. Physiol.* 272: E255-E261; 1997.

31. Terjung, R.L. Muscle fiber involvement during training of different intensities and durations. *Am. J. Physiol.* 230: 946-950; 1976.

32. Terjung, R.L., and Hood, D.A. Biochemical adaptations in skeletal muscle induced by exercise training. In *Nutrition and Aerobic Exercise.* Ed. D.K. Layman, 8-27. Washington: American Chemical Society; 1986.

33. Termin, A., and Pette, D. Changes in myosin heavy-chain isoform synthesis of chronically stimulated rat fast-twitch muscle. *Eur. J. Biochem.* 204: 569-573; 1992.

34. van Deursen, J., Heerschap, A., Oerlemans, F., Ruitenbeek, W., Jap, P., ter Laak, H., and Wieringa, B. Skeletal muscles of mice deficient in muscle creatine kinase lack burst activity. *Cell* 74: 621-631; 1993.

35. Williams, R. S. Mitochondrial gene expression in mammalian striated muscle. Evidence that variation in gene dosage is the major regulatory event. *J. Biol. Chem.* 261: 12,390-12,394; 1986.

36. Williams, R.S., and Neufer, P.D. Regulation of gene expression in skeletal muscle by contractile activity. In *Handbook of Physiology. Section 12: Exercise: Regulation and Integration of Multiple Systems.* Ed. L.B. Rowell and J.T. Shepherd. New York: Oxford Press; American Physiological Society; 1996.

CHAPTER 23

Vascular Remodeling Induced by Muscle Activity

Ronald L. Terjung, H.T. Yang, Hiroshi Itoh*, Michael R. Deschenes**, and Robert W. Ogilvie***

Biomedical Sciences, College of Veterinary Medicine, University of Missouri, Columbia, MO, USA
*Department of Physical Education, Nagoya Institute of Technology, Nagoya, Japan
**Department of Kinesiology, College of William and Mary, Williamsburg, VA, USA
***Department of Cell Biology and Anatomy, Medical University of South Carolina, Charleston, SC, USA

Vascular changes induced by physical activity can be divided into several fundamental aspects, those related to: alterations in the central cardiovascular control system; changes in vascular responsiveness of the peripheral vessels; and structural changes to the existing system. While important alterations in the central control (43) and peripheral vasomotor modulation (35) of the cardiovascular system are induced by physical exercise, the focus of the present chapter is to consider recent developments in structural changes that involve expansion of the vascular network. This can occur by de novo expansion of the capillary bed, a process termed angiogenesis, and by enlargement of existing vessels, often referred to as neovascular development. For the purposes of this discussion, structural changes will be divided arbitrarily into two aspects related to function: 1) the microvascular, which serves to support blood/tissue nutrient exchange; and 2) the conduit vessels, whose function establishes flow delivery and bulk flow capacity to a tissue.

Microvasculature

A cardinal feature of training adaptations, established by endurance-type exercise, is an increase in capillarity within the active muscles (27). This increase in capillary density, often characterized as capillary-to-fiber ratio or capillary contacts per fiber, establishes a larger capillary volume within the muscle. This lengthens the average red blood cell transit time for the same high blood flow, increases the surface area for nutrient exchange with the tissue, and should reduce the average diffusion path length within the muscle. All these factors should enhance the diffusion capacity of the tissue and could account for the long-recognized greater peak oxygen extraction across the active limbs of trained subjects (35, 44). However, a redistribution of blood flow within the limb could enhance blood flow to specific critical areas of the active muscle and lead to the observed lower venous oxygen content that effluxes the limb after training. Relevant evidence now is available in experiments where blood flow and therefore oxygen delivery was controlled to be the same in the trained and untrained muscle (54). The peak oxygen consumption obtained by sequentially increasing the muscle contractile demands was ~30% greater for the trained muscle, as compared with the sedentary control muscle (cf. table 23.1). This was apparent even though oxygen delivery to the contracting muscle was virtually identical between the groups. Muscle capillarity (capillary contacts per fiber) was ~20% greater in the trained muscle. A similar training-induced increase in oxygen consumption, observed at similar blood flows, was reported by Bebout et al. (9) for dog muscle. Thus, aerobic-type training remodels muscle to increase its capacity for oxygen exchange and thereby better utilize the oxygen delivered in the blood flow. Interestingly, this adaptation is just as robust in the aged animal (54). The elevation in muscle capillarity evident in trained elderly humans (13) implies that the capacity for angiogenesis, and its functional outcome of enhanced blood/tissue exchange, is fully responsive to exercise stimuli throughout life span.

While the greater capillarity of trained muscle likely contributes to the increase in peak oxygen uptake observed for trained muscle, it may not be the sole determining factor. For example, as the site of oxygen utilization, mitochondrial content, and distribution within the active fiber could impact the muscle's diffusive exchange properties for oxygen. In fact, in the general case there is a fairly good match between capillarity and mitochondrial content among the different skeletal muscle fiber types (44). As expected, the trained muscle that exhibited the increase in capillarity and the increase in peak oxygen consumption also possessed an increase in mitochondrial content (52, 54). The activity of citrate synthase, a marker enzyme

Table 23.1   Training Increases Muscle Capillarity and Peak Oxygen Consumption

	Oxygen Delivery	Peak Oxygen Consumption		Capillary Contacts	Citrate Synthase
	(μmol/min/g)	(μmol/min/g)	(ml/min/kg)	(#/fiber)	(μmol/min/g)
Sedentary	9.53	4.34	97.2	4.68	18.1
(n = 10)	± 1.3	± 0.29	± 6.5	± 0.11	± 1.64
Trained	9.46	5.68	127	5.70	38.4
(n = 12)	± 0.6	± 0.34	± 7.6	± 0.10	± 0.98

Data adapted from Yang et al. (54).

of the mitochondrion, was increased ~100%. This raises the possibility that mito-chondrial density changes also contributed to the greater peak oxygen consumption. This possibility is further strengthened by the demonstration that experimental removal of the elevated mitochondrial capacity resulted in loss of the training-induced improvement in peak oxygen consumption (42). Thus, as illustrated in figure 23.1, the elevated mitochondrial content was essential to realize the increase in peak oxygen consumption. Alternatively, the increase in capillarity of the trained muscle appeared insufficient, in itself, to support an elevation in peak oxygen consumption when mitochondrial electron flux capacity was no greater than in the sedentary control muscle. These findings imply that both adaptations, a high muscle capillarity and a high mitochondrial content, represent a coordinate response to exercise training that serves to enhance muscle peak aerobic capacity. To date, it has not been possible to eliminate the increase in capillarity, while keeping the increase in mitochondrial content that occurs with training, in order to experimentally evalu-ate the specific impact of the change in mitochondrial content.

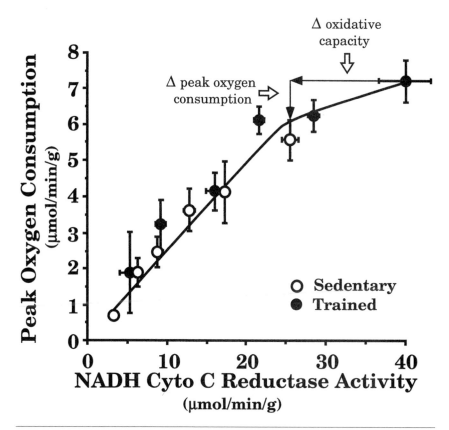

Figure 23.1.   Loss of the training-induced increase in mitochondrial capacity results in a loss of the increased peak oxygen consumption in trained muscle. Adapted from Robinson et al. (42).

Important new insights have been gained into the process of de novo capillary development. Essential elements of this angiogenic process must include a prompting stimulus, capillary wall disruption, endothelial cell proliferation, migration, and tube formation, and finally closure, probably established by contact inhibition (6, 38). It has long been recognized that biophysical and metabolic factors can be important determinants prompting angiogenesis and vascular remodeling. Relevant biophysical factors suspected of stimulating angiogenesis include (27): an increase in vessel wall tension, established by an increase in vessel diameter or pressure; an increase in shear stress within the vessel, usually established by a critical increase in flow velocity; and changes in vessel wall integrity, established for example by injury, disease, and/or endothelial/extracellular matrix disruption. An increase in vessel wall tension has been implicated in the vascular expansion that occurs with growth and development. Expansion of the vascular tree appears to occur with invasion of the capillary network by vascular smooth muscle originating from the terminal arterioles adjacent to the capillaries (41). Proliferating smooth muscle cells envelop the endothelial layer of the capillary, which is normally devoid of smooth muscle, to form new arterioles and arcade arterioles, the vessels interconnecting vascular branches to the capillary network. This "arteriolization of capillaries" may be the process which supports expansion of the whole arterial network of vessels as the muscle enlarges. Experimentally increasing flow through the vascular circuit of a muscle leads to an increase in the number of proliferating terminal arterioles and arcade arterioles (40). Interestingly, this arteriolar smooth muscle/capillary interface is thought to remain quite plastic, responsive to vascular expansion and contraction depending on the prevailing stimulus (40). While it may be easy to imagine how similar biophysical factors established by high flow conditions could be the critical stimulus prompting the enhanced capillarity within trained muscle, definitive evidence is not available. Thus, this is an important area for future inquiry.

In addition to biophysical factors, a number of other factors generally grouped as "metabolic factors" are important modulators of the angiogenic process. Events that lead to hypoxia, release of certain cytokines, and/or interactions with heparin or heparin sulfates of the extracellular matrix have been implicated in angiogenesis and neovascular development (19, 27). Exciting developments in the recent past have identified several families of heparin-binding cytokines produced by endothelial cells and myocytes that are potent angiogenic growth factors (cf. table 23.2). The acidic and basic fibroblast growth factors are potent mitogens for all three cell types that comprise the vasculature: endothelial cells, smooth muscle cells, and fibroblasts. In contrast, the vascular endothelial growth factor family of mitogens are specific for endothelial cells. VEGF is the same peptide as Vascular Permeability Factor (VPF) that causes increased permeability in the microvasculature. Each of these growth factors cause cultured endothelial cells to proliferate, migrate into the collagen substratum, and form tube-like cylinders (3, 38), similar to that characterized in de novo capillary formation (27). Unlike the fibroblast growth factors, the VEGF transcript contains a signal sequence typical for producing protein that is secreted from cells. All of these full-size growth factors contain a binding domain that binds to heparin or heparin-like compounds. Thus, once positioned extracellularly, these growth factors are retained in the extracellular matrix via their binding to the heparin sulfates. This extracellular pool of angio-

Table 23.2   Heparin-Binding Angiogenic Growth Factors

Factor	Abbrev.	Responsive Cells	Ref.
acidic Fibroblast Growth Factor	aFGF	Endothelial cells Fibroblasts	(5, 20, 24, 32, 48)
basic Fibroblast Growth Factor	bFGF	Myocytes	
Vascular Endothelial Growth Factor	VEGF (VPF)	Endothelial cells	(11, 33)

genic growth factor probably represents a potential source of angiogenic effectors available to respond to initiating stimuli that leads to vascular expansion and remodeling. A number of excellent reviews are available synthesizing the considerable information regarding the genes, chemistry, receptors, and biological actions of these growth factors (5, 6, 24, 32).

It is highly probable that these angiogenic growth factors are the agents that ultimately effect an angiogenic response initiated by some other biophysical and/or metabolic stimulus. In fact, there is good reason to expect the involvement of VEGF in the angiogenic response to hypoxia. The VEGF gene possesses a response element, the same as the control element on the erythropoietin gene, that when activated prompts an increase in VEGF mRNA (23). Decreasing the pO_2 of endothelial cells in culture increases the expression of VEGF mRNA; conversely, returning the pO_2 to its elevated value decreases VEGF mRNA production (46). The greater the abundance of VEGF message, the greater should be the growth factor protein production. Thus, there is an established relationship between a "metabolic" angiogenic stimulus, in the general case, and the response of the angiogenic effector VEGF in the specific case. It is probable that this response is physiologically relevant. An approximate threefold increase in VEGF mRNA was observed in the ischemic region of the heart within just two hours following coronary artery occlusion (25). Ischemic exercise, established by walking rats who have an inadequate calf muscle flow capacity, results in a similar outcome (unpublished observations). It is possible, however, that ischemia is not an obligatory requirement to realize the increase in VEGF mRNA abundance in exercising muscle. Breen et al. (10) observed this increase in VEGF mRNA abundance in the muscles of normal rats while running on the treadmill. In their study the running speed was fairly modest, well within the running capacity of the rat. Most importantly the speed required a blood flow well below that measured at higher steady-state treadmill speeds. This implies that flow and therefore oxygen delivery was adequate and that hypoxie conditions should not be found in the muscle. Thus, exercise per se may be an important stimulus prompting up-regulation of VEGF. Future developments in this important area of research are likely to provide fundamental insights into the process of angiogenesis in skeletal muscle.

Conduit Vessels

There is some evidence to indicate that the conduit vessels of normal animals increase after exercise training to establish a higher flow capacity to muscle (35). Laughlin and Ripperger (34) found that flow to the papaverine-dilated isolated perfused hindquarters of the exercise trained rat was greater, as compared with nontrained controls. Mackie and Terjung (36) reported a significantly greater peak flow during contractions in situ to the muscle of trained rats; however, this was observed only in the low flow capacity, fast-twitch white region of the gastrocnemius and not in the high flow capacity, slow-twitch red and fast-twitch red muscle sections. While a high flow capacity is an essential element for high-oxidative muscle, even for trained muscle of normal individuals, it is likely that vascular adaptations to increase flow capacity are of modest impact physiologically. Flow capacity of normal muscle of untrained individuals is already so extraordinarily high, especially compared with that typically commanded during whole-body exercise. This probably accounts for the modest muscle blood flow increase (~10%) during maximal treadmill running in highly aerobic hound dogs after training (39). In a landmark study, Andersen and Saltin (2) demonstrated that flow capacity to human quadriceps muscle is far in excess of that ever commanded by the muscle during even maximal whole-body exercise. Thus, cardiac output is the primary determinant of whole-body maximal oxygen consumption and not flow capacity of the active muscles. This indicates that further enhancement of flow capacity to trained muscle is probably superfluous. The possible exception might be training-induced increases in flow capacity to the relatively low flow muscle regions of nonprimate mammals that contain predominantly fast-twitch white fibers. The increase in biochemical aerobic capacity of these fibers with training possibly could benefit from an increase in blood flow capacity. The greater vascular conductance is expected to permit this region of muscle to better "compete" for flow (and gain a larger fraction of cardiac output) to support the enhanced performance capabilities of these motor units after training.

While vascular expansion of the conduit vessels to muscle may not be a major expectation with training in normal individuals, clinical conditions where flow capacity is abnormally low, e.g., caused by peripheral vascular disease, are very different. It is well-recognized that physical activity is an important treatment in the management of patients with peripheral arterial insufficiency (15, 16, 26, 30). Patients who become physically active uniformly exhibit an improvement in exercise tolerance (15, 16, 26, 30). While a major reason for this improvement is the peripheral adaptation(s) in the distal muscles (49, 52, 54, 55, 56), an increase in collateral-dependent blood flow is commonly, but not always, observed (1, 14, 16, 17, 47, 49, 56). An increase in collateral blood flow is made possible by an enlargement of the collateral vessels and a corresponding decrease in vascular resistance. This can occur by dilation of existing collateral vessels, as it is now evident that collateral conduits are responsive to vasomotor control (18). For example, we have found that existing collateral vessels are responsive to angiotensin converting enzyme (ACE) inhibition; the involvement of angiotensin was evident by a significant improvement in collateral blood flow with ACE inhibition following acute occlusion of a large peripheral vessel (53). Similarly, using a nexopamil (L-type Ca^{++} channel blocker and $5HT_2$ receptor antagonist 31) leads to a reduction of collateral resistance and resultant increase in calf muscle blood flow following acute occlusion of the femoral artery of rats (unpublished observations). As illustrated by these two examples, modulation of vascular tone to relax smooth muscle can improve collateral-dependent blood flow. However, this modulation of vessel diameter exhibits a limit established by the "size" of the dilated conduit vessel. Thus, a true expansion of the vascular tree to increase the caliber of the vessels is the only means to establish an actual improvement in collateral blood flow capacity. There is now good evidence that exercise can expand the collateral vasculature by neovascularization.

Fujita et al. reported the first evidence to implicate involvement of the heparin-binding angiogenic growth factors in the enhanced collateral blood flow established by exercise in patients with coronary insufficiency. They found that daily heparin injections, shortly before exercise, resulted in a marked improvement in exercise tolerance, an improvement in clinical indices of coronary ischemia, and angiographic evidence of enhanced collateral function (21). Similar patients who exercised but did not receive daily heparin injections did not exhibit the same improved collateral function; or, patients who received heparin but did not exercise did not realize any benefit. This specific influence of heparin and exercise implies the involvement of the angiogenic growth factors in the vascular expansion stimulated by exercise. Similar results have been observed in trained rats with peripheral arterial insufficiency caused by femoral artery occlusion (55). As shown in figure 23.2, animals that were trained by walking at a well-tolerated modest speed on a treadmill for ~6 weeks exhibited some evidence of an improvement in collateral-dependent flow capacity to the calf muscles. This response was significantly enhanced by daily heparin injections, sufficient to double clotting time, ~30 min prior to the exercise. Although the heparin-administered group was capable of a much better exercise tolerance, the exercise intensity and duration were kept identical to the modest-intensity, saline-trained group in order to establish the same exercise "stimulus." In addition, a second pair of exercised groups were included; in this case the exercise "stimulus" was optimized by running the animals at progressively increased speeds, as their exercise tolerance improved. The marked improvement in collateral-dependent blood flow of the high-intensity, saline-administered group was further enhanced in the heparin injected group (cf. figure 23.2). Angiographic evaluation of the vasculature revealed enlargement of existing conduit vessels in the upper thigh. These vessels probably served as the collateral channels to permit the greater blood flow down stream to the lower leg muscles. Interestingly, as with the work in the heart patients (22), heparin administration in the absence of exercise did not increase collateral blood flow. Thus, exercise imparted the essential stimulus. However, heparin served to interact with the exercise "stimulus" to greatly enrich the vascular expansion. These findings implicate the angiogenic growth factors in the biological adaptions induced by aerobic exercise.

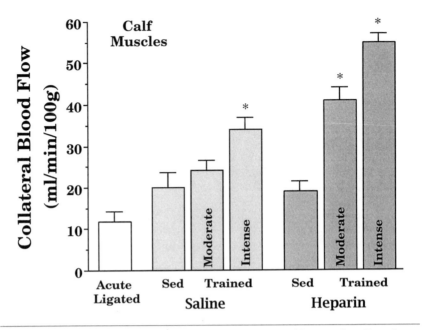

Figure 23.2.   The influence of heparin on the training-induced increase in collateral blood flow. Adapted from Yang et al. (55).

The potent vascular effects of the angiogenic growth factors has prompted considerable interest in their potential therapeutic use in managing patients with peripheral arterial insufficiency (28). Preclinical evidence in animals is encouraging (4, 7, 8, 12, 50). We have observed significant improvements in collateral-dependent blood flow in rats following femoral artery occlusion (51). Infusion of bFGF that was complexed with heparin into the iliac artery, for delivery near the source of the collateral vessels, caused marked improvement in calf muscle blood flow during the first two weeks of delivery but not thereafter (cf. figure 23.3). Infusion of heparin or saline as the control did not appreciably increase collateral-dependent blood flow. Not withstanding potential contraindications, these findings suggest the utility of angiogenic growth factor administration in the management of patients with peripheral arterial insufficiency, especially those incapable of physical activity.

At present, it is unclear whether biophysical and/or metabolic factors are the prompting stimuli established by exercise that initiates the neovascular expansion of the collateral vessels. An increase in wall stress and/or shear stress, established by flow increases during exercise, are likely candidates (29, 45). For example, increasing shear stress to cultured endothelial cells prompts up regulation of growth factor mRNA expression (37). While metabolic factors, established by the active muscle during exercise, are implicated in the angiogenic process, their involvement as the dominant stimulus for collateral expansion may be suspect; this is especially true when the site for collateral development is quite distant from the ischemic muscle that is active during exercise (29). This can be the case in peripheral arterial insufficiency where the tissue most at risk of ischemia is typically in the distal limb;

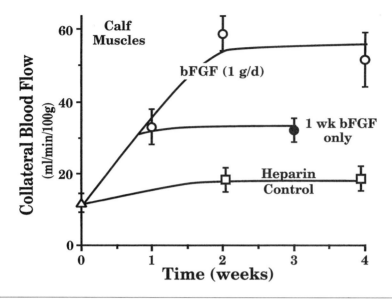

Figure 23.3.   The influence of bFGF infusion on collateral blood flow in rats with peripheral arterial insufficiency. Adapted from Yang et al. (51).

whereas, the occlusion is often associated with proximal large-vessel obstruction(s) in the iliac and/or femoral arteries. This distance seems to preclude a local direct effect of metabolic factors influencing neovascular collateral development. Thus, biophysical influences are the focus of attention (29). Consistent with this focus, we have observed a further increase in collateral blood flow induced by bFGF administration when the animals were exercised by daily treadmill running (unpublished observations). These results would be expected, if flow increases through the collateral vessel were important in establishing the vasculature's responsiveness to growth factor-induced expansion.

Important vascular remodeling that is induced by exercise is becoming better understood. It is well recognized that physical activity imparts a critical influence to maintain and optimize the cardiovascular system and its function. The complexity of events operating physiologically in the process of training makes it extremely difficult to interpret relevant experimental evidence unambiguously. However, important new insights should be forthcoming as investigations into the interactions between exercise, vascular remodeling, and the angiogenic growth factors are undertaken.

Acknowledgements

Work in the authors' laboratory has been supported by NIH grants HL R01-37387 and AR R37-21617, and by a grant from the American Heart Association New York affiliate; H. Itoh was supported by a Fellowship from the Japan Ministry of Education; M. Deschenes was supported by an NIH NRSA HL 09178.

References

1. Alpert, J., Larsen, O.A., and Lassen, N.A. Exercise and intermittent claudication. Blood flow in the calf muscle during walking studied by the xenon-133 clearance method. *Circulation* 39: 353-359; 1969.
2. Andersen, P., and Saltin, B. Maximal perfusion of skeletal muscle in man. *J. Physiol. (London)* 366: 233-249; 1985.
3. Auerbach, R., Auerbach, W., and Polakowski, I. Assays for angiogenesis: a review. *Pharmacol. and Thera.* 51: 1-11; 1991.
4. Baffour, R., Berman, J., Garb, J.L., Rhee, S.W., Kaufman, J., and Friedmann, P. Enhanced angiogenesis and growth of collaterals by in vivo administration of recombinant basic fibroblast growth factor in a rabbit model of acute lower limb ischemia: dose-response effect of basic fibroblast growth factor. *J. Vasc. Surg.* 16: 181-191; 1992.
5. Basilico, C., and Moscatelli, D. The FGF family of growth factors and oncogenes. *Adv. in Cancer Res.* 59: 115-165; 1992.
6. Battegay, E.J. Angiogenesis: mechansitic insights, neovascular diseases, and therapeutic prospects. *J. Mol. Med.* 73: 333-346; 1995.
7. Bauters, C., Asahara, T., Zheng, L.P., Takeshita, S., Bunting, S., Ferrara, N., Symes, J. F., and Isner, J.M. Physiological assessment of augmented vascularity induced by VEGF in ischemic rabbit hindlimb. *Am. J. Physiol.* 267: H1263-H1271; 1994.
8. Bauters, C., Asahara, T., Zheng, L.P., Takeshita, S., Bunting, S., Ferrara, N., Symes, J.F., and Isner, J.M. Site-specific therapeutic angiogenesis after systemic administration of vascular endothelial growth factor. *J. Vasc. Surg.* 21: 314-325; 1995.
9. Bebout, D.E., Hogan, M.C., Hempleman, S.C., and Wagner, P.D. Effects of training and immobilization on VO2 and DO2 in dog gastrocnemius muscle in situ. *J. Appl. Physiol.* 74: 1697-1703; 1993.
10. Breen, E.C., Johnson, E.C., Wagner, H., Tseng, H.M., Sung, L.A., and Wagner, P.D. Angiogenic growth factor mRNA responses in muscle to a single bout of exercise. *J. Appl. Physiol.* 81: 355-361; 1996.
11. Brown, L.F., Detmar, M., Claffey, K., Nagy, J.A., Feng, D., Dvorak, A.M., and Dvorak, H.F. Vascular permeability factor/vascular endothelial growth factor: a multifunctional angiogenic cytokine. *EXS* 79: 233-269; 1997.
12. Chleboun, J.O., and Martins, R.N. The development and enhancement of the collateral circulation in an animal model of lower limb ischaemia. *Aust. N.Z. J. Surg.* 64: 202-207; 1994.
13. Coggan, A.R., Kohrt, W.M., Spina, R.J., Bier, D.M., and Holloszy, J.O. Endurance training decreases plasma glucose turnover and oxidation during moderate intensity exercise in men. *J. Appl. Physiol.* 68: 990-996; 1990.
14. Dahllof, A.G., Bjorntorp, P., Holm, J., and Schersten, T. Metabolic activity of skeletal muscle in patients with peripheral arterial insufficiency. *Eur. J. Clin. Invest.* 4: 9-15; 1974.
15. Dahllof, A.G., Holm, J., Schersten, T., and Sivertsson, R. Peripheral arterial insufficiency, effect of physical training on walking tolerance, calf blood flow, and blood flow resistance. *Scand. J. Rehab. Med.* 8: 640-643; 1976.

16. Ekroth, R., Dahllof, A.G., Gundevall, B., Holm, J., and Schersten, T. Physical training of patients with intermittent claudication: indications, methods, and results. *Surgery* 84: 640-643; 1978.

17. Ericsson, B., Haeger, K., and Lindell, S.E. Effect of physical training of intermittent claudication. *Angiology* 21: 188-192; 1970.

18. Feldman, R.D., Christy, J.P., Paul, S.T., and Harrison, D.G. Beta-adrenergic receptors on canine coronary collateral vessels: characterization and function. *Am. J. Physiol.* 257: H1634-H1639; 1989.

19. Folkman, J., and Shing, Y. Control of angiogenesis by heparin and other sulfated polysaccharides. *Adv. Exptl. Med. Biol.* 313: 355-364; 1992.

20. Friesel, R.E., and Maciag, T. Molecular mechanisms of angiogenesis: fibroblast growth factor signal transduction. *FASEB Journal* 9: 919-925; 1995.

21. Fujita, M., Sasayama, S., Asanoi, H., Nakajima, H., Sakai, O., and Ohno, A. Improvement of treadmill capacity and collateral circulation as a result of exercise with heparin pretreatment in patients with effort angina. *Circulation* 77: 1022-1029; 1988.

22. Fujita, M., Yamanishi, K., Hirai, T., Ohno, A., Miwa, K., and Sasayama, S. Comparative effect of heparin treatment with and without strenuous exercise on treadmill capacity in patients with stable effort angina. *Am. Heart J.* 122: 453-457; 1991.

23. Goldberg, M.A., and Schneider, T.J. Similarities between the oxygen-sensing mechanisms regulating the expression of vascular endothelial growth factor and erythropoietin. *J. Biol. Chem.* 269: 4355-4359; 1994.

24. Gospodarowicz, D. Expression and control of vascular endothelial cells: proliferation and differentiation by fibroblast growth factors. *J. Invest. Dermatol.* 93: 39S-47S; 1989.

25. Hashimoto, E., Ogita, T., Nakaoka, T., Matsuoka, R., Takao, A., and Kira, Y. Rapid induction of vascular endothelial growth factor expression by transient ischemia in rat heart. *Am. J. Physiol.* 267: H1948-H1954; 1994.

26. Hiatt, W.R., Wolfel, E.E., Meier, R.H., and Regensteiner, J.G. Superiority of treadmill walking exercise versus strength training for patients with peripheral arterial disease. *Circulation* 90: 1866-1874; 1994.

27. Hudlicka, O., Brown, M., and Egginton, S. Angiogenesis in skeletal and cardiac muscle. *Physiol. Rev.* 72: 369-417; 1992.

28. Isner, J.M. The role of angiogenic cytokines in cardiovascular disease. *Clin. Immuno. Immunopath.* 80: 82S-91S; 1996.

29. Ito, W.D., Arras, M., Scholz, D., Winkler, B., Htun, P., and Schaper, W. Angiogenesis but not collateral growth is associated with ischemia after femoral artery occlusion. *Am. J. Physiol.* 273: H1255-H1265; 1997.

30. Jonason, T., and Ringqvist, I. Prediction of the effect of training on the walking tolerance in patients with intermittent claudication. *Scand. J. Rehab. Med.* 19: 47-50; 1987.

31. Kirchengast, M. Nexopamil (LU 49938), a combined Ca^{2+} channel blocker and $5HT_2$ receptor antagonist. *Cardiovasc. Drug Rev.* 12: 195-207; 1994.

32. Klagsbrun, M. The fibroblast growth factor family: structural and biological properties. *Prog. Growth Factor Res.* 1: 207-235; 1989.

33. Klagsbrun, M., and D'Amore, P.A. Vascular endothelial growth factor and its receptors. *Cytokine & Growth Factor Rev.* 7: 259-270; 1996.

34. Laughlin, M.H., and Ripperger, J. Vascular transport capacity of hindlimb muscles of exercise trained rats. *J. Appl. Physiol.* 62: 438-443; 1987.

35. Laughlin, M.L., Korthuis, R.J., Duncker, D.J., and Bache, R.J. Control of blood flow to cardiac and skeletal muscle during exercise. In *Handbook of Physiology. Section 12: Exercise: Regulation and Integration of Multiple Systems.* New York: Oxford Press; American Physiological Society. 705-769; 1996.

36. Mackie, B.G., and Terjung, R.L. Influence of training on blood flow to different skeletal muscle fiber types. *J. Appl. Physiol.* 55: 1072-1078; 1983.

37. Malek, A.M., Gibbons, G.H., Dzau, V.J., and Izumo, S. Fluid shear stress differentially modulates expression of genes encoding basic fibroblast growth factor and platelet-derived growth factor B chain in vascular endothelium. *J. Clin. Invest.* 92: 2013-2021; 1993.

38. Montesano, R. 1992 Mack Forster Award Lecture: Review: Regulation of angiogenesis in vitro. *Eur. J. Clin. Invest.* 22: 504-515; 1992.

39. Musch, T.I., Haidet, G.C., Ordway, G.A., Longhurst, J.C., and Mitchell, J.H. Training effects on regional blood flow response to maximal exercise in foxhounds. *J. Appl. Physiol.* 62: 1724-1732; 1987.

40. Price, R.J., and Skalak, T.C. Chronic alpha1-adrenergic blockade stimulates terminal and arcade arteriolar development. *Am. J. Physiol.* 271: H752-H759; 1996.

41. Price, R.J., Owens, G.K., and Skalak, T.C. Immunohistochemical identification of arteriolar development using markers of smooth muscle differentiation. *Circ. Res.* 75: 520-527; 1994.

42. Robinson, D.M., Ogilvie, R.W., Tullson, P.C., and Terjung, R.L. Increased peak oxygen consumption of trained muscle requires increased electron flux capacity. *J. Appl. Physiol.* 77: 1941-1952; 1994.

43. Rowell, L.B. *Human Cardiovascular Control.* New York: Oxford Press; 1993.

44. Saltin, B., and Gollnick P.D. Skeletal muscle adaptability: significance for metabolism and performance. In *Handbook of Physiology.* New York: Oxford Press; American Physiological Society. 555-631; 1983.

45. Schaper, W., and Ito, W.D. Molecular mechanisms of coronary collateral vessel growth. *Circ. Res.* 79: 911-919; 1996.

46. Schweiki, D., Itin, A., Soffer, D., and Keshet, E. Vascular endothelial growth factor induced by hypoxia may mediate hypoxia-initiated angiogenesis. *Nature* 359: 843-845; 1992.

47. Skinner, J.S., and Strandness, D.E. Exercise and intermittent claudication. II. Effect of physical training. *Circulation* 36: 23-29; 1967.

48. Slavin, J. Fibroblast growth factors: at the heart of angiogenesis. *Cell Biol. Intl.* 19: 431-444; 1995.

49. Sorlie, D., and Myhre, K. Effects of physical training in intermittent claudication. *Scand. J. Clin. Lab. Invest.* 38: 217-222; 1978.

50. Takeshita, S., Zheng, L. P., Brogi, E., Kearney, M., Pu, L.Q., Bunting, S., Ferrara, N., Symes, J.F., and Isner, J.M. Therapeutic angiogenesis. A single intraarterial bolus of vascular endothelial growth factor augments revascularization in a rabbit ischemic hind limb model. *J. Clin. Invest.* 93: 662-670; 1994.

51. Yang, H.T., Deschenes, M.R., Ogilvie, R.W., and Terjung, R. L. Basic fibroblast growth factor increases collateral blood flow in rats with femoral arterial ligation. *Circ. Res.* 76: 62-69; 1996.
52. Yang, H.T., Ogilvie, R.W., and Terjung, R.L. Low-intensity training produces muscle adaptations in rats with femoral artery stenosis. *J. Appl. Physiol.* 71: 1822-1829; 1991.
53. Yang, H.T., and Terjung, R.L. Angiotensin-converting enzyme inhibition increases collateral-dependent muscle blood flow. *J. Appl. Physiol.* 75: 452-457; 1993.
54. Yang, H.T., Ogilvie, R.W., and Terjung, R.L. Peripheral adaptations in trained aged rats with femoral artery stenosis. *Circ. Res.* 74: 235-243; 1994.
55. Yang, H.T., Ogilvie, R.W., and Terjung, R.L. Heparin increases exercise-induced collateral blood flow in rats with femoral artery ligation. *Circ. Res.* 76: 448-456; 1995.
56. Zetterquist, S. The effect of active training on the nutritive blood flow in exercising ischemic legs. *Scand. J. Clin. Lab. Invest.* 25: 101-111; 1970.

CHAPTER 24

Contributions of Muscle Activity Patterns and Neutral Influences on Regulating Glucose Transporters in Skeletal Muscles

Arend Bonen

Department of Kinesiology, University of Waterloo, Waterloo, Ontario, Canada

Glucose enters the cell via a facilitative transport system. It is now known that there is a family of glucose transport proteins (GLUT1–GLUT7) that are expressed and regulated in a tissue-specific manner (1, 23). GLUT1 is ubiquitously expressed and is responsible for basal glucose transport in skeletal muscle. GLUT4 is one of the most important glucose transporters. It is expressed in insulin-sensitive tissues such as adipocytes, the heart, and skeletal muscles. Insulin and contraction promote glucose uptake in skeletal muscle by translocating GLUT4 from intracellular sites to the plasma membrane (22) and T-tubules (10, 31). Because the effects of insulin and contraction are addictive on GLUT4 translocation, it has been proposed that there are separate insulin-recruitable and contraction-recruitable GLUT4 pools in muscle. There is now some evidence for this, since insulin and contraction stimulate GLUT4 translocation in an additive manner to plasma membrane (13).

The available GLUT4 pool within muscle can be increased with exercise training, resulting in improved insulin sensitivity (29, 30). This suggests that muscle activity patterns are key to maintaining the GLUT4 levels within skeletal muscles. Therefore, over the past number of years we have begun to examine how normal and increased muscle activity, as well as decreased muscle activity influence GLUT1 and GLUT4, as well as basal and insulin-stimulated glucose transport. While muscle activity is an important factor in regulating GLUT4 and GLUT1, an unexpected observation has been the apparent contribution of a nerve-derived factor(s) that may affect GLUT4 expression, whereas contractility affects GLUT4 signaling.

GLUT4 and Insulin-Stimulated Glucose Transport in Rat Hindlimb Muscles

Rat skeletal muscles are extremely heterogeneous with respect to their metabolic capacities. Early studies also show this to be so for glucose uptake and insulin binding, with greater insulin-stimulated glucose transport occurring in an oxidative muscle such as soleus when compared with more glycolytic types of muscles such as the EDL and plantaris (4, 5). With the discovery that GLUT4 content was greater in red than in white muscles (14, 17, 21) we began to examine the relationship between glucose transport and GLUT4 expression among rat hindlimb muscles. In addition, in these studies, as well as in most others reported below, we always compared GLUT4 with insulin-stimulated glucose transport in perfused rat muscles. There were two reasons for doing so. First, since we measured total GLUT4 in all of our studies the insulin stimulation provided an indirect index of the translocated GLUT4, which is highly correlated with the total GLUT4 in all of our studies (this also suggests that insulin translocates a relatively constant proportion of the GLUT4 pool). Second, in some studies, changes in glucose transport have not been associated with concurrent change in total GLUT4 (28), implying that signaling mechanisms designed to translocate GLUT4 may be regulated independently from GLUT4 expression. Thus, it was important in all our work to compare changes in GLUT4 with changes in insulin-stimulated glucose transport.

When we examined GLUT4 expression in rat hindlimb muscles there was a high correlation between GLUT4 and the oxidative capacities of the muscles, as measured by the sum of the oxidative muscle fibers (% SO + % FOG) (figure 24.1). In addition, insulin-stimulated glucose transport was highly correlated with GLUT4 content of the rat muscles (figure 24.1) (25).

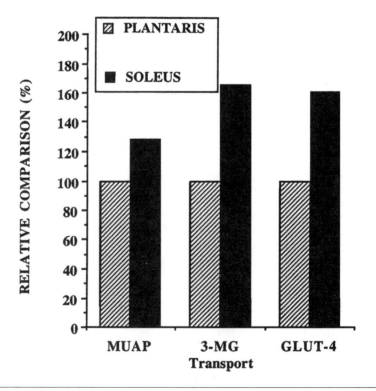

Figure 24.1. Comparison of motor unit action potentials (MUAPs) (six day average), GLUT4 expression, and insulin-stimulated glucose transport in soleus and plantaris muscles.

Relationship Between Normal and Increased Muscle Activity and GLUT4 and GLUT1

Since muscle metabolic capacities are related to their natural innervation patterns it was of interest to determine the normal activities of muscle related to the GLUT4 expression. Therefore, we compared the motor unit action potentials (MUAPs) received by soleus and plantaris muscles over a 6-day period (sampling rates 10 sec. every 2 min., 24 h/day for 6 days) in freely moving, caged rats (3). When averaged, the soleus muscle exhibited about 30% more MUAPs than the plantaris. This corresponded almost identically to the 30% difference in GLUT4 content between these two muscles, suggesting that muscle activity patterns were related closely to GLUT4 expression (figure 24.2a). Moreover, insulin-stimulated glucose transport also differed by about the same magnitude (figure 24.2b) (25).

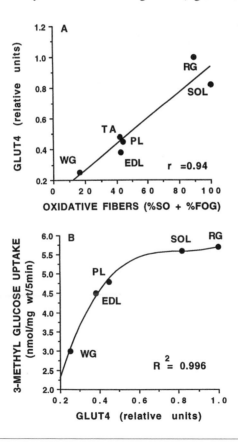

Figure 24.2.   Comparison of muscle fiber composition and GLUT4 expression (A) and GLUT4 expression and insulin-stimulated glucose transport in perfused rat hindlimb muscles (B).

Reproduced, by permission, from Megeney 1993.

To examine further the relationship between muscle activity and glucose transporters we used a chronic stimulation model to increase muscle activity. In this model wire electrodes were implanted next to the peroneal nerve and after one week of recovery from surgery, muscle in one hindlimb was stimulated (10Hz, 50μsec., 24 h/day for 6 days). The contralateral muscles in the same animals served as control. In this model we can stimulate an oxidative muscle (red tibialis anterior [RTA; 69% FOG fibers, 8% SO fibers, 23% FG fibers]), a glycolytic muscle (white tibialis anterior [WTA; 20% FOG, 0% SO, 80% FG]), and a mixed muscle (extensor digitorum longus [EDL; 39% FOG, 4% SO, 57% FG]). It was possible to increase GLUT4 and GLUT1 expression in both glycolytic and oxidative types of muscles, although the relative (%) increase was greater in the glycolytic muscles (figure 24.3). Interestingly, the observed change in GLUT4 occurred independently of changes in muscle fiber composition (i.e., no fiber type changes occurred in 6 days), although there was an increase in the muscles' oxidative capacities (19). More recently, we also have been able to demonstrate in human muscle that the increase in GLUT4 and GLUT1 are due to an increased muscle activity per se independent of changes in the muscles' oxidative capacities (27).

The functional significance of increase in GLUT4 is that the muscle becomes more sensitive to insulin, presumably because more GLUT4 can be translocated (figure 24.3). However, for unknown reasons this was not the case for contraction-induced glucose transport, since there was an unexpected decrease in contraction-stimulated glucose transport despite the increase in total GLUT4 (figure 24.3) (19).

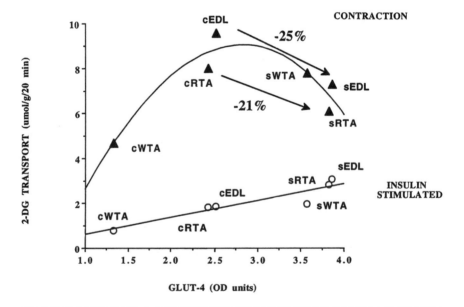

Figure 24.3. Comparison between insulin-stimulated and contraction-stimulated glucose transport in control and chronically stimulated rat muscle (c = control; s = stimulated; RTA = red tibialis anterior; WTA = white tibialis anterior; EDL = extensor digitorum longus).

Effects of Reducing Muscle Activity on GLUT1, GLUT4, and Glucose Transport

To examine the role of muscle activity on GLUT4 further, a number of studies have been performed in which the muscle activity was either eliminated or changed. The models used in these studies, as in the chronic stimulation model above, provide important experimental controls, in that the muscle activities in one leg can be altered while the contralateral leg muscles serve as control. Thus, both the experimental and control muscles are exposed to the same systemic milieu.

Tenotomy

Tenotomizing a group of muscles accomplishes the following: the muscle is unable to exert force, since the tendon is no longer attached to the bone while at the same time the neural innervation to the muscle is not disrupted. Indeed, in preliminary studies, we observed that the MUAP patterns in tenotomized soleus and EDL were barely altered. In this model of muscle disuse (3 days) we observed remarkably little change in GLUT1, GLUT4, or insulin-stimulated glucose transport (unpublished data). Thus, this provides strong evidence that muscle innervation is important to maintain the glucose transport system in muscle, but it would seem that the ability to generate force is not particularly important.

Denervation

By sectioning the sciatic nerve, muscle activity is instantly eliminated. This also results in insulin resistance of the denervated muscles (35). When we examined the effects of 3 days of denervation on GLUT4 it was observed that there was a decrement in the transport protein. However, the decrement was not uniform. The decrements were greatest in the more oxidative muscles, while only a very slight decrease occurred in the glycolytic muscles, for reasons that are not entirely clear, except to note that the loss of muscle innervation is greater in oxidative muscles. The decrease in muscle GLUT4 was highly correlated with a decrease in insulin-stimulated glucose transport (figure 24.4).

Interestingly, we also observed that GLUT1 was increased with denervation, resulting in specific increases in GLUT1 in the muscles' plasma membranes (15). Indeed, there was rather substantial increase in GLUT1-mediated basal glucose transport, and one that was inversely related to the decrease in GLUT4-mediated insulin-stimulated glucose transport (figure 24.5) (15).

We recognized that by severing the sciatic nerve that not only had the innervation been removed but also the axoplasmic flow to the muscle. In some studies this axoplasmic flow has been identified as an important parameter in muscle atrophy (8), and several nerve-derived proteins with the potential to influence muscle gene expression have been identified (11, 16). Therefore, studies were undertaken to identify the possible role of axoplasmic flow on GLUT4 expression.

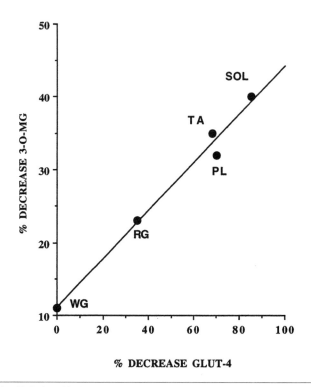

Figure 24.4.   Relationship between decrease in GLUT4 and decrease in insulin-stimulated glucose transport in denervated rat muscles.

Reprinted, by permission, from Megeney et al. 1993.

A very simple procedure for altering the axoplasmic flow to muscle is to section the sciatic nerve as close as possible (short stump) or as far away as possible from the muscle (long stump). This leaves different lengths of nerve attached to the muscle, and this enables the axoplasmic flow to the muscle to be maintained longer from the long nerve stump than from the short nerve stump. With these experiments we examined GLUT4 in red gastrocnemius (RG) and white gastrocnemius (WG) over a 3-day period. These studies showed that when axoplasmic flow was maintained longer (long stump) in RG, the GLUT4 decrease was much shorter than when less axoplasmic flow was maintained (short stump) (figure 24.6a). This however was not observed in WG in which GLUT4 decrements were also much less (figure 24.6a) (26).

An interesting facet of these studies was that despite the different rate of GLUT4 decrease, the insulin-stimulated glucose uptake decrease was identical (figure 24.6b). Similar dissociations between decrease in GLUT4 and insulin-stimulated glucose transport have been observed by others in denervated muscles (18). This began to suggest the possibility that muscle activity was associated more closely with maintaining glucose transport, whereas nerve-derived factors may be more critical for maintaining GLUT4 content of muscles.

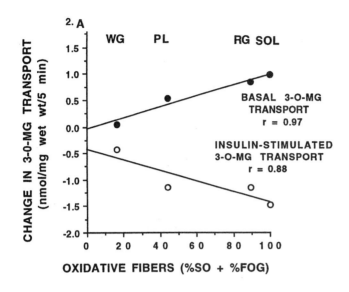

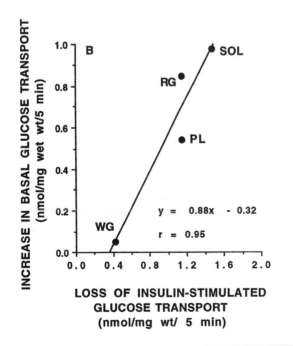

Figure 24.5.   Effects of denervation on basal and insulin-stimulated glucose transport (A) and the relationship between the decrease in insulin-stimulated glucose transport and the increase in basal glucose transport (B).

Reprinted, by permission, from Handberg et al. 1996.

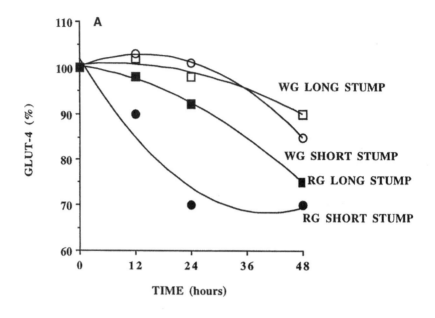

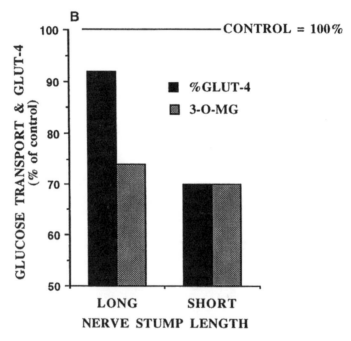

Figure 24.6.   Effects of nerve stump length on GLUT4 (A) and on insulin-stimulated glucose transport after 24 h (B).

Reprinted, by permission, from Megeney et al. 1994.

Tetrodotoxin

By using the Na⁺ channel blocker tetrodotoxin (TTX) it is possible to eliminate muscle activity while maintaining axoplasmic flow. This also provides an interesting comparison model with the denervation model, in which both axoplasmic flow and muscle activity are affected. It was found that TTX resulted in large decreases in insulin-stimulated glucose transport while only minimally altering GLUT4 content. In contrast, the same decrease in insulin-stimulated glucose transport in denervated muscles was accompanied by large decrements in GLUT4 (figure 24.7). These studies point out several important features. First, when axoplasmic flow and muscle activity are removed, as with denervation, both GLUT4 and insulin-stimulated glucose transport are reduced (24). Second, when only muscle activity is eliminated, as with TTX, then only insulin-stimulated glucose transport was reduced with almost no changes in GLUT4. These results can be taken to indicate that muscle activity is critical for maintaining the insulin signaling system whereas axoplasmic flow is critical for maintaining GLUT4.

The criteria for a trophic substance involves the following: a) the delivery of a substance by the nerve, b) the release of the substance from the nerve, and

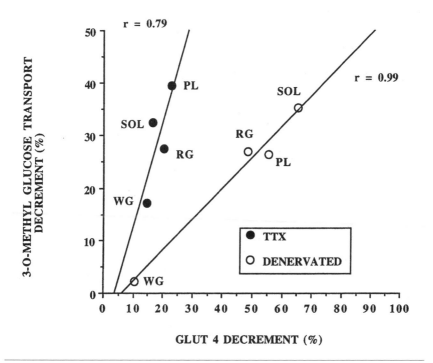

Figure 24.7.   Comparison of alternatives in GLUT4 and insulin-stimulated glucose transport in 3 day TTX-treated and 3 day denervated rat muscles.
Reprinted, by permission, from Megeney et al. 1995.

c) action(s) of the released substance at a step beyond the synapse, possibly via a receptor mediated mechanism. Ciliary neurotrophic factor (CNTF) a protein related to the cytokine class of molecules satisfies the requirements of being an important trophic factor for muscle. CNTF has a highly specific receptor expressed exclusively within skeletal muscle and the nervous system (7). Interestingly, CNTF content is decreased in the distal portion of an injured nerve (12, 33), and CNTF expression in the peripheral nerve is only evident post-natally (34). These trends parallel closely both the decrements in GLUT4 expression in denervated muscles (present study and 2, 18, 25) and the post-natal increments in the expression of GLUT4 in muscle (32, 36). Whether CNTF is the trophic factor that is involved with the regulation of GLUT4 expression is not known. However, CNTF administration has been shown to partially inhibit muscle atrophy and protein loss associated with denervation (9, 16). Moreover, CNTF-like activity is reduced in sciatic nerves of diabetic animals (6), suggesting a possible role for this trophic substance in the maintenance of glucose transporter expression.

Cross-Reinnervation

A final test of the importance of neural innervation patterns on GLUT4 was provided by experiments in which the soleus and EDL muscles were cross reinnervated (20). In these studies the nerve normally innervating the soleus muscle was cut, and the nerve that normally innervates the EDL was now attached to the soleus (cross-reinnervated soleus [X-soleus]). The net result is that the spontaneous neural innervation pattern that is normally delivered to the EDL is now being delivered to the X-soleus muscle. In addition, the converse experiment also was conducted, in which the nerve normally innervating the soleus was surgically attached to sectioned EDL motor neurons, thereby providing the X-EDL with a soleus-type innervation pattern. After surgery, the animals were permitted to recover for 3 months to reestablish complete reinnervation with the "foreign" nerves.

Over this 3-month period there was a remarkable shift in muscle fiber composition so that compared with the normal EDL (3.8% SO, 39.2% FOG, 57% FG) the X-EDL (70% SO, 26.2% FOG, 3.3% FG) began to resemble the normal soleus muscle (90% SO, 10% FOG, 0% FG), but the X-soleus (21.4% SO, 78.2% FOG, 0% FG) did not entirely resemble the normal EDL. Nevertheless, from these results it was evident that changing the innervation patterns to the muscle remodeled their fiber composition. When we summed the oxidative characteristics (% SO + % FOG), the X-EDL and X-soleus, as well as normal soleus, consisted of 96.7%, 100%, and 100% oxidative fibers, respectively. These innervation-induced oxidative characteristics also produced similar GLUT4 characteristics (figure 24.8), although GLUT1 was increased much more in the X-EDL. Thus, it is evident that innervation patterns can remodel the muscles' biochemical characteristics, and the GLUT4 expression is related to the oxidative nature of these remodeled muscles.

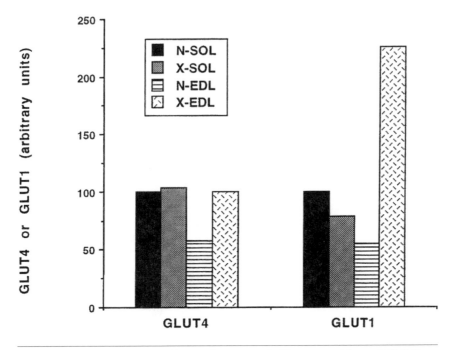

Figure 24.8. Effects of cross-reinnervation on muscle fiber composition and GLUT4 and GLUT1. (N = normally innervated; X = cross-reinnervated)

Summary

In summary, the studies from the past few years have shown that muscle activity patterns are critically important for maintaining insulin-stimulated glucose transport. It appears that nerve-derived factors may regulate GLUT4 expression whereas the contractile activity may influence the insulin signaling pathways that are critical for GLUT4 translocation to the muscles' surface. It is clear, therefore, to recognize not only the role of muscle activity maintaining the integrity of the glucose transport system, but also the role of possible nerve-derived factors in this process.

Acknowledgments

Work in A. Bonen's laboratory has been supported by the Canadian Diabetes Association and the Natural Sciences and Engineering Research Council of Canada.

References

1. Bell, G.I., Kayano, T., Buse, J.B., Burant, C.F., Takeda, J., Lin, D., Fukumoto, H., and Seino, S. Molecular biology of mammalian glucose transporters. *Diabetes Care* 13: 198-208; 1990.

2. Block, N.E., Menick, D.R., Robinson, K.A., and Buse, M.G. Effects of denervation on the expression of two glucose transporter isoforms in rat hindlimb muscles. *J. Clin. Invest.* 88: 546-552; 1991.

3. Bonen, A., Blewett, C., McDermott, J.C., and Elder, G.C.B. A model for non-exercising hindlimb muscles in exercising animals. *Can. J. Physiol. Pharm.* 68: 914-921; 1990.

4. Bonen, A., Tan, M.H., and Watson-Wright, W.M. Insulin binding and glucose uptake differences in rodent skeletal muscles. *Diabetes* 30: 702-704; 1981.

5. Bonen, A., Tan, M.H., and Watson-Wright, W.M. Effects of exercise on insulin binding and glucose metabolism in muscle. *Can. J. Physiol. Pharm.* 62: 1500-1504; 1984.

6. Calcutt, N.A., Muir, D., Powell, H.C., and Mizisin, A.P. Reduced ciliary neurotrophic factor-like activity in the nerves from diabetic or galactose-fed rats. *Brain Res.* 575: 320-324; 1992.

7. Davies, S., Aldrich, T.H., Valenzuela, D.M., Wong, V., Furth, M.E., Squinto, S.P., and Yancopoulus, G.D. The receptor for ciliary neurotrophic factor. *Science* 253: 59-63; 1991.

8. Davis, H.L. Trophic influences of neurogenic substances on adult skeletal muscle in vivo. In *Nerve-Muscle Cell Trophic Communication.* Ed. H.L. Fernandez and J.A. Donoso. Boca Raton, FL: CRC Press; 1988.

9. DiStefano, P.S., Yancopoulos, G.D., and Squinto, S.P. Ciliary neurotrophic factor (CNTF) prevents denervation induced atrophy of rat skeletal muscle. *Neurology* (Suppl 3) 42: 336; 1992.

10. Dombrowski, L., Roy, D., Marcotte, B., and Marette, A. A new procedure for the isolation of plasma membranes, T tubules and internal membranes from skeletal muscles. *Am. J. Physiol.* (Endocrinol. Metab.) 270: E667-E676; 1996.

11. Falls, D.L., Ropsen, K.M., Corfas, G., Lane, W.S., and Fishbach, G.D. ARIA, a protein that stimulates acetylcholine receptor synthesis, is a member of the Neu ligand family. *Cell* 72: 801-815; 1993.

12. Friedman, B., Schere, Rudge, J.S., Helgren, M., Morrisey, D., McClain, J., Wang, D.W., Wiegand, S.J., Furth, M.E., Lindsay, R.M., and Ip, N.Y. Regulation of ciliary neurotrophic factor expression in myelin-related Schwann cells in vivo. *Neuron* 9: 295-305; 1992.

13. Gao, J., Ren, J., Gulve, E.A., and Holoszy, J.O. Additive effects of contractions and insulin on GLUT-4 translocation into the sarcolemma. *J. Appl. Physiol.* 77: 1597-1601; 1994.

14. Goodyear, L.J., Hirshman, M.F., Smith, R.J., and Horton, E.S. Glucose transporter number, activity and isoform content in plasma membranes of red and white skeletal muscle. *Am. J. Physiol.* (Endocrinol. Metab.) 261: E556-E561; 1991.

15. Handberg, A., Megeney, L.A., McCullagh, K.J.A., Kayser, L., and Bonen, A. Reciprocal GLUT1 and GLUT4 expression and glucose transport in denervated muscles. *Am. J. Physiol.* 271: E50-E57; 1996.

16. Helgren, M.E., Squinto, S.P., Davis, H.L., Parry, D.J., Boulton, T.G., Hecks, C.S., Zhu, Y., and Yancopoulus, G.D. Trophic effect of ciliary neurotrophic factor on denervated skeletal muscle. *Cell* 76: 493-504; 1994.

17. Henriksen, E.J., Bourney, R.E., Rodnick, K.J., Koranyi, L., Permutt, M.A., and Holloszy, J.O. Glucose transporter protein content and glucose transport capacity in rat skeletal muscles. *Am. J. Physiol.* (Endocrinol. Metab.) 259: E593-E598; 1990.

18. Henriksen, E.J., Rodnick, K.J., Mondon, C.E., James, D.E., and Holloszy, J.O. Effect of denervation or unweighting on GLUT-4 protein in rat soleus muscle. *J. Appl. Physiol.* 70: 2322-2327; 1991.

19. Johannsson, E., Jensen, J., Gunderson, K., Dahl, H.A., and Bonen, A. Effects of electrical stimulation patterns on glucose transport in rat muscles. *Am. J. Physiol.* (Regulatory Integrative Comp. Physiol.) 271: R426-R431; 1996.

20. Johannsson, E., Waerhaug, O., and Bonen, A. Effect of cross-reinnervation on the expression of GLUT-4 and GLUT-1 in slow and fast rat muscles. *Am. J. Physiol.* (Regulatory Integrative Comp. Physiol.) 270: R1355-R1360; 1996.

21. Kern, M., Wells, J.A., Stephens, J.M., Elton, C.W., Friedman, J., Tapscott, E.B., Pekala, P.H., and Dohm, G.L. Insulin responsiveness in skeletal muscle is determined by glucose transporter (GLUT4) protein level. *Biochem. J.* 270: 397-400; 1990.

22. Klip, A., and Marette, A. Acute and chronic signals controlling glucose transport in muscle. *J. Cell. Biochem.* 48: 51-60; 1992.

23. Klip, A., Tsakiridis, T., Marette, A., and Ortiz, P.A. Regulation of expression of glucose transporters by glucose: a review of studies in vivo and in cell cultures. *FASEB J.* 8: 45-93; 1994.

24. Megeney, L.A., Michel, R.N., Boudreau, C.S., Fernando, P.K., Prasad, M., Tan, M.H., and Bonen, A. Regulation of muscle glucose transport and GLUT-4 by nerve-derived factors and activity-related processes. *Am. J. Physiol.* (Regulatory Integrative Comp. Physiol.) 269: R1148-R1153; 1995.

25. Megeney, L.A., Neufer, P.D., Dohm, G.L., Tan, M.H., Blewett, C.A., Elder, G.C.B., and Bonen, A. Effects of muscle activity and fiber composition on glucose transport and GLUT4. *Am. J. Physiol.* (Endocrinol. Metab.) 264: E583-E593; 1993.

26. Megeney, L.A., Prasad, M., Tan, M.H., and Bonen, A. Expression of the insulin-regulatable transporter GLUT-4 is influenced by neurogenic factors. *Am. J. Physiol.* (Endocrinol. Metab.) 266: E813-E816; 1994.

27. Phillips, S., Han, X.-X., Green, H.J., and Bonen, A. Increments in skeletal muscle GLUT-1 and GLUT-4 following endurance training in humans. *Am. J. Physiol.* (Endocrinol. Metab.) 270: E546-E562; 1996.

28. Richardson, J.M., Balon, T.W., Treadway, J.L., and Pessin, J.E. Differential regulation of glucose transporter activity and expression in red and white skeletal muscle. *J. Biol. Chem.* 266: 12,690-12,694; 1991.

29. Rodnick, K.J., Henrikson, E.J., James, D.E., and Holloszy, J.O. Exercise training, glucose transporters and glucose transport in rat skeletal muscles. *Am. J. Physiol.* (Cell Physiol.) 262: C9-C14; 1992.

30. Rodnick, K.J., Holloszy, J.O., Mondon, C.E., and James, D.E. Effects of exercise training on insulin-regulatable glucose-transporter protein levels in rat skeletal muscle. *Diabetes* 39: 1425-1429; 1990.

31. Roy, D., Johannsson, E., Bonen, A., and Marette, A. Electrical stimulation induces fiber types specific translocation of GLUT4 to transverse tubules in rat skeletal muscle. *Am. J. Physiol.* (Endocrinol. Metab.). In press; 1997.

32. Santalucia, P.S., Camps, M., Castello, A., Munoz, P., Nuel, A., Testar, Z., Palacin, M., and Zorzano, A. Developmental regulation of GLUT-1 (erythroid/Hep G2) and GLUT4 (muscle/fat) glucose transporter expression in rat heart, skeletal muscle, and brown adipose tissue. *Endocrinology* 130: 837-846; 1992.

33. Sendtner, M., Stockli, K.A., and Thoenen, H. Synthesis and localization of ciliary neurotrophic factor in the sciatic nerve of the adult rat after lesion and during regeneration. *J. Cell. Biol.* 118: 139-148; 1992.

34. Stockli, K.A., Lottspeich, F., Sendtner, M., Masiakowski, P., Carroll, P., Gotz, R., Lindholm, D., and Thoenen, H. Molecular cloning, expression and regional distribution of the rat ciliary neurotrophic factor. *Nature* 342: 920-923; 1989.

35. Turinsky, J. Dynamics of insulin resistance in denervated slow and fast-twitch muscles in vivo. *Am. J. Physiol.* (Regulatory Integrative Comp. Physiol.) 252: R531-R537; 1987.

36. Wallace, S., Campbell, G., Knott, R., Gould, G.W., and Hesketh, J. Development of insulin sensitivity in rat skeletal muscle. Studies of glucose transporter and insulin receptor mRNA levels. *FEBS Lett.* 301: 69-72; 1992.

CHAPTER 25

Stress Proteins and the Adaptive Response to Muscle Loading

Earl G. Noble and F. Kris Aubrey

School of Kinesiology, The University of Western Ontario,
London, Ontario, Canada

Both chronic and acute physical exercise are stressors that ultimately necessitate an adaptive response in order for cellular homeostasis to be maintained (6). Adaptive responses to muscle loading are not confined to whole organisms, but are also evident in the differential capacities of muscle fibers that are recruited to greater or lesser degrees (35). Several recent articles suggest that a class of proteins known as heat shock or stress proteins (SPs) may play an important role in cellular response to and survival from stress (for reviews, see 19, 27, 45). This report provides a brief overview of the stress response to exercise and differences in muscle specific expression of stress proteins. It offers some preliminary observations on factors, which might be involved in these differences.

The Stress Response

Most organisms exhibit a highly conserved response to stress that includes a rapid, preferential synthesis of a group of proteins known as heat shock (HSPs) or stress proteins (SPs). Of the SPs studied to date, the most characterized are those of the 70 kDa SP family. Two primary members of this family are a 73 kDa protein (SP73) that is present in most tissues under control or unstressed conditions (constitutively expressed) and a 72 kDa stress inducible isoform (SP72). Although SP72 once was believed to be induced primarily during times of stress (45), it is now clear that SP72 is constitutively expressed at low levels in a variety of tissues (41) and probably performs a routine although presently unknown function.

SP72 is believed to control its own cellular level through a feedback system involving an interaction with a heat shock transcription factor (HSF) (1, 3, 29). Normally, HSF exists in the cytoplasm in a complex with members of the SP70 family. When bound in this fashion HSF is unavailable to activate a stress response. During times of stress, however, SP70s may interact with damaged or denatured proteins and shift the equilibrium away from binding with HSF. At such times, three HSF monomers may combine to form a trimer that interacts with heat shock elements

(HSE) on the SP70 gene, thereby promoting its transcription. When levels of SP70s subsequently rise, they again bind HSF, thereby self-regulating their own synthesis.

Exercise as an Inducer of the Stress Response

Acute exercise is a sufficient stressor to induce the synthesis of a number of SPs (15, 20, 22, 38, 39). In fact, in both cardiac tissue (22) and skeletal muscle (37, 38; unpublished observations) exercise causes a rapid increase in the levels of activated HSF, SP72 mRNA, and ultimately SP72 protein. In response to treadmill running at approximately 70% maximal oxygen uptake (24 m/min, 0% grade), 40% of the exercising rats exhibit HSF activation after only 20 minutes of running and SP72 mRNA shortly thereafter. These numbers increase to 100% at exhaustion (22). Hence, as with heat shock (22), exercise loading of skeletal muscle results in a very rapid and substantial stress response. With chronic exercise training (15) or electrical stimulation (30), this response is not diminished, as the muscle adapts to the exercise load. This finding suggests that the increased synthesis of SP72 observed in skeletal muscle after a single bout of acute exercise may not represent simply an acute stress response but may be associated ultimately with some more long-term, adaptive process.

While the exact factor(s) responsible for the exercise-induced stress response remains unknown, it is clear that many conditions potentially associated with exercise could be involved. These include altered ATP status (5, 26, 43), changes in calcium homeostasis (28) and pH (34), hypoxia (4), generation of oxygen free radicals (38), elevated temperature (22, 27, 39), altered protein synthesis rates (17, 26), and any proteotoxic condition that could lead to cell damage (45). Elevated temperature is a known and well-characterized inducer of the stress response (22, 27, 39). Skidmore et al. (38) employed exercise in a cool environment as a means of separating other potential SP-inducing effects of exercise from that of elevated temperature. In this study, an accumulation of SP72 was detected in several muscles of treadmill-running rats, despite their having a core temperature that was not different from baseline. More recently, Neufer et al. (30) were able to detect elevated levels of SP72 mRNA and protein in electrically stimulated (6–10 Hz continuously) rabbit skeletal muscle despite the fact that the temperature in the contracting muscle was elevated less than $0.5°$ C ($39°$ C to $39.4°$ C). Hence, factors other than elevated temperature are involved in the initiation of the stress response to exercise.

The ultimate effect of exercise-induced increases in SP72 is unclear. However, one possible role is that of cellular protection against proteotoxic insults. It is known that heat-induced elevation of SP72 content in various cells or tissues helps maintain cellular viability and function (19, 27, 45). In cardiac tissue of transgenic animals, overexpression of SP72 is associated with reduced muscle damage and enhanced metabolic recovery from ischemia and subsequent reperfusion (7, 25, 36). Similarly, when SP72 levels in rat heart were elevated in response to three bouts of treadmill running (60 min/day for 3 days at 30 m/min) their functional recovery from ischemia was enhanced tremendously (24). Although few studies have examined the potential protective role of SP72 in skeletal muscle, as in cardiac muscle, elevated levels of this protein may help preserve metabolic and structural homeostasis following stress (10).

Muscle Specific Expression of SP72

One interesting observation is that skeletal muscle that has a high proportion of type I muscle fibers tends to exhibit an elevated constitutive expression of the inducible SP70 isoform, SP72 (15, 21, 23, 33). Since type I fibers tend to be recruited more frequently for ambulation and postural maintenance, this stress of "muscle loading" could account for the elevated constitutive expression of SP72. Consistent with this conjecture is the observation that when loading of the rat plantaris muscle is dramatically increased by removal of its synergists (9) the expression of SP72 also is elevated (16, 18). In addition, maintenance of this "muscle loading" stress could account for the elevated levels of SP72 detected in skeletal muscle that had undergone eight weeks of exercise training (15) or which had been subjected to chronic electrical stimulation (30, 33). In contrast, when muscle loading is reduced, such as occurs with hindlimb unweighting, whole-tissue SP70 (both SP73 and SP72) levels decrease within 18 hours in the soleus (17). Whether SP72 levels are altered by this treatment is unclear, but it is likely that they decline in response to the reduced muscle load.

Another possibility, which could explain the constitutive expression of SP72 in muscles with a high proportion of type I fibers, is that SP72 may play a specific, routine role in type I fibers and be expressed in co-ordination with another protein(s) in these fibers. For example, the increased expression of SP72 in muscle undergoing compensatory hypertrophy is accompanied by an increase in type I myosin heavy chain (MHC) content (18). Similarly, when conducted at sufficient intensity and for prolonged periods of time, both exercise training (11) and chronic nerve stimulation (35) have been observed to increase the type I MHC content in skeletal muscle, as well as SP72 (15, 30, 33). When skeletal muscle is unloaded, via hindlimb unweighting, a procedure that results in decreased SP72 expression, a decrease in the expression of type I MHC also is ultimately observed (42).

Effect of Tri-Iodothyronine Treatment and Compensatory Overload on SP72 Levels

One of the problems in attempting to ascertain whether the constitutive expression of SP72 in skeletal muscle rich in type I MHC is a direct response to changes in muscle load, or is more closely linked to expression of other proteins in type I muscle fibers, is that many of the experimental models employed to study this question simultaneously affect a number of factors that are known inducers of the stress response. Two such factors are the production of abnormal proteins accompanying muscle damage and altered protein synthetic rates accompanying growth or atrophy (19). For example, exercise training and chronic electrical stimulation may result in muscle damage and, like hindlimb unweighting, alter protein synthesis (6, 17). Similarly, compensatory overload of the rat plantaris leads not only to an increase in the content of type I muscle fibers but also results in early muscle damage (2) and a chronic elevation in protein synthetic rate (32).

Previous studies (14, 40) have demonstrated that the increase in type I MHC normally accompanying compensatory overload may be inhibited by tri-iodothyronine (T3) administration. Therefore, in order to determine whether the elevated constitutive expression of SP72 accompanying compensatory overload is co-ordinated with that of type I MHC or is simply a response to the stress of increased muscle loading, rats were subjected to compensatory overload of the left plantaris (13) either with (T150) or without (TS) administration of 150 µg/kg T3, every second day, for 40 days. As indicated in table 25.1, this dosage was sufficient to significantly elevate T3 levels in the blood, raise body temperature, and reduce body weight.

As anticipated (16, 18), in the absence of T3 treatment, surgical removal of the ipsilateral gastrocnemius and the resultant compensatory overload of the rat plantaris was accompanied by increased expression of type I MHC and SP72 (figures 25.1 and 25.2). Also, as previously demonstrated (14, 40), T3 treatment was effective in abolishing the overload-induced increase in type I MHC content (figure 25.1). The novel observation was that hyperthyroidism also inhibited the normal increase in SP72 expression accompanying overload (figure 25.2). This is particularly interesting, since the overloaded muscles of T3-treated animals exhibited a similar relative growth to those in the TS group (table 25.1), suggesting a similar increase in loading. In addition, plantaris muscles from the T3 group probably experienced a slightly elevated temperature (table 25.1) and mitochondrial proliferation (18). As SP72 may be induced by elevated temperatures and is thought to act as a cytoplasmic chaperone during mitochondrial biogenesis (27), T3 treatment would have been expected to augment the SP72 induction that normally accompanies muscle overload.

The observation that the normal overload-induced elevation in the expression of SP72 and type I MHC was inhibited in the plantaris muscle of T3-treated rats,

Table 25.1   Serum Thyroid Hormone Level, Temperature, Body and Plantaris Weights Following 40 days of Compensatory Overload and Thyroid Hormone Injection

		TS	T150
Serum T3 (ng/dL)		48.9 ± 3.7	120.0 ± 15.4*
Body weight (g)		451.7 ± 8.1	400.1 ± 11.8*
Rectal temperature (°C)		37.0 ± 0.1	37.5 ± 0.1*
Plantaris weight (mg)	C	434 ± 6.8	385 ± 14.6
	OV	727 ± 22.5[†]	654 ± 17.5[†]
	% change	67.5	69.9

Rats were injected with carrier solution (TS) or 150 (T150) µg/kg body weight of tri-iodothyronine (T3) every second day over a 40-day period. The left plantaris was overloaded (OV) by removal of the ipsilateral gastrocnemius, the right plantaris served as a contralateral control (C). Values are expressed as means ± SEM.
* significantly different from TS (p ≤ 0.05), [†] OV significantly different from C (p ≤ 0.05).

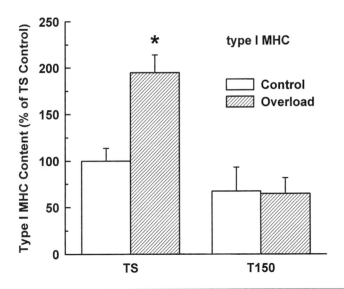

Figure 25.1.   T3 treatment is able to suppress the increase in type I MHC in the plantaris muscle of rats following 40 days of compensatory overload. Values are expressed as a percentage (mean ± SEM) of the value for the unloaded plantaris of the sham-treated animals when this value was set to 100%. N = 4-6. *Significantly different from all other conditions.

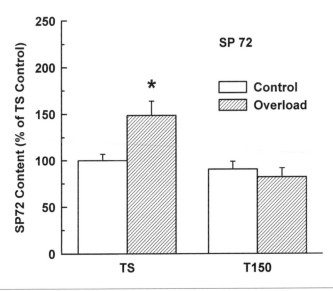

Figure 25.2.   T3 treatment is able to suppress the increase in SP72 in the plantaris muscle of rats following 40 days of compensatory overload. Values are expressed as a percentage (mean ± SEM) of the value for the unloaded plantaris of the sham-treated animals when this value was set to 100%. N = 4-6. *Significantly different from all other conditions.

despite factors associated with these treatments that might be expected to induce a stress response, suggests that under some circumstances these proteins may come under co-ordinate control. Whether this is due to a direct effect of thyroid hormone on both the type I MHC and SP72 genes or whether it is mediated through transcription factors that are in turn regulated by T3, remains to be determined. As no thyroid hormone response elements have been detected in the promoter regions of SP70 genes, it is tempting to speculate the latter. Myogenic regulatory factors have been suggested to be responsible for the distinctive pattern of protein expression in various muscle fiber types (8, 12, 44). When alterations in muscle fiber type are induced by T3 administration, levels of these regulatory factors shift in a corresponding manner (12). Hence, regulation by specific myogenic factors could explain the constitutive expression of SP72 in muscles rich in type I MHC as well as the T3-induced inhibition of the normal SP72 response to compensatory overload. Such a concept is supported by the observation that the small SP, αB-crystallin, which demonstrates a fiber-specific distribution in control muscle, may be upregulated coincident with changes in members of the MyoD family in response to chronic contractile activity (31).

Summary

From the foregoing, it is tempting to speculate that in skeletal muscle more than one pathway is involved in regulating SP72 expression. Clearly, exercise and muscle activity are sufficient stressors to activate the usual cascade of events leading to increased expression of SP72. However, the observation of fiber-type specific constitutive expression of this protein, coupled with an ability of T3 to inhibit the normal increase in SP72 in overloaded muscle, suggests that other pathways, possibly those associated with muscle specific transcription factors, also may be important. If so, this could suggest a differential role for SP72 during normal muscle function as opposed to that during a response to stress.

Acknowledgments

This work was supported by a grant from the National Sciences and Engineering Research Council of Canada.

References

1. Abravaya, K.A., Phillips, B., and Morimoto, R.I. Attenuation of the heat shock response in HeLa cells is mediated by the release of bound heat shock transcription factor and is modulated by changes in growth and in heat shock temperatures. *Genes and Development* 5: 2117-2127; 1991.

2. Armstrong, R.B., Marum, P., Tullson, P., and Saubert , C.W., IV. Acute hypertrophic response of skeletal muscle to the removal of synergists. *Journal of Applied Physiology* 46: 835-842; 1979.

3. Baler, R., Welch, W.J., and Voellmy, R. Heat shock gene regulation by nascent polypeptides and denatured proteins: hsp70 as a potential autoregulatory factor. *The Journal of Cell Biology* 117: 1151-1159; 1992.

4. Benjamin, I.J., Kroger, B., and Williams, R.S. Activation of the heat shock transcription factor by hypoxia in mammalian cells. *Proceedings of the National Academy of Sciences USA* 87: 6263-6267; 1990.

5. Benjamin, I.J., Horie, S., Greenberg, M.L., Alpern, R.J., and Williams, R.S. Induction of stress proteins in cultured myogenic cells. *Journal of Clinical Investigation* 89: 1685-1689; 1992.

6. Booth, F.W., and Thomason, D.B. Molecular and cellular adaptation of muscle in response to exercise: perspectives of various models. *Physiological Reviews* 71: 541-585; 1991.

7. Bradford, N.B., Fina, M., Benjamin, I.J., Moreadith, R.W., Graves, K.H., Zhao, P., Gavva, S., Wiethoff, A., Sherry, A.D., Malloy, C.R., and Williams, R.S. Cardioprotective effects of 70 kDa heat shock protein in transgenic mice. *Proceedings of the National Academy of Sciences USA* 93: 2339-2342; 1996.

8. Buonanno, A., and Rosenthal, N. Molecular control of muscle diversity and plasticity. *Developmental Genetics* 19: 95-107; 1996.

9. Gardiner, P., Michel, R., Browman, C., and Noble, E. Increased EMG of rat plantaris during locomotion following surgical removal of its synergists. *Brain Research* 380: 114-121; 1986.

10. Garramone, R.R., Jr., Winters, R.M., Das, D.K., and Deckers, P.J. Reduction of skeletal muscle injury through stress conditioning using the heat shock response. *Plastic and Reconstructive Surgery* 93: 1242-1247; 1994.

11. Green, H.J., Klug, G.A., Reichmann, H., Seedorf, U., Wiehrer, W., and Pette, D. Exercise-induced fibre type transitions with regard to myosin, parvalbumin, and sarcoplasmic reticulum in muscles of the rat. *Pflügers Archiv: European Journal of Physiology* 400: 432-438; 1984.

12. Hughes, S.M., Taylor, J.M., Tapscott, S.J., Gurley, C.M., Carter, W.J., and Peterson, C.A. Selective accumulation of MyoD and myogenin in fast and slow adult skeletal muscle is controlled by innervation and hormones. *Development* 118: 1137-1147; 1993.

13. Ianuzzo, C.D., and Chen, V. Compensatory hypertrophy of skeletal muscle: contractile characteristics. *Physiology Teacher* 6: 4-7; 1977.

14. Ianuzzo, D., Hamilton, N., and Li, B. Competitive control of myosin expression—hypertrophy vs hyperthyroidism. *Journal of Applied Physiology* 70: 2328-2330; 1991.

15. Kelly, D.A., Tiidus, P.M., Houston, M.E., and Noble, E.G. Effect of vitamin E deprivation and exercise training on induction of HSP70. *Journal of Applied Physiology* 81: 2379-2385; 1996.

16. Kilgore, J.L., Timson, B.F., Saunders, D.K., Kraemer, R.R., Klemm, R.D., and Ross, C.R. Stress protein induction in skeletal muscle: comparison of laboratory models to naturally occurring hypertrophy. *Journal of Applied Physiology* 76: 598-601; 1994.

17. Ku, Z., Yang, J., Menon, V., and Thomason, D.B. Decreased polysomal HSP-70 may slow polypeptide elongation during skeletal muscle atrophy. *American Journal of Physiology* 268: C1369-C1374; 1995.

18. Locke, M., Atkinson, B.G., Tanguay, R.M., and Noble, E.G. Shifts in type I fiber proportion in rat hindlimb muscle are accompanied by changes in HSP72 content. *American Journal of Physiology* 266: C1240-C1246; 1994.

19. Locke, M., and Noble, E.G. Stress proteins: the exercise response. *Canadian Journal of Applied Physiology* 20: 155-167; 1995.

20. Locke, M., Noble, E.G., and Atkinson, B.G. Exercising mammals synthesize stress proteins. *American Journal of Physiology* 258: C723-C729; 1990.

21. Locke, M., Noble, E.G., and Atkinson, B.G. Inducible isoform of HSP70 is constitutively expressed in a muscle fibre type specific pattern. *American Journal of Physiology* 261: C774-C779; 1991.

22. Locke, M., Noble, E.G., Tanguay, R.M., Field, M.R., Ianuzzo, S.E., and Ianuzzo, C.D. Activation of heat-shock transcription factor in rat heart after heat shock and exercise. *American Journal of Physiology* 268: C1387-C1394; 1995.

23. Locke, M., and Tanguay, R.M. Increased HSF activation in muscles with a high constitutive Hsp70 expression. *Cell Stress and Chaperones* 1: 189-196; 1996.

24. Locke, M., Tanguay, R.M., Klaburde, R.E., and Ianuzzo, C.D. Enhanced postischemic myocardial recovery following exercise induction of HSP72. *American Journal of Physiology* 269: H320-H325; 1995.

25. Marber, M.S., Mestril, R., Chi, S.H., Sayen, M.R., Yellon, D.M., and Dillman, W.H. Overexpression of the rat inducible 70-kD heat stress protein in a transgenic mouse increases the resistance of the heart to ischemic injury. *Journal of Clinical Investigation* 95: 446-456; 1995.

26. Menon, V., Yang, J., Ku, Z., and Thomason, D.B. Decrease in heart peptide initiation during head-down tilt may be modulated by HSP-70. *American Journal of Physiology* 268: C1375-C1380; 1995.

27. Morimoto, R.I., Tissieres, A., and Georgopolis, C., ed. *The biology of heat shock proteins and molecular chaperones.* Cold Spring Harbor, NY: Laboratory Press; 1994.

28. Mosser, D.D., Kotzbauer, P.T., Sarge, K.D., and Morimoto, R.I. *In vitro* activation of heat-shock transcription factor DNA-binding by calcium and biochemical conditions that affect protein conformation. *Proceedings of the National Academy of Sciences USA* 87: 3748-3752; 1990.

29. Mosser, D.D., Duchaine, J., and Massie, B. The DNA-binding activity of the human heat shock transcription factor in regulated *in vivo* by Hsp70. *Molecular and Cellular Biology* 13: 5427-5438; 1993.

30. Neufer, P.D., Ordway, G.A., Hand, G.A., Shelton, J.M., Richardson, J.A., Benjamin, I.J., and Williams, R.S. Continuous contractile activity induces fiber type specific expression of HSP70 in skeletal muscle. *American Journal of Physiology* 271: C1828-C1837; 1996.

31. Neufer, P.D., and Benjamin, I.J. Differential expression of αB-crystallin and Hsp27 in skeletal muscle during continuous contractile activity. *Journal of Biological Chemistry* 271: 24,089-24,095; 1996.

32. Noble, E.G., Tang, Q., and Taylor, P.B. Protein synthesis in compensatory hypertrophy of rat plantaris. *Canadian Journal of Physiology and Pharmacology* 62: 1178-1182; 1984.

33. Ornatsky, O.I., Connor, M.K., and Hood, D.A. Expression of stress proteins and mitochondrial chaperonins in chronically stimulated skeletal muscle. *Biochemical Journal* 311: 119-123; 1995.
34. Petronini, P.G., Alfieri, R., Campanini, C., and Borghetti, A.F. Effect of alkaline shift on induction of the heat shock response in human fibroblasts. *Journal of Cell Physiology* 162: 322-329; 1995.
35. Pette, D., and Staron, R.S. Mammalian skeletal muscle fiber type transitions. *International Review of Cytology* 170: 143-223; 1997.
36. Plumier, J.-C., Ross, B.M., Currie, R.W., Angelidis, C.E., Kazlaris, H., Kollias, G., and Pagoulatos, G.N. Transgenic mice expressing the human heat shock protein 70 have improved post-ischemic myocardial recovery. *Journal of Clinical Investigation* 95: 1854-1860; 1995.
37. Puntschart, A., Vogt, M., Widmer, H.R., Hoppeler, H., and Billeter, R. Hsp 70 expression in human skeletal muscle after exercise. *Acta Physiolgica Scandinavica* 157: 411-417; 1996.
38. Salo, D.C., Donovan, C.M., and Davies, K.J.A. HSP70 and other possible heat-shock or oxidative stress proteins are induced in skeletal muscle, heart, and liver during exercise. *Free Radical Biology and Medicine* 11: 239-246; 1991.
39. Skidmore, R., Gutierrez, J.A., Guerriero, J.V., and Kregel, K.C. HSP70 induction during exercise and heat stress in rats: role of internal temperature. *American Journal of Physiology* 268: R92-R97; 1995.
40. Swoap, S.J., Haddad, F., Caiozzo, V.J., Herrick, R.E., McCue, S.A., and Baldwin, K.M. Interaction of thyroid hormone and functional overload on skeletal muscle isomyosin expression. *Journal of Applied Physiology* 77: 621-629; 1994.
41. Tanguay, R.M., Wu, Y., and Khandjian, E.W. Tissue-specific expression of heat shock proteins of the mouse in the absence of stress. *Developmental Genetics* 14: 112-118; 1993.
42. Tsika, R.W., Herrick, R.E., and Baldwin, K.M. Interaction of compensatory overload and hindlimb suspension on myosin isoform expression. *Journal of Applied Physiology* 62: 2180-2186; 1987.
43. Van Why, S.K., Mann, A.S., Thulin, G., Zhu, X.-H., Kashgarian, M., and Siegel, N.J. Activation of heat-shock transcription factor by graded reductions in renal ATP, in vivo, in the rat. *Journal of Clinical Investigation* 94: 1518-1523; 1994.
44. Voytik, S.L., Przyborski, M., Badylak, S.F., and Konieczny, S. Differential expression of muscle regulatory factor genes in normal and denervated adult rat hindlimb muscles. *Developmental Dynamics* 98: 214-224; 1993.
45. Welch, W.J. Mammalian stress response: Cell physiology, structure/function of stress proteins, and implications for medicine and disease. *Physiological Reviews* 72: 1063-1081; 1992.

Related Books from Human Kinetics

Biochemistry of Exercise IX
Ronald J. Maughan, PhD, and Susan M. Shirreffs, BSc, editors
1996 • Hardcover • 608 pp • Item BMAU0486   ISBN 0-88011-486-X • $69.00 ($103.50 Canadian)

Exercise Metabolism
Mark Hargreaves, PhD, Editor
1995 • Hardcover • 272 pp • Item BHAR0453   ISBN 0-87322-453-1 • $39.00 ($58.50 Canadian)

Antioxidants and Exercise
Jan Karlsson, PhD
1996 • Hardcover • 224 pages • Item BKAR0896   ISBN 0-87322-896-0 • $39.00 ($58.50 Canadian)

Exercise and Immunology
[Current Issues in Exercise Science, Monograph Number 2]
Laurel T. Mackinnon, PhD
1992 • Paper • 128 pp • Item BMAC0347   ISBN 0-87322-347-0 • $22.00 ($32.95 Canadian)

Biochemistry Primer for Exercise Science
Michael E. Houston, PhD
1995 • Paper • 144 pp • Item BHOU0577   ISBN 0-87322-577-5 • $22.00 ($32.95 Canadian)

The Blood Lactate Response to Exercise
[Current Issues in Exercise Science, Monograph Number 4]
Arthur Weltman, PhD, FACSM
1995 • Paper • 128 pp • Item BWEL0769   ISBN 0-87322-769-7 • $22.00 ($32.95 Canadian)

Related Journal from Human Kinetics

Exercise Immunology Review
Editor: Hinnak Northoff, PhD
Frequency: Annual
First Issue: March 1995
Subscription Rates (including shipping):

	Individual	Institution	Student
U.S.	$20.00	$40.00	$12.00
Canada—surface	30.00	57.00	19.00
Canada—air	34.00	61.00	23.00

ISSN: 1077-5552 • Item: JEIR

To request more information or to order, U.S. customers call 1-800-747-4457, e-mail us at humank@hkusa.com or visit our Web site at www.humankinetics.com. Persons outside the U.S. can contact us via our Web site or use the appropriate telephone number, postal address, or e-mail address shown in the front of this book.

HUMAN KINETICS
The Information Leader in Physical Activity
P.O. Box 5076, Champaign, IL 61825-5076